Humanities, Science, Scimat

From Two Cultures to Bettering Humanity

L O V E
WISDOM
L O V E
FREEDOM

WHEN HERE DWELL POETICALLY

Science Matters Series

Lui Lam
Founder and Editor

Scimat (Science Matters) is the new multidiscipline that treats all human matters as part of science, wherein, humans (the material system of *Homo sapiens*) are studied scientifically from the perspective of complex systems. That *Everything in Nature Is Part of Science* was well recognized by Aristotle and da Vinci and many others. Yet, it is only recently, with the advent of modern science and experiences gathered in the study of evolutionary and cognitive sciences, neuroscience, statistical physics, complex systems, and other disciplines, that we know how the human-related disciplines can be studied scientifically. Science Matters Series covers new developments in all the topics in the humanities and social science from the scimat perspective, with emphasis on the humanities.

1. *Science Matters: Humanities as Complex Systems*
   M. Burguete & L. Lam, editors
2. *Arts: A Science Matter*
   M. Burguete & L. Lam, editors
3. *About Science: Philosophy, History, Sociology & Communication*
   M. Burguete & L. Lam, editors
4. *Humanities, Science, Scimat: From Two Cultures to Bettering Humanity*
   L. Lam
5. *Scimat Anthology: Histophysics, Art, Philosophy, Science*
   L. Lam

Science Matters Series | No. 4

Humanities, Science, Scimat

From Two Cultures to Bettering Humanity

Lui Lam
San Jose State University, USA

NEW JERSEY • LONDON • SINGAPORE • BEIJING • SHANGHAI • HONG KONG • TAIPEI • CHENNAI • TOKYO

Published by

World Scientific Publishing Co. Pte. Ltd.
5 Toh Tuck Link, Singapore 596224
USA office: 27 Warren Street, Suite 401-402, Hackensack, NJ 07601
UK office: 57 Shelton Street, Covent Garden, London WC2H 9HE

Library of Congress Control Number: 2024935050

British Library Cataloguing-in-Publication Data
A catalogue record for this book is available from the British Library.

Cover design: Lui Lam

Science Matters Series — Vol. 4
HUMANITIES, SCIENCE, SCIMAT
From Two Cultures to Bettering Humanity

ISBN 978-981-12-8439-7 (hardcover)
ISBN 978-981-12-8440-3 (ebook for institutions)
ISBN 978-981-12-8441-0 (ebook for individuals)

For any available supplementary material, please visit
https://www.worldscientific.com/worldscibooks/10.1142/13627#t=suppl

Contents

PART II SCINECE AND SCIENTIST

PART III ARTICLES

About the Author

Lui Lam, humanist and physicist, is professor emeritus at San Jose State University, California; recipient of SJSU's 2017 Distinguished Service Award; guest professor at Chinese Academy of Sciences and China Association for Science and Technology Education; BS (First Class Honors), University of Hong Kong; MS, University of British Columbia; PhD, Columbia University, thesis mentor: Philip Platzman, Bell Labs.

Apart from the United States, Lam had worked in Europe (Belgium and West Germany) and Beijing (Institute of Physics, CAS, 1978-1983); worked 30 years in natural science (physics, chemistry, complex systems), then 20 years in the humanities (history, art, philosophy).

Lam invented Bowlics (1982), one of three existing types of liquid crystals; Active Walk (1992), a new paradigm in complex systems; and two new multidisciplines: Histophysics (2002) and Scimat (Science Matters, 2007). He published over 180 papers and 24 books, including *Nonlinear Physics for Beginners* (1998), *Arts* (2011), and *All About Science* (2014).

Founder of International Liquid Crystal Society (1990); founder and editor of two book series: Science Matters (World Scientific) and Partially Ordered Systems (Springer). Current research: philosophy, humanities-science synthesis, and innovation. (For more, see [A1].)

Email: lui2002lam@icloud.com
Website: www.sjsu.edu/people/lui.lam/scimat
More: https://www.researchgate.net/profile/Lui-Lam/research

Preface

The humanities (and social science) are the disciplines that study human, which are essential in helping us to understand ourselves and others and the world around us. Since science is the study of everything in the universe and human is a material system consisting of the same atoms that make up other nonhuman systems, humanities are part of science. Thus, understanding correctly what science is about will be helpful, if not a prerequisite, in making progress in the humanities.

Historically, the humanities and natural science (of nonhuman systems) resulted from the fragmentation of science that happened only few hundred years ago, giving rise to the so-called "two cultures" as noted by C. P. Snow. To patch up the gap between these two cultures and to help both the humanities and natural science to grow healthily, the best way is to recognize the common root of these two branches of knowledge. That is, through humanities-science synthesis, as advocated by Scimat, the new multidiscipline I proposed in 2007.

Furthermore, the humanities are more important than natural science for two reasons. First, the humanities include art. It is art that makes life worth living while natural science only makes life more comfortable. Second, all great tragedies in history (like the six million jews terminated in WWII) are human-made—due to disastrous human decisions, and the study of decision making belongs to the humanities. Thus, the way to make the world better is to raise the scientific level of the humanities.

These ideas were put forward in the first scimat book *Science Matters* (2008) and elaborated in the third scimat book *All About Science* (2014). Since then I have been trying to explain them to a lot of people, through lectures and writings, in both English and Chinese, including the WeChat groups (loosely equivalent to Facebook) I joined. Most of these explanations—together with new insights on history, art, philosophy, science, and scientist from the scimat perspective—are grouped together here in this book, *Humanities, Science, Scimat.*

The book consists of three parts. Part I is about Scimat and the new humanities (history, philosophy, art). Part II is on the origin and nature of science, observations on the life and work of selected scientists, some thoughts on science communication/popularization, and case examples of science innovation. While Parts I and II are short essays with no references (with rare exceptions), Part III are longer articles with full references that supplement Parts I and II. Original papers and book chapters backing up the subjects of this book are provided in the accompanying volume *Scimat Anthology*.

Each essay/article starts with a color picture. They are all easy to read—nothing technical—by students and researchers alike. The contents of the book are summarized in How to Read this Book. My personal journey leading to this book is presented in the Epilogue.

Note that social science is not included in this book because its importance is well recognized while the humanities are not—not yet. Hopefully, this book will help to change that.

In short, this book contains the basic knowledge about the humanities and science that everyone should know. The aimed readership is anyone, from high school students and laypeople to professors, who are interested in what the humanities and science are about, and how we can work together to achieve a better humanity.

Lui Lam

August 11, 2023
San Jose, California

About Chinese Names

There is no perfect way to write Chinese names in English. The spelling and ordering conventions of a Chinese name's characters are different in different geological areas—mainland China, Hong Kong, Taiwan, and United States. The conventions adopted in this book are as follows.

All Chinese names in text and references are written with family name *first*, with first name's characters connected by a hyphen.

All Chinese names from mainland China are spelled out in pinyin.

For those who made their career in the US, whether they settled later in mainland China or not, their name's old spelling is adopted, i.e., *not* in pinyin. For example, Chen-Ning Yang in this book is Chen Ning Yang in the US (which would be Yang Zheng-Ning if he made his career in mainland China but not in the US).

Lui Lam made his career in both places, outside and inside China. The name Lui Lam and his pinyin name Lin Lei appear both in text and references. (His family name, Lin in pinyin, is Lam in Cantonese.)

How to Read this Book

This book contains 100 essays and 10 articles. They can be read randomly—the best way to read, in fact. In the text, [2.4] and [17], say, mean Sec. 2.4 and Chap. 17, respectively, which are more for further reading than prerequisite reading.

The core, must-read part is Chap. 1 (Human; A Brief History), Sec. 2.1 (Origin of disciplines), Sec. 3.3 (Scimat), and Sec. 8.1 (Defining science).

Here is a guide to the contents, for those who want to know before reading the book. Otherwise, the following can be skipped or read after finishing a chapter.

Chapters 1-13 are short essays without references (with rare exceptions).

Chap. 1: Human. Secs. 1.1-1.3 contains three basic scientific information about where human came from and where to go after death. The atoms that made up any human body were generated after the big bang 13.7 billion years ago and will spread to anywhere on Earth after death, which could be absorbed by other living beings. Furthermore, all humans are descendants of fish. Therefore, all humans, dead or living, are relatives that share the same family tree. Sec. 1.4 explains why the West and China are different in culture today after sharing many similarities 2,600 years ago.

Chap. 2: Academic Disciplines. Sec. 2.1 explains how the academic disciplines evolved from a single discipline called Philosophy since the ancient Greek time. Sec. 2.2 introduces the Knowscape, a pictorial way to view all the knowledge in the universe, including those discovered or not yet discovered. Secs. 2.3 and 2.4 are brief introduction to complex systems and basic probability concepts, respectively. Sec. 2.5 provides a useful classification of all disciplines according to simple/complex and deterministic/probabilistic systems. That mathematics is *not* part of science is also explained here.

Chap. 3: Scimat. Sec. 3.1 gives a brief history of the search for a science of human after human settled down to form villages 10,000 years ago. Sec. 3.2 shows Alfred Nobel's dream of achieving world peace. Sec. 3.3 introduces Scimat, a new multidiscipline proposed in 2007, which is the present effort to develop a science of human and achieve world peace. Secs. 3.4 and 3.5 are two interesting consequences that follow from the definition of Scimat.

Chap. 4: On Humanities. The split of the humanities and (natural) science happened about 200 years ago, due to specialization through the introduction of graduate schools, hindering the progress in the humanities—the most important knowledge affecting human's future. To patch up this deficiency, humanities-science synthesis was proposed and general education was implemented in 1918 by Cai Yuan-Pei, the president of Peking University, ahead of others in the world, as discussed in Secs. 4.1 and 4.2. Sec 4.3 shows the three research levels in any discipline, which is well known in physics but less known in the humanities. Sec. 4.4 highlights examples of the humanities-science approach in the humanities. Sec. 4.5 points out what George Sarton failed to do as the father of history of science, and its dire consequences, followed by a new direction in science history in Sec. 4.6.

Chap. 5: History. That history is part of science is explained in Sec. 5.1. Two quantitative laws in history are described in Secs. 5.2 and 5.3. The invention of the new discipline Histophysics (physics of history) in 2002 is recalled in Sec. 5.4. Secs. 5.5 and 5.6 are about the characteristics of modern civilizations.

Chap. 6: Philosophy. The content of Philosophy shrinks since ancient Greek time, as pointed out and emphasized in Sec. 6.1. Secs. 6.2 and 6.3 compare the Western and Chinese philosophies and explain why they are so different from each other, with the first Chinese philosopher Guanzi introduced in Sec. 6.2. Sec. 6.4 gives a brief description of Confucious' life and career, and compares the schools set up by Confucious and Plato while Sec. 6.5 explains why there was Confucious in the first place. Sec. 6.6 introduces the most distinctive Chinese philosopher Mozi, also a pioneering scientist and engineer—and a man of action. Sec. 6.7 explains why most of Kant's philosophy are wrong. The relationship of philosophy and science and philosophy's future are discussed in Secs. 6.8 and 6.9, respectively.

Chap. 7: Art. A viable answer to the origin and nature of art is given in Secs. 7.1 and 7.2 while the relationship between art and science is given in Sec. 7.3. The three important artists, Su Dong-Po, Cézanne, and Da Vinci, are covered in Secs. 7.4-7.6. Secs. 7.7 and 7.8 contain insights on art history and art's relationship to aesthetics. Secs. 7.9 and 7.10 present the work of two gifted Chinese women artists: Li Qing-Zhao, the Song dynasty's poet, and the contemporary sculptor Luo Li-Rong.

Chap. 8: Science Basics. Science is universally understood to be human's effort to understand nature *without* supernatural considerations. Sec. 8.1 traces this modern definition of Science to 1867 and explains how science and scientists before 1867 are identified. Confusion about the definition of science persists today among experts and laypeople alike, leading to many unnecessary debates. Secs. 8.2-8.5 discusses some common misunderstandings of science. In particular, the Reality Check in science, often ignored by historians and philosophers of science, is discussed in Sec. 8.3.

Chap. 9: Scientific Confusion. Secs. 9.1 and 9.2 present the world's *first* scientific theory—Guanzi's "all things originate from water." Sec. 9.3 explains why Einstein's letter of 1953—often invoked to explain why *modern* science did not appear in China (the Needham Question)—is misunderstood. Sec. 9.4 discusses why traditional Chinese medicine is a science, contrary to the view held by many people. Finally, in Sec. 9.5, we are able to give a new answer to the famous Needham Question. Secs. 9.6 and 9.7 present the new discipline *Sciphilogy*—the human science of philosophers of science, and explain why science philosophers like Mach, Popper, and Kuhn could be so wrong in their works.

Chap. 10: On Science. Secs. 10.1-10.7 present further discussion on the nature of science, covering science and religion, pseudoscience, antiscience, folk science, and science's relationship with technology. Sec. 10.8 points out when and how people should trust science and scientists, including the example of climate change. The important characteristics of science are summarized in Sec. 10.9.

Chap. 11: Scientist. Scientists are humans. The human side of Newton, Einstein, and Gödel are presented in Secs. 10.1.10.3. Rabi and Oppenheimer played pivotal roles in the development of modern physics in the United States while Yan Ji-Ci, the one in China. The three all went to Europe to study. Their stories are given in Secs. 11.4 and 11.5. Secs. 11.6-11.9 provide perspectives on Madame Wu's work on parity nonconservation and the reasons why she did not win the Nobel Prize. Human story of the famous Feynman is given in Sec. 11.10, followed by that of three less known but accomplished theoretical physicists, Platzman, Hohenberg, and Lax (with them I interacted personally), in Secs. 11.11-11.13. In particular, the three's contact with China are told here for the first time. Similarly, Anderson's story is presented in Sec. 11.14. Sec. 11.15 describes the history of a group of Chinese scientists who returned to settle and work in their motherland around 1980—the so-called China's second generation of returnees. Secs. 11.16-11.18 are observations on the roles of scientists.

Chap. 12: Popsci. Secs. 12.1-12.5 present some insights on popular science (popsci). In particular, a brief introduction to the most successful folk scientist Ovshinsky in given in Sec. 12.3. The two popsci books correcting some science myths in Sec. 12.4 are highly recommended. Sec. 12.6 is about my involvement in getting Shermer's popsci book translated and published in China while Sec. 12.7 is my personal journey in doing popsci, pretty interesting and hopefully will inspire others to do the same. Secs. 12.8 and 12.9 is the proposal to do popsci based on martial arts. The importance of science appreciation (instead of science explanation) in popsci, often overlooked by the professionals, is pointed out and emphasized in Sec. 12.10.

Chap. 13: Personal. Secs. 13.1-13.4 present examples of how innovative research in science come about, drawing from my own experience of working in liquid crystals and complex systems. The key lies in picking topics by not following others. Sec. 13.5 is the firsthand story of how the Chinese Liquid Crystal Society (1980) and the International Liquid Crystal Society (1990) were established. Sec. 13.6 tells how I came up with the important idea that art's origin and nature are linked to killing time (see Secs. 7.1 and 7.2)—my favorite story.

Chapters 14-23 are long articles with references.

Chaps. 14-16 are about *Histophysics* (physics of history), a new discipline I proposed in 2002. Chap. 14 tells the birth and subsequent development of Histophysics. My paper on self-organization (2000) that leads to the Histophysics idea is reproduced in Chap. 15 while the content of Histophysics is introduced in Chap. 16.

Chaps. 17-18 are about *Scimat*, the new multidiscipline I introduced in 2007. Chap. 17 traces the development of Scimat and the international scimat program. My paper on the two cultures (2006) provides the foundation of scimat and is reproduced in Chap. 18.

Chap. 19 is a piece of firsthand science history, an account of China's *first* non-government visiting scholars traveling in 1979 from Beijing to Evanston, Illinois, USA, and their two-year stay at Northwestern University there. The delegation was from the Institute of Physics, Chinese Academy of Sciences. The visit changed the academic life of the delegates involved.

Chap. 20 is my paper on what science-and-art message to send to the aliens, published in *Leonardo* (2004). It is a work exemplifying the scimat approach of using humanities-science synthesis in humanities research. It is my favorite short paper.

Chaps. 21-23 (together with Chap. 16) is the slightly updated version of my first popsci book *This Pale Blue Dot* (2004). It is based on my three talks given at Tamkang University, covering Histophysics, complex systems, and science and religion.

PART I

SCIMAT AND HUMANITIES

1

Human: A Brief History

Big bang

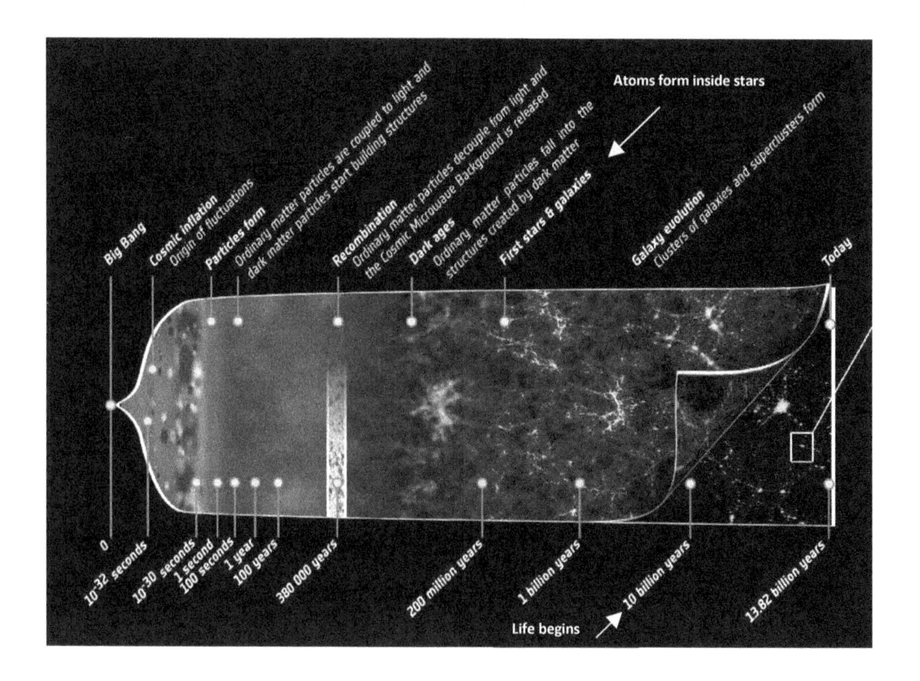

Everything in the universe originated from the big bang 13.7 billion years ago. Afterwards, atoms were formed, and part of it coalesced into the solar system 4.7 billion years ago, one of which was Earth. One billion years later, life appeared on Earth.

The theoretical framework of the big bang is based on Einstein's theory of general relativity, supported by two important astronomical observations:

1. In 1929, Edwin Hubble (1889-1953) found through observations that all distant galaxies and galaxy clusters are moving away from us, and the farther the distance, the greater the speed, thus deriving the view of the expansion of the universe.

2. The cosmic microwave background radiation discovered by Bell Labs in 1964 (Nobel Prize in 1978).

The origin of the big bang is currently a frontier topic in physics.

It is a *confirmed* scientific fact:

> All things **originated from the big bang**.

This is the *first* basic knowledge that everyone should know.

1.2 Descendants of fish

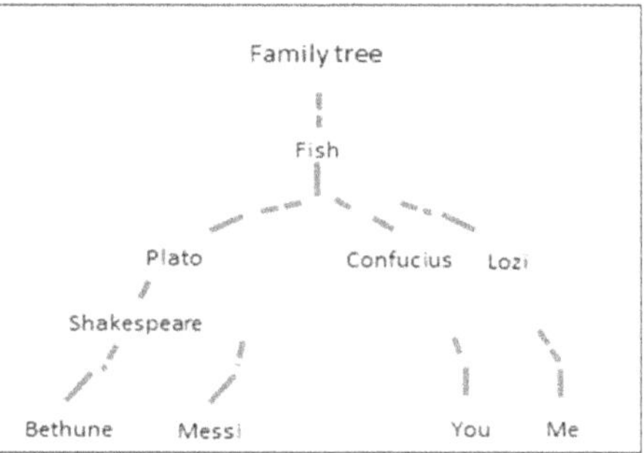

Life appeared on Earth 3,700 million years ago. We don't know how life began. It could come from outer space or originate on Earth itself. (The remote possibility of Aliens brought it to Earth is not ruled out by science.) However, through fossil and DNA research, we know roughly the evolutionary history of life after its emergence. The guiding theory behind its research is Darwin's *On the Origin of Species by Means of Natural Selection*, published in 1859 by Charles Darwin (1809-1882).

After the beginning, life generally evolves from simple to complex. About 375 million years ago, the first fish climbed onto land; the front fins of this fish, called Tiktaalik, had enough strength to do push-ups and get themselves out of the water. Our arms, legs, necks, and lungs can be traced back to fish (supported by DNA tracking). And so, everyone in the world is a descendant of fish. (The book/video of *Your Inner Fish* won the 2009 National Academy of Sciences Award and the 2014 AAAS Award.)

We even know who our fish *ancestor* was (before any fish went ashore): a fish called *Microbrachius*. Specimens of Microbrachius have been found in Scotland, Belarus, Estonia, and China. This fish is 8 cm long and lived 400 million years ago. It is divided into male and female according to its own sexual organs. Their method of egg fertilization is like that of humans, not like that of the current fish (spraying eggs in water) [*Nature* **517**, 196-199 (2015), https://doi.org/10.1038/nature 13825.]

Later, after a long evolution, humans and chimpanzees appeared, both descendants of fish. The two separated 6 million years ago. Humans have undergone many evolutionary stages such as *Homo erectus*, and finally evolved into *Homo sapiens* (modern humans) about 195,000 years ago.

In other words, everyone in the world, ancient and modern, is a family, either distant relatives or close relatives, sharing a family tree, and the common ancestor is fish. There is a Chinese saying: Blood is thicker than water. Thus, be kind to anyone in the world.

It is a *confirmed* scientific fact:

We are all **descendants of fish**.

This is the *second* basic knowledge that everyone should know.

1.3 Recycled stardust

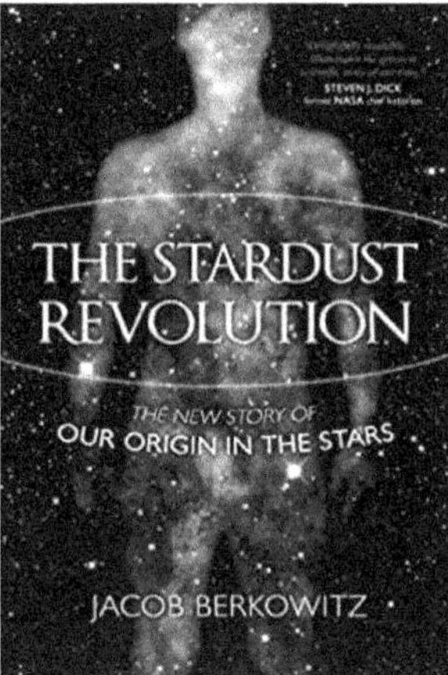

All material systems on Earth, including humans, are composed of atoms. Where did the atoms come from?

Light atoms such as hydrogen (H), helium (He), and lithium (Li) were formed shortly after the big bang. Heavier atoms all come from stars, but silver (Ag) and gold (Au) are created when neutron stars merge. When a star's internal atomic furnace goes out, it collapses and then explodes, causing the atoms inside to scatter into space.

The atoms floating in the cold dark space are quite lonely. A few of them meet by chance, attracted to each other due to gravitational force, and form a big family of the solar system, one member of which is called Earth. According to this, the atoms on Earth all come from outer space, and almost all of them come from the stars. Therefore, our bodies are all *stardust*.

Except for the occasional new atoms brought in by meteorites, the overall number of atoms on Earth is limited and unchanged. So, *every* atom in *any* living person is *recycled* from somewhere else, possibly from someone else's body, dead or alive. You never know.

In other words, any two persons on earth, at a material if not spiritual level, could be *you in me* and *me in you*, whether you like it or not. Treating other people's body kindly is treating your own body kindly.

Therefore, all people in ancient and modern times, in any country or continent, are not only related by blood (being the descendants of fish) but also related to each other in terms of bodily material because we are all recycled stardust.

It is worth noting that the above-mentioned two conclusions about our body—"descendants of fish" and "recycled stardust"—are scientific *facts* that can only be established after the appearance of Darwin's evolutionary theory (1859) and Einstein's (indirect) proof of the existence of atoms (through his theory of Brownian motion, 1905). The consequence is that all human-related research written 160 years ago, including the works of science history and science philosophy, must be read with care. Of course, one should also be careful when reading books written in the last 160 years which ignore these two scientific facts.

It is a *confirmed* scientific fact:

Our bodies are **recycled stardust**.

This is the *third* basic knowledge that everyone should know.

Summary. Three basic scientific facts related to humans:

- Historical: All things originated from the big bang.
- Biological: We are all descendants of fish.
- Physical: Our bodies are recycled stardust.

Lesson. The recorded human history is only a few thousand years long, which, when put in the background of billions years of the universe, is less than a flash of time. All matters, dividual or national, should be considered in this perspective. Live peacefully with others who shared this Earth. And be forgiving and humble.

1.4 The world today

Years ago	Evolution	Migration	Lifestyle	Art related
6 million	Human and chimpanzee lineages split			
3.5-1.8 million			First hominids move from forest to savannah; meat eating begins	
2 million	*Homo erectus* appears; brain enlarged; mimesis capabilities			
1.8 million		First wave of migration out of Africa begins		
1.6 million			First use of fire; more complex stone tools created; art could begin	
400,000			Earliest evidence of cooking	
195,000	*Homo sapiens* (early modern humans) appears.			
150,000				Language begins
60,000		Second wave of migration out of Africa		
50,000				Cultural revolution: ritualistic burials, clothes-making, invention of complex hunting techniques
35,000				Cave art (in France, Spain)
10,000			Agriculture begins; first villages appear	
5,500				Bronze Age begins
5,000				Earliest known writing

Out of Africa

After the human and chimpanzee linages split from each other 6 million years ago humans developed through several stages of evolution. Two million years ago *Homo erectus* appeared. Modern humans, called *Homo sapiens*, appeared 195,000 years ago. In between and afterwards, due to climate change in Africa, Africans migrated to different parts of Africa but also went to the Middle East through the northeastern link between the two continents—Africa and (western) Asia. This happened two times at least: The first time was 1.8 million years ago and the second, 60,000 years ago. Later, some of them migrated west to Europe and some towards east, to Asia and then the Americans.

Common Ancestor

In fact, like it or not, by chance, all humans, dead or alive, shared the same ancestor who originated in Africa, and before that, the fish called *Microbrachius* [1.2]. The DNA in all humans is 99.9% identical. An important consequence is that humans are pretty homologous, bearing almost identical genes and having similar intelligence: No race is more intelligent than the rest, on average. The skin color is due to how much sunlight there are and is controlled by only three genes.

It was merely 10,000 years ago that humans settled down to form villages, resulting in the emergence of civilizations of different degrees of maturity. Note that the earliest-known writing, essential to recorded history, was about 5,000 years ago. Thus, accurate human history is extremely short compared to the universe's 13.7 billion-year history starting from the big bang.

The World Today: What Happened

How did the world—the two major ones, West and East, say—end up like this, as we know it today? As it turned out, it happened through three similarities between the history of ancient China and ancient Greece, which occurred at about the same time:

1. In China, the Qin dynasty ended the Warring States in 221 BC; in Europe, Rome destroyed the Athenian city-state in 146 BC (ending the philosophical classical period in both China and Europe).
2. In China, the Qin dynasty unified the central plains in 221 BC; in Europe, Rome unified a large part of Europe in 117 AD (resulting in a scarcity of new ideas/ideologies for a long time, in both East and West).
3. In China, the Han dynasty worshiped exclusively *Confucianism* in 134 AD; in Europe, Rome recognized *Christianity* in 313 AD and established it as the state religion 10 years later.

However, Christianity and Confucianism are two very different systems of thought. Christianity is adaptive or forced to be adaptive in later history, which emphases rule of law and allows rational thinking as long as the basic premise of God's existence is not challenged, ushering in *modern* science 400 years ago. In contrast, Confucianism remains rigid, emphasizing absolute obedience and discourage innovations. And so, here we are today.

Two paths, two thousand years, two results.

2

Academic Disciplines

2.1 Origin of disciplines

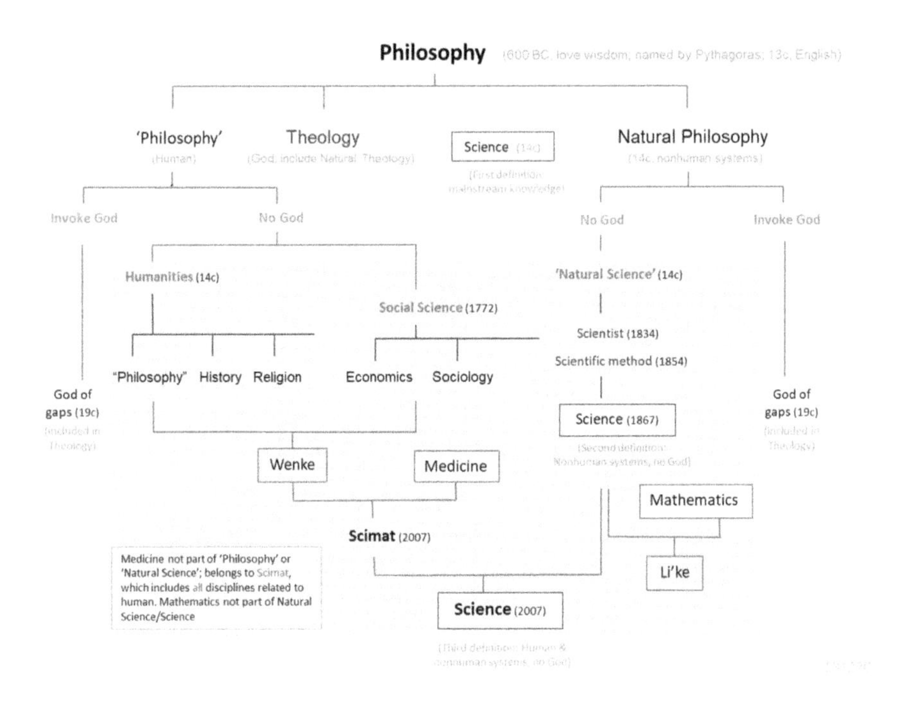

2,600 years ago, Thales (c.624-c.546 BC) of ancient Greece broke away from Greek mythology and started Philosophy. Philosophy was the study of *everything* in the universe, including human and nonhuman systems, and that was the *only* discipline at that time.

In addition, the Zhou dynasty (1046-256 BC) in China had *Six Arts* (rites, music, shooting, royalty, calligraphy, and numeracy), and Europe from ancient Greece to the Middle Ages, there were *Seven Arts* (grammar, logic, rhetoric, arithmetic, geometry, music, astronomy; or, called liberal-arts education). While these are related to modern disciplines, they are *not* the direct source of the latter.

Historical Development

An important turning point in the classification of disciplines occurred in the 14th century when ancient Greek Philosophy was divided into three parts: 'Philosophy' (single quotes), Natural Philosophy, and Theology. 'Philosophy' is the study of the *human* system; Natural philosophy is the study of *nonhuman* systems; Theology is the study of everything related to God under the premise of God's (then supernatural) existence. The content of 'Philosophy' and Natural Philosophy contains two parts: "no God" and "invoke God."

Subsequently, the no-God part of 'philosophy' was divided into the *humanities* and *social science*, and the no-God part of natural philosophy (also known as "*natural science*") was formally defined as Science in 1867 [8.1]. The invoke-God parts in 'philosophy' and natural philosophy are incorporated into theology. The humanities contain the "philosophy" (double quotes)—the kind of philosophy currently found in universities' philosophy department [6.1].

In Chinese vocabulary there are two terms, *Wenke* (文科) and *Li'ke* (理科), that have no equivalents in English. Specifically, Li'ke = Humanities + Social Science; Wenke = (nonhuman-systems) Natural Science + Mathematics. So, Li'ke = no-God part of 'Philosophy' [2.5])

However, Wenke and Li'ke do not include medicine: an important discipline on human physiology and health, which obviously is part of science. Hence entered *Scimat* in 2007: Scimat = Humanities + Social Science + Medicine. These three disciplines are all research on human, and from the perspective of disciplinary development, there is a need for mutual support and penetration.

Also, the second definition of Science in 1867 only includes nonhuman systems, which is no longer in line with the current understanding of nature (e.g., human is a kind of animal [1]). Thus, there is a third definition of Science in 2007, the Scimat definition: *Science* = Scimat + Science of nonhuman systems [8.1, 3.3]. According to this, Wenke and Li'ke (except Mathematics) are part of science. The reason that Mathematics is *not* part of science is explained in [2.5].

Conclusion

The classification of disciplines is for administrative convenience, not based on the nature of knowledge or the disciplines. All disciplines have the same purpose: to understand the workings and laws of nature. They are split aways from the single discipline of ancient Greek's Philosophy.

All disciplines complement each other, need to be mutually supportive, and cross discipline in their development. In particular, the synthesis of humanities and "natural science" is conducive to the development of both [4.1].

Disciplinary separation/classification is historical, not a top-down design, and a bit annoying. But not paying attention to these details can lead to communication difficulties and even misrepresentations [6.1].

2.2 Knowscape

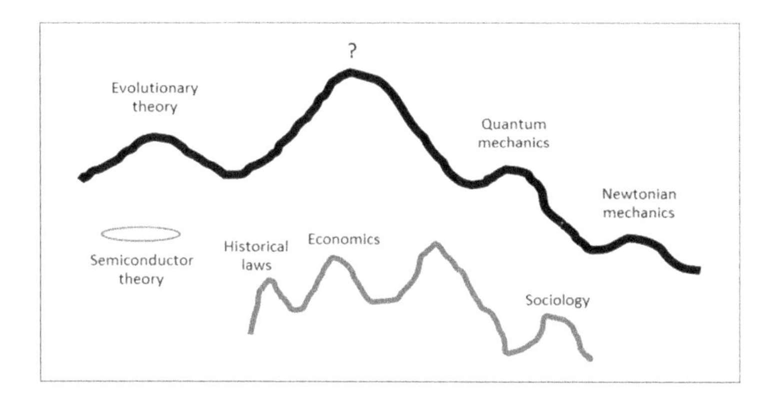

Knowscape is a new word I coined in 2014, combining the words knowledge and landscape (see Chapter 1 of my 2014 book *All About Science*).

Knowscape is an abstract landscape in which all (known and unknown) knowledge are distributed. Doing research, the search for knowledge, is to do exploration on the surface of the knowscape. An important discovery is to climb up to a peak in the knowscape that no one has ever reached before. Innovation is to climb up a rather high peak.

Knowscape as the landscape of knowledge is a metaphor. As a metaphor, it is far from perfect and has its limitations—not everything in the figure should be taken literally. There are hills/mountains, valleys and plateaus in the landscape, interconnected in a vast terrain, and, in addition, some *man-made* lakes. Each summit represents a bright spot in human knowledge. The height of the mountain corresponds to the difficulty of reaching it. Furthermore, just like in real mountaineering, an explorer usually has to pass through the lower hills before they can reach the higher ones.

However, there are two categories of mountains on the knowscape:

1. One category is human-*independent*. That is, if there were no humans in the universe, it would still exist there (upper curve in the figure). In other words, they could even be found by nonhumans—aliens, say.
2. The other category is human-*dependent*. This category includes knowledge of the humanities, social science, and medicine (lower curve). There are also man-made nonhuman systems (such as semiconductors, computers, artificial life), which also depend on human existence and are represented by isolated lakes (ellipses) in the knowscape.

Conclusion

Researchers in *all* disciplines are explorers and adventurers of the knowscape.

To be a first-rate scholar or a master in research is to aim at and try to find the peaks in the knowscape to climb. Within one's own capacity, the higher the peak, the better.

2.3 Complex system

Following the success of nonlinear science and chaos research in the 1970s, the importance of complex systems as a *multidiscipline* was recognized in the early 1980s (exemplified, for example, by the establishment of the Santa Fe Institute in 1984). The application of complex systems to different disciplines, spanning the humanities to social science to natural science, is therefore noteworthy.

Due to the difficulty of defining *complexity*, there is no strict definition of complex systems [20]. In general, a *complex system* usually consists of many interacting components, which can be of the same type or of different types. Each component can have several or more internal states and is adaptive in its behavior.

The *weakness* of this definition is obvious. For example, a system may seem complex simply because we do not yet understand it. Once understood, it becomes a *simple* system. In addition, whether a system is complex or not may depend on which aspect of it we want to study. If we want to know the internal structure and formation mechanisms of a rock, the rock may be a complex system. But if we just want to know how a rock moves when kicked, then using Newtonian mechanics will do; in this case, the rock is just a simple system. The lack of a strict definition and the ambiguity of simple/complex concepts make this definition unsuitable for use as a demarcation tool for anything.

In the absence of a strict definition, a simple *working* definition can be adopted: With the exception of traditional courses in physics, chemistry, and engineering departments, almost all disciplines covered at university fall into the field of complex systems. In other words, biological systems, including humans, and all topics covered by the humanities and social science fall under the category of complex systems. Most of the rest are simple systems.

At present, there is no basic theory that unifies complex systems. As a science, complex systems are more like a banner, bringing together scientists who do not want to do simple systems. But in individual subfields (such as chaos and network systems), there are some useful theories.

Discussion

1. A *single* human being is the most complex system in the whole universe. There is some saying that females are more complex than males but there is no scientific data to support this claim.

2. Complex systems may also include nonliving systems (e.g., climate change [10.8]).

3. “Simple” does not mean “easy”—these are two very different concepts.

4. Theoretical methods for studying simple systems have been and will also be used in complex systems, but the latter is too complex after all, and many times has to rely on computers to simulate simplified models. Thus, the study of complex systems did not flourish until the popularization of personal computers in the 1980s.

5. The simulation tools in complex systems falls into two broad categories: active walk and agent-based. *Active walk* was invented by Lui Lam in 1992: Each particle (the walker) operates in a classical continuum field representing the environment, where the particle interacts with the environment and interacts indirectly with each other through the shared environment (not excluding direct interaction between the particles). *Agent-based* generally allows the particles to interact directly with each other. Both have a wide range of applications, especially in the humanities and social science. In fact, agent-based, a latecomer, is just a kind of active walks with a different name.

2.4 Probabilistic life

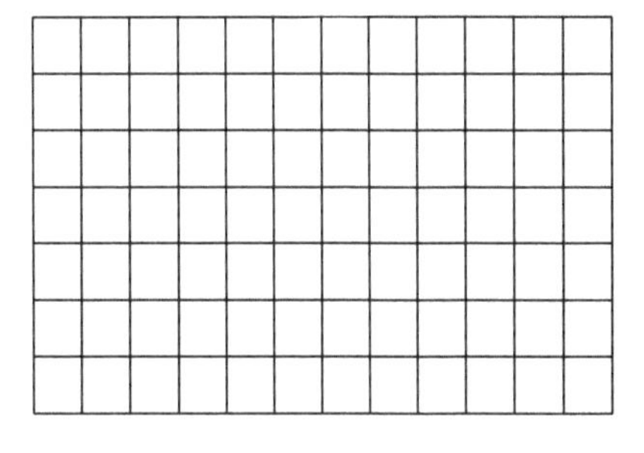

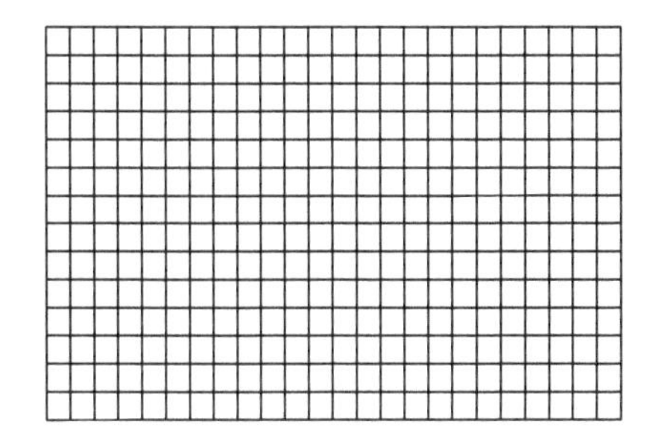

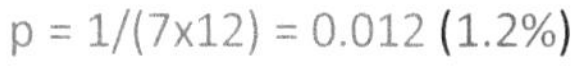

p = 1/(7x12) = 0.012 (1.2%)

p = 1/(7x2x12x2) = 0.003 (0.3%)

Every time we cross the street, there is a danger of being hit by a car, although the probability is low, but not zero. In San Francisco's Chinatown, there are several cases of elderly people being hit and killed by cars when crossing the street every year. Similarly, driving is probabilistic. No matter how careful you are, there is always a chance that you will be hit by a lousy driver. Therefore, every day when you go home safely, you should thank God for blessing, or thank yourself for being careful and lucky.

The human world is a probabilistic world: Nothing can be 100% certain in advance. For example, you cannot be sure that the person you call dad every day is really your dad, because DNA identification is only 99.9% accurate to determine that he is your dad (100% accuracy to exclude dad though). It is a pity that such an important knowledge as probability is usually not taught in school, and when it is taught, not teach well enough—concentrating too much in the mathematical details, say.

One of the consequences is that many people think that something with a very low probability (called "black swan") will not happen, and they are not mentally prepared (see Remarks). Another consequence is that university teaching is so focused on deterministic systems that engineering graduates come out to work and realize that what they learn is often not useful, because the actual problems are probabilistic (such as climate change). For how to avoid bad things in the probabilistic world, see [10.8].

Therefore, when your mother quietly tells you one day that your father is someone else, you should be mentally prepared and calmly say: Mom, it is okay. Then just go walk the dog, or do whatever waiting to be done.

In life, in addition to being *humble*, should also be *calm*.

Remarks

1. The novel about probabilistic life is *The Plague*, by Albert Camus (1913-1960, Nobel Prize in Literature 1957).

2. Probabilistic is important and universal. Accordingly, university courses should include content of *probabilistic* systems in addition to the emphasis on deterministic systems. For example, *physics* courses should include an introduction to *Random Walk*. And popular science should be more cooperative by popularizing knowledge and content about probability.

In fact, random walk is not only the simplest probabilistic system, but also plays an extremely important role in the history of science:

- Einstein's 1905 theory to explain Brownian motion (and thus indirectly prove the existence of atoms) is random walk.
- Five years earlier, in 1900, Louis Bachelier (1870-1946), a doctoral student of the French mathematician Henri Poincaré (1854-1912), first used random walk to explain the rise and fall of stocks.

3. No matter how low the probability of an event, it could actually happen. The explanation is as follows: Suppose a single (zero-size) raindrop will fall on the graph paper (figure, left); p = probability that a cell will be hit. Now by subdividing the cells, p can become super small (figure, right). However, there is always a small cell that will be hit by the raindrop. See?

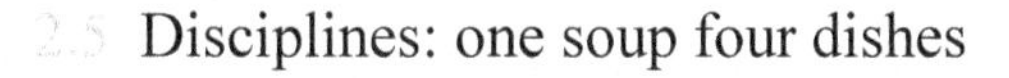

2.5 Disciplines: one soup four dishes

The object of scientific research is all things in nature. According to the two properties of simple/complex, deterministic/probabilistic, they can be divided into four classes. As shown in the Table (with examples), the four categories of A, B, C and D belong to "Natural Science" (*li'ke*), and the human system belong to Liberal Arts (*wenke*). The ancient Greeks (such as Aristotle) did not distinguish between these four categories and treated them equally. Note that the beginning of modern science 400 years ago came from Galileo's breakthrough in the study of class A.

	Simple	Complex
Deterministic	A Free fall, projectile path	C Chaos (three-body problem), fractal
Probabilistic	B Random walk (Brownian motion), elementary particle (probabilistic because of quantum mechanics)	D Part of "Natural Science" (meteorology, climate change, nonhuman biological systems) Liberal Arts (Humanities, Social Science)

Class A is the simplest, and the use of mathematics is relatively easy and productive, so much so that Galileo (1564-1642) claimed that the laws of nature are mathematical. In fact, Galileo merely had experience in class A, and he was referring to situations in class A only (which he himself may not have realized). Unfortunately, later generations mistook that he was referring to all four classes. Whether all four classes of systems can eventually be mathematicized cannot be predicted; the answer is "yes" only when the mathematization is carried out successfully. Naturally, mathematization as an ultimate goal is quite fascinating. (See Mitchell Waldrop's 1992 popular science book *Complexity*.)

The current situation is that mathematics works well for simple systems (classes A and B) but is ineffective or not useful for complex systems. For example, chaos in class C was popular about 50 years ago, mainly because of computers, and fractals are not even mathematically possible (although some people have tried). As for meteorology and climate research in class D, there are equations but too numerous to solve by hand, and can only be calculated by supercomputers. Mathematics is rare in biology. In social science, only economics uses more mathematics. In other disciplines and the humanities, mathematics is not common. Complex systems are indeed complex.

To understand the relationship between science and mathematics let us look at the very nature of mathematics as it evolved in history.

Mathematics

Early mathematics was connected to reality. For example, numbers were invented because people had to count apples. After the abstraction and axiomatization of modern mathematics, mathematics is not necessarily related to reality, even though axioms are sometimes inspired by facts in the real world (such as Riemannian geometry). Mathematics is *not* part of science, because science is the study of nature, and mathematics is the study of abstract axiomatic systems, not part of nature.

On the other hand, some mathematics can be used in physics and other disciplines, becoming one of the tools of science. The reason is that mathematicians start from a few basic axioms to see what system can be deduced. Later, when there are many such different mathematical systems, it is not surprising that a few of them are useful in science.

Because of the success of mathematics in class A in science, people, by generalization, overestimate the necessity of mathematics. In fact, in physics, neither Newton's third law of motion nor the third law of thermodynamics can be expressed mathematically, but only in words. Similarly, evolutionary theory has no mathematics, but it liberates man from the dogma that God creates human—a great and influential achievement.

Mathematics and Science

Mathematics is not part of science. What then is the relationship between mathematics and science?

Answer: It is "one soup and four dishes." Mathematics is *soup*, and the four *dishes* are the four classes of science in the Table. Eating, it is best to have soup, but not necessarily.

Conclusion

- Science includes simple systems and complex systems.
- Mathematics is only a signature of simple systems, not a signature of complex systems.
- Moreover, basically, mathematics is not part of science and definitely not a signature of science.

3
Scimat

Science of human

Everything in the universe originated with the big bang 13.7 billion years ago. Later, atoms formed, part of which coalesced into the solar system 4.7 billion years ago, one of which is *Earth*. After 1 billion years, *life* appeared on Earth and roughly evolved from simple to complex. About 375 million years ago, the first fish climbed onto land and evolved into a variety of animals, including *human*. So, everyone in the world is a descendant of fish. After human and chimpanzee separated from each other 6 million years ago, human underwent several evolutions such as *Homo erectus*, and finally evolved into modern human (*Homo sapiens*) 195,000 years ago. Then 10,000 years ago, humans settled down to form villages, and with writing invented 5,000 years ago, written history began.

Human's Interest in Understanding Human

One of the factors that humans have survived for 6 million years is that humans are curious animals. Initially the curiosity was about the surroundings of the environment: Does the small grass movements indicate that some beast is nearby? Or is it just the wind blowing the grass? Being able to identify situations allowed our ancestors to sleep well and live a few more years, which was conducive to the survival of the fittest in harsh environments.

After *Homo sapiens* settled down to farm 10,000 years ago, the population density increased greatly compared to before, and human relationships suddenly became important. Understanding neighbors and fellow villagers can help reduce friction and increase community stability. In addition, an understanding of the relationship between human and nature contributes to the development of agricultural economies. All this understanding, in addition to helping to live practically, is driven by people's *curiosity* about everything.

Ancient Research on Human

The recorded history of humankind is only about 5,000 years, and the academic research on human is even shorter, about 3,000 years. In ancient China, Guanzi and Confucius had a lot of discussion on political philosophy, and so did Mozi. But Mozi was also an innovator of science and technology, which was unique among ancient Chinese philosophers [6.6]. All these philosophers were interested in personal cultivation, governance, social system, etc., but Zhuangzi had written more on the relationship between human and nature. All of this is academic research on human.

In ancient Greece, academic research on human was representative of Plato and Aristotle. But Aristotle was not only interested in human, but also studied nonhuman systems (such as plants and nonhuman animals). In Europe, the focus of philosophers of different eras were also mainly human beings.

Scientific Research on Human

The premise of *science* as we currently understand it is that neither data collection and analysis nor theoretical interpretation can involve God/ supernatural, i.e., neither the process nor the results of science can contain the supernatural. Where does this concept that people take for granted come from? In fact, it came from the *second* definition of the word Science in 1867, which is to strip away the "invoke-God" part of

Natural Philosophy at that time, for a better understanding of nature. Specifically, the 1867 definition of science says: Science is the study of *nonhuman* systems in nature without introducing God/supernatural considerations [8.1].

The restriction of nonhuman systems was removed in 2007 because it was no longer in line with later scientific understanding (e.g., evolutionary theory that humans are a type of animal). Specially, the *third* definition, the 2007 *Scimat* definition of science says: Science is the study of *all* systems in nature without introducing God/supernatural considerations. Here, all systems mean both human and nonhuman systems.

Before the word science appeared, it does not mean that there was no scientific research, just as there were humans before the word human. Therefore, retrospectively, in Newton's book *Principia* published in 1687, the "no-God" part in it was posthumously considered science.

The following year, inspired by Newton's *Principia*, the 101-year long *Enlightenment* (1688-1789) swept through Europe, with the goal of imitating Newtonian mechanics and creating a "Science of Man," but failed. Why did it fail? They did not notice, or probably did not know, that human things are *probabilistic* complex systems, and that Newtonian mechanics is successful in dealing with deterministic simple systems. A wrong understanding of the nature of the system, coupled with the wrong use of tools, can only end in failure.

But the Enlightenment did not fail completely, because after all, the first discipline of social science appeared: *Economics*—founded in 1776 by the Scotsman Adam Smith (1723-1790). The Enlightenment was the first large-scale movement in history to pursue the *science* of human, and its characteristics were premised on the freedom from religion/supernatural, a category that none of the previous efforts of Plato and Aristotle belonged.

Forty-nine years after the Enlightenment, the French philosopher Auguste **Comte** (1798-1857) coined the term sociology in 1838, opening the second discipline of social science: *Sociology*. Six years later, in

1844, he published *On the Spirit of Positivism*, arguing that there were three stages in the development of human speculation: theology, metaphysics, and positivism. He emphasized that human society, like physics, can be understood through rational research. Although Comte's theory was wrong, it was a *milestone* in the science of human after the Enlightenment, influencing the development of this area in the second half of the 19th century and a large number of people, including Karl **Marx** (1818-1883).

Marx's "Science of Man"

Comte was 20 years older than Marx and lived in the late stages of the Industrial Revolution (1760-1820/40) while Marx grew up at the end of the Industrial Revolution and beyond, which influenced their different views and ideas on social reform. Moreover, Marx lived at a time that spanned almost the entire 19th century, a period when modern science advanced by leaps and bounds.

In 1844, at the age of 26, Marx produced a manuscript on the relationship between natural science and human science, and finally concluded that human science and natural science would merge into a *whole* science. (See Marx's *Economic and Philosophic Manuscripts of 1844*.)

Unfortunately, Marx's wish seems to have been forgotten, at least not taken seriously, and the multidiscipline *Scimat* that emerged in 2007 is precisely towards this goal.

Scimat

Scimat is the scientific study of humans. In terms of content, scimat is a collective term for the combination of humanities, social science, and medicine. Scimat suggests that when studying human problems, one can and should consult the scientific findings from these three disciplines. This research method or approach usually points to the synthesis of the humanities and natural science.

Scimat points out that both the humanities and natural science are the study of systems in nature, with different focus on the human system and nonhuman systems, respectively. Since all nature's systems are composed of atoms, the humanities and natural science have a lot in common. Thus, humanities-science synthesis is the most natural tool of research in these two fields. Furthermore, scimat provides the theoretical basis for humanities-science synthesis.

Scimat is the "science of human"—the goal that the Enlightenment wanted but did not achieve. Scimat advocates a rational discussion of any question about human from a scientific point of view and cognition—all questions, big or small [3.3].

Conclusion

People's interest in understanding humans began millions of years ago, born of curiosity and the need for survival. More than 2,000 years ago, philosophers studied humans, especially the Chinese.

More than 300 years ago, the European Enlightenment was the first large-scale movement aimed at creating a human science, followed by Comte and Marx who made important contributions. Marx even predicted that human science and natural science would merge into a whole science in the future.

Today, **Scimat**—the current science of human—is working towards this goal.

3.2 Nobel's dream

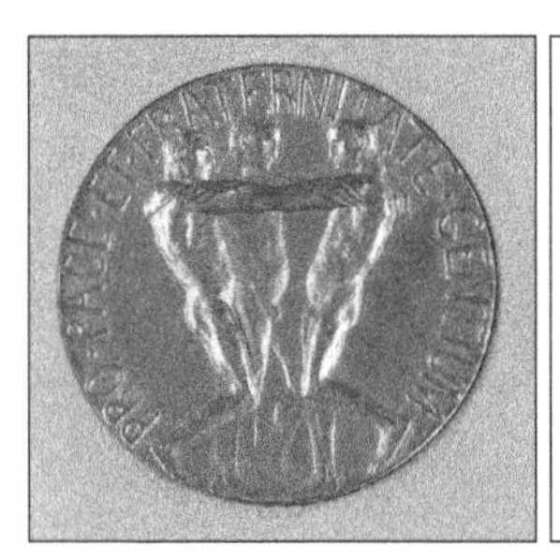

Alfred Nobel (1833-1896), 19th-century Swedish chemist and inventor, rich by improving gunpowder, never married, childless, lifelong advocate of pacifism. He made a will before his death:

> All of my remaining realisable assets are to be disbursed as follows: the capital, converted to safe securities by my executors, is to constitute a fund, the interest on which is to be distributed annually as prizes to those who, during the preceding year, have conferred the greatest benefit to humankind. The interest is to be divided into five equal parts and distributed as follows: one part to the person who made the most important discovery or invention in the field of physics; one part to the person who made the most important chemical discovery or improvement; one part to the person who made the most important discovery within the domain of physiology or medicine; one part to the person who, in the field of literature, produced the most outstanding work in an idealistic direction; and one part to the person who has done the most or best to advance fellowship among nations, the abolition or reduction of standing armies, and the establishment and promotion of peace congresses… It is my express wish that when awarding the prizes, no consideration be given to nationality, but that the prize be awarded to the worthiest person, whether or not they are Scandinavian.

According to this will, starting in 1901, five international Nobel Prizes were created: physics, chemistry, physiology or medicine, literature, and *peace*.

The Peace Prize

Nobel speaks six languages (Swedish, French, Russian, English, German, Italian) and has no formal secondary or higher education. He was a folk scientist.

He wrote poetry, but his literary talent was not high. When he was dying, his only play was printed, which was considered "slanderous and blasphemous"; almost all of them were destroyed after his death, with three copies survived. In 2003, the first surviving edition was published in Sweden.

Nobel was in love three times. His marriage proposal to the first one was rejected. His second love was with Bertha von Suttner (1843-1914). And the third was with an 18-year old through long-distance exchange of letters (at that time there was no Internet and the woman was far away in Vienna).

Born in Prague, Bertha was a novelist and radical pacifist who became Nobel's secretary in Paris in 1876. But she left shortly thereafter to marry someone else. Since then, the two have kept in contact. In 1889, two years after Paris' Eiffel Tower was built, Bertha made a name for herself with the book *Lay Down Your Arms*. Because of Bertha, Nobel set up the Peace Prize. Bertha deservedly took the 1905 one. The influence of women should not be underestimated. Thanks Bertha!

World Peace

World peace is a universal aspiration. There are many approaches. Each has its own shortcomings, and it is difficult to implement. Thus, the Nobel Peace Prize in previous years has been awarded to people or organizations, instead of a single person, that solve partially the peace problem.

By any measure, the Peace Prize that can *truly* solve global peace, should be the most difficult award among the Nobel Prizes and the greatest contribution to humankind.

Nobel's dream of world peace has yet to be realized.

Remarks

The award money offered by the Nobel Prize is not low, but it is not the highest in the world. Still, it remains the most prestigious award in the academic world.

The reason is that, apart from being the oldest award of its kind, the nomination and review processes are carried out carefully by four high-level professional organizations (Royal Swedish Academy of Sciences, Karolinska Institute, Swedish Academy, Nobel Committee of the Norwegian Parliament), a full-time operation throughout the year.

This is the secret of the Nobel Prize, which makes it invincible and cannot be surpassed.

3.3 Scimat

Scimat (science of human) is a new multidiscipline created by Lui Lam in 2007 [17].

The founding article "Science Matters: A unified perspective" was published in *Science Matters: Humanities as Complex Systems* (World Scientific, 2008), the first scimat book. The first Chinese article "Science Matters: The latest and largest interdiscipline" was published in *China Interdisciplinary Science* (Science Press, 2008). The Chinese translation of the first book is *Renke: Humanities as Complex Systems* (China Renmin University Press, 2013). Renke is pinyin of 人科 (human science). The term Science Matters emphasizes that the humanities/social science—the disciplines studying human—are part of science.

Scimat is a *new* thinking based on *current* scientific cognition.

Scientific Cognition

1. All things originated from the big bang [1.1].
2. We are all descendants of fish [1.2].
3. Our bodies are recycled stardust [1.3].

New Thinking

1. The 1867 definition of Science [8.1], which studies only nonhuman systems, is outdated and should be expanded to study both human and nonhuman systems.
2. We agree that the decoupling of science from religion as the main thrust of the definition is a good thing and should be retained.
3. All research disciplines are complementary to each other, should be mutually supportive, and considered together.

Explanation: **Thinking 1**. Darwin's theory of evolution has been published for more than 160 years, and it is no longer disputed that human is an animal (even the Pope has acknowledged it). Like other animals (such as bees), everything about humans should be the object of scientific study. The similarities and differences between humans and other animals can be clarified through research.

Thinking 2. The object of scientific study is things within nature, and religion by definition necessarily involves the supernatural (e.g., God, reincarnation), so it cannot be discussed in the context of science.

Thinking 3. Bee research includes bee anatomy, flight dynamics, social histology, etc. To fully understand bees, we should combine the results of these sub-disciplines. The same is the study on humans.

Definition of Scimat

1. *Scimat* = Humanities + Social Science + Medicine
2. *Science* consists of the process and results of human's effort to understand all things in nature without introducing any supernatural considerations.

3. A *scientist* is a person who works in science.

Explanation: 1. **Scimat**. "Scimat" is a new term, which advocates the unification of the three groups of disciplines (Humanities, Social Science, and Medicine) that study human. But each discipline should retain its existing identity, and carry on with what each has been doing while looking out for others are doing.

2. **Science**. "Understand nature" means "serious and honest study of nature" whereas, "serious study" means *Reality Check* (including experiments) [8.3]; "honest study" means not deceiving yourself or others. "All things in nature" include both human system and all nonhuman systems. "*Process* and *results*" refer to scientific processes and scientific results, respectively. "Supernatural" includes God and the heavens who can detect human cheatings and suffering. Science = Scimat + Science of nonhuman systems. Science consists of two parts: scientific process and scientific results.

3. **Scientist**. Scientists or researchers include professional and non-professional, full-time and part-time people. "Scientist" is not an official title and does not need to be verified or recognized by others. Besides, what scientists say is not necessarily correct except that they usually know a little more than the average person, in the field they are familiar with. Note that this extra knowledge often does not stand the test of time [8.4]. Scientist is the name of a profession, which is only a division of labor with artists and Internet celebrities, say.

Why Scimat Is Important

Nobel laureate Friedrich von Hayek (1899-1992) said something like this: "In the long run, it is ideas, and therefore the people who disseminate new ideas, who dominate the course of historical development." For example: "Man is born free" abolished slavery; "All men created equal" eliminated aristocracy and totalitarianism; "Gender equality" removes restrictions on women's rights in education, employment and voting. Similarly, scimat is a new concept and equally important (see below).

"Human science and natural science will merge into a whole science" is Marx's prophecy and vision already put forward in 1844 [3.1]. Scimat provides the rationale for this prophecy. Note that Marx's prophecy was 4 years *after* William Whewell proposed the word Scientist and 23 years *before* the second definition of the word Science, which decoupled it from religion. Did Marx have this prediction because he was an atheist?

Four Goals

1. Straighten out the humanities and natural science.
2. Enhance the humanities.
3. Unify definitions.
4. Achieve world peace.

Explanation: 1. **Straighten out** the humanities and natural science. Both the humanities and natural science (not mathematics) are about the study of nature. The two share same origin since both human and nonhuman systems are made up of atoms. Scimat recognize this and is the only way and the most reasonable approach to smooth the relationship between the two.

2. **Enhance** the humanities. The number of humanities graduates is decreasing year by year, and employment is relatively difficult because humanities research has long stagnated at the empirical level [4.3], resulting in multiple theories that address the same problem and the correctness of the theories impossible to be ascertained. This is a desperate situation for the humanists and, worse, making their graduates lacking skills needed by the rapidly changing market. In this regard, scimat advocates humanities-science synthesis [4.1], and encourages interaction and cooperation between humanists (including scholars and students) and scientists, which will raise the scientific level of humanities research and open up new research areas for science [4.4, 5.4]. In the *long* run, the fundamental solution to the two-culture problem is to tackle it from the ground up by recognizing the humanities as part of science [18]. In the *short* term, it is to clarify the problem through general education and pass on the known results of humanities-science synthesis to the students.

Why are the humanities particularly important? Because it is human decision-makings that affect the wellbeing of individuals or society, and the science of decision-making belongs to the humanities.

3. **Unify** definitions. Science history, science philosophy and science communication are new disciplines established in the last century, and they belong to the humanities. From the perspective of content, science history and science philosophy are branches of history and philosophy, respectively. In the current situation where there is no consensus on the definition of science, when making reports in seminars or conferences, unless everyone explains their own definition of science first, they cannot effectively discuss the issues; if they do, the discussion will take a long time, wasting everybody's time and energy. The same goes when writing papers. The lack of a unified definition of science will also hinder the compilation of a general history of science because the extent of the contents will be unclear: Should it be confined to natural science (and medicine and mathematics as it is done now) or should it also cover the history of humanities/social science? Scimat's definition of science is the only one that fits current scientific cognition and the historical evolution of the term Science.

4. **World peace**. World peace is a universal aspiration. The ways to achieve this is divided into two categories: secular and religious. The *secular* approach often conflicts with human nature while the *religious* approach requires the abandonment of independent thinking. Both approaches are difficult to implement universally, and the world is still restless [3.2]. Fundamentally, all human injustice and war result from someone's decision making, which is based on the decision maker's worldview. Therefore, if all decision-makers have a peaceful, life-respecting, and humble attitude to life, the world will be peaceful.

The question is: How to make this happen? Because it is impossible to predict who the future decision-makers will be, it is necessary to see everyone as a possible decision-maker and train them from an early age. The method is to teach the three major scientific understandings related to human [1.1, 1.2, 1.3] from the first grade on, which can be included in science or humanities courses. Of course, it can also be talked about in university general-education courses. Thus, when everyone looks at everything from the long history since the big bang, they know that they share one ancestor, belong to the same family by *blood*, and all the things on Earth (plants, water, air, humans, dogs, fish, etc.) are made up of stardust by exchanging atoms. Hopefully, they will become humble, mutually accommodating, and love each other. And we will achieve world peace, which is scimat's most important goal.

1-2-3 Insight

One culture, two systems, three levels!

1. Humanities and science belong to *one* culture: scientific culture as defined by scimat.
2. All systems fall into *two* categories: simple systems and complex systems (the two are different) [2.3].
3. There are *three* levels of study in any discipline: empirical, phenomenology, and bottom-up [4.3].

Conclusion

1. Everything in nature, including humans, is the object of scientific study.
2. The correct image of science: two connected animals (one controlling the other).

3. Scimat breaks down the barriers between the humanities and science, promotes world peace, and increases human happiness.

What better expresses this meaning and aim of scimat are two posters (see figures at beginning of article).

3.4 Scimat rules

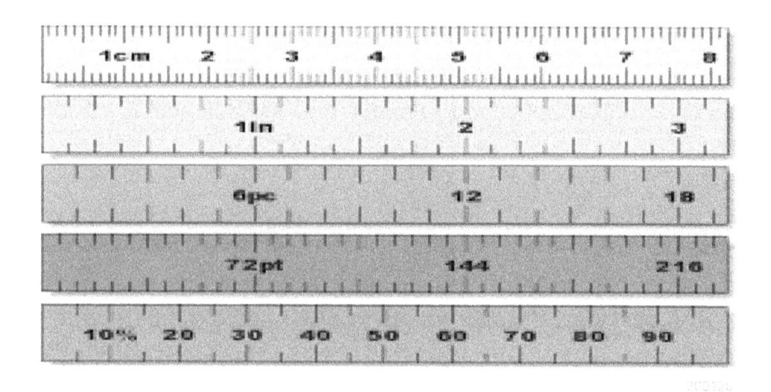

There are five scimat rules for everybody:

1. We will be honest with the reader and ourselves and present our findings in clear writings, and will not try to hide our relevant thinking.
2. We will *not* quote anyone's writing to support our own argument.
3. We will not be ashamed to admit our own mistakes in our findings and correct them as soon as possible.
4. A conjecture or hypothesis becomes a (temporary) theory only *after* it is confirmed by experiments or by practices in the real world.
5. We will abandon (or revise) the theory if it does not agree with confirmed and irrefutable evidence.

Explanations for these five rules are in order.

Rule 1. Rule 1 is the basic ethic of any honest researcher or knowledge seeker, but is not always practiced by everyone. Yes, we understand that complex systems such as those studied in humanities/social science are very complicated, and one does not always have a clear idea of what one's thinking really is. If that is the case, please tell it to your reader which part is clear to you, which part is not, and mark your work as "work in progress." Better still, present your ideas like these in a seminar or cocktail party but not in a conference. If every paper was written clearly and findings/results were presented as "objectively" as possible, and if the paper always ended with a section of Discussion/Conclusion in which the author presented what lessons they have learned, perhaps the Sokal hoax would never have to happen and the Science Wars could be avoided.

Rule 2. Quoting others to support one's argument is a common practice in non-physical disciplines. But this is completely useless. For example, while Einstein was proven right in his many writings such as the two theories of relativity, he could not always be right, and he did not. We all know about this; that is why Rule 2. To argue that your point of view is correct, just talk about the matter, lay out the evidence, and explain the results clearly, without involving others (neither Plato nor Einstein).

Rule 3. One should not be ashamed of making mistakes when doing complex-system studies since the job itself is so difficult. What one should do is simply admit their mistakes once recognized, and correct them as soon as possible. In this regard, good economists are real scientists who know their limits and act accordingly. They keep on adjusting their predictions of the stock index or the gross domestic product (GDP) and should be respected for doing that. That explains our Rule 3.

Rule 4. If anyone put out an educated guess (called *Hypothesis*), this guess has to be confirmed before it can be called a *theory*. Common sense, right?! Rule 4 is copied from the practice in physics and other "natural sciences." We just want to unify our terminology in communicating to each other, since in scimat we very likely are coming from different disciplines with different training and background.

Let me emphasize this: We do not mean that physical science is superior to humanities/social science. It is not. In fact, the opposite could be true. Humanists and social scientists are tackling very complex systems, while most physical scientists are dealing with simple systems. Those dealing with complex problems could be more courageous and should be admired.

In fact, to be a good artist is more difficult than being a good physicist. In physics, there are rules to be obeyed and experience to follow and the choices in solving any physical problem are more restricted than what is available to a painter who wants to create something new. The painter has infinite possibilities and really needs imagination and talent. That is why there are more good physicists than good painters in the world.

With Rule 4 in place, no societal theory in the form of political ideology of any kind could be validated, since it is unethical to try experiments on living humans, especially in large numbers. Political leaders are advised to try their "experiments" with computer simulations and be prepared to adjust their policies frequent enough.

Rule 5. Rule 5 is obvious. Finally, it follows from the spirit of this rule that we will adopt better rules if that become available.

3.5 Names incorrect

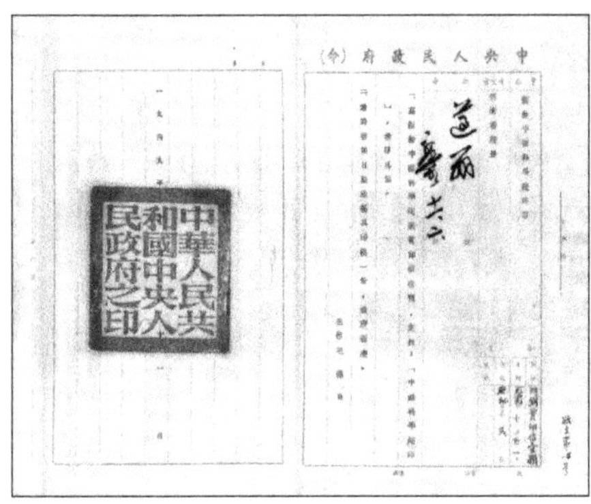

Left: Decree of the Chinese Academy of Sciences to President Guo Mo-Ruo (October 31, 1949). *Right*: Chinese Academy of Sciences headquarters, Sanlihe, Beijing.

The *Analects of Confucius, Zilu*:

> If the name is not correct, the words will not make sense, and if the words does not make sense, things cannot be done.

In 1949, the Chinese Academy of Sciences (CAS) was founded to include the humanities, social science, and natural science, which is justified because the humanities and social science are part of science [3.3].

In 1977, Philosophy and Social Science were detached from CAS to form the Chinese Academy of Social Sciences (CASS), marking the beginning of humanities-science separation, and making the CAS and CASS both misnomers. Why misnomers? Because the remaining CAS has only natural science while the new CASS, apart from social science, does include the humanities. In other words, the CASS is actually a liberal-arts (called *Wenke* 文科 in Chinese [2.1]) academy. Therefore, presently, the names of CAS and CASS are both incorrect—being inconsistent with their contents.

Recommended Name Changes

Suggested *name changes*:

1. Chinese Academy of Sciences → Chinese Academy of Nonhuman Sciences (CANS)
2. Chinese Academy of Social Sciences → Chinese Academy of Human Sciences (CAHS; or, better, replace Human Sciences by Scimat).

And let the two combine to create a new Chinese Academy of Sciences (CAS), viz., CAS = CANS + CAHS.

The *new* Chinese Academy of Sciences could be organized either way:

1. One leader: There is a new president, and two vice presidents served concurrently by the presidents of CANS and CAHS.
2. Two leaders: There are two presidents, served concurrently by the presidents of CANS and CAHS.

The new CAS is merely a formal organization to coordinate cooperations between the two constitutive academies; an office and one secretary are sufficient.

There are seven *benefits* to rebranding and integration:

1. It helps to dispel the current misunderstandings about the separation of liberal arts and natural science among everyone (scholars, students, and the masses).
2. It is conducive to the humanities-science synthesis [4.1], can promote the construction of new interdiscipline and the *new* humanities [4.4], and promote *innovation*.
3. Promoting new humanities can help enhance *soft* power.
4. Promoting innovation helps technology become *self-reliant*.
5. Demonstrate the implementation of "seeking truth from facts" and drive the whole country to attach importance to the scientific spirit.
6. Implement Confucius' teaching that "names must be correct" and demonstrate cultural self-confidence.

7. Officially announce that Marx's last wish (to merge human science and natural science [3.1]) has been realized in China.

More Recommendation

Finally, it is proposed to withdraw health-related research institutes from the Chinese Academy of Sciences and merge them with other relevant research institutes in the country into a new Chinese Academy of Health Sciences, similar to the National Institutes of Health (NIH) in the United States. The reasons:

1. Health and medical issues are super-important, life-stake, and require leaders with different expertise, while the CAS mainly studies nonhuman systems.
2. The funding of the new Chinese Academy of Health Sciences will be much larger than that of the CAS, and it is not suitable to be placed under the jurisdiction of the CAS. (For example, the NIH funding in the United States is *five* times that of NSF, the National Science Foundation.)

4

On Humanities

4.1 Humanities-science synthesis

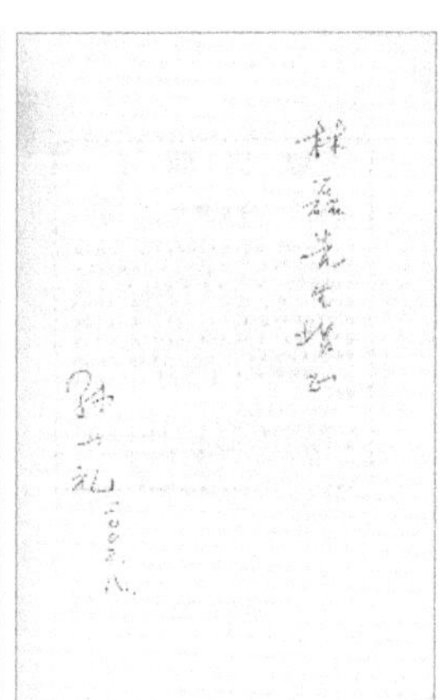

The Chinese word *Wen* (文) or *Wenke* (文科) refers to the humanities plus social science while *Li* (理) or *Li'ke* (理科), mathematics plus natural science (of nonhuman systems) [2.1]. Since mathematics is not part of science li'ke is not equivalent to science [2.5]. The word *Ke* here means discipline/subject. The two words wenke and li'ke exemplify the elegance and beauty of the Chinese language and have no counterparts in English.

Wen-Li Separation

The Industrial Revolution (1760-1820/1840) that began in Britain and swept across Europe spanned the 18th and 19th centuries, boosting economic development, creating a middle class, and prospering education. In the 19th century, the Prussian Minister of Education, Wilhelm von Humboldt (1767-1835), founded the *research*-oriented University of Berlin (renamed Humboldt University in 1948) in 1810, which advocated equal emphasis on teaching and research, and students learn through research.

Later, German universities invented the graduate-school system (then called Seminars, not today's seminars) to systematically train *graduate* students, which was introduced to the United States in 1876 when Johns

Hopkins University opened. In other words, the undergraduate program of American universities inherits the educational tradition of British universities, while graduate schools inherit the educational tradition of German universities.

Importantly, the Faculty of Philosophy of German universities was split into liberal arts and sciences (i.e., wen and li) in the second half of the 19th century. Unavoidably, this practice spread from graduate schools to undergraduate and secondary schools, and from Germany to other parts of the world.

The wen-li separation was designed to serve the needs of the industrial society but it also destroys the common root and unified foundation of all the academic disciplines [2.1]. And there are consequences, one of which is the slow or lack of progress of the humanities, making the world a less peaceful place to live [3.3]—the motivation for earthlings to migrate to Mars, say, not to mention the increase of military conflicts here and there.

Cai Yuan-Pei

Cai Yuan-Pei (蔡元培, 1868-1940), born in the Qing dynasty, was a modern revolutionist, educator, and politician. At the age of 17, he was admitted as a *xiucai* scholar (秀才), and at 25, he was selected as a *jinshi* scholar (进士) in the Imperial Academy. After the failure of the 1898 reform movement, in the autumn of that year, he left the Imperial Academy and went south to devote himself to education and advocate "new learning" (新学). In 1904, he established the Restoration Association (光复会) in Shanghai and served as its president. The following year, he joined the Alliance (同盟会) established by Sun Yat-Sen (孙中山) and was appointed as the head of the Shanghai Branch of the league.

In 1907, at the age of 40, he joined the tide of students studying abroad and went to Berlin, Germany, to study German for one year. The next year, he studied philosophy, aesthetics, psychology, and ethnology at the University of Leipzig for three years. He returned to China near the end of 1911.

In January 1912, Sun Yat-Sen formed the Nanjing interim government and appointed Cai as Minister of Education on the third day. Cai was president of Peking University from 1916 to 1927. He innovated Peking University's educational system and started the tradition of "scholarship" (学术) and "academic freedom" in the university.

Wen-Li Blending

As a foreign student in Germany for four years, Cai Yuan-Pei saw the short comings of wen-li separation firsthand. Accordingly, at Peking University, he advocated *blending wenli* (融通文理, i.e., roughly speaking, integrating the humanities and natural science), believing that *wen* and *li* cannot be separated from each other because there is wen in li and vice versa. He said: "Philosophy in li'ke must be based on natural science, and the final assumptions of wenke scholars often involve philosophy. In the past, psychology was attached to philosophy, but now experimental method is used, and psychology should be included in li'ke."

In 2002, Lui Lam (林磊) created the new discipline called *Histophysics* through wen-li blending [14]. In 2003, Sun Xiao-Li (孙小礼), director of the Science and Social Research Center of Peking University, published the book *Wen-Li Blending* (文理交融). In 2007, the new multidiscipline *Scimat* (science of human), founded by Lam, provided the theoretical basis for wen-li blending [3.3].

4.2 General education

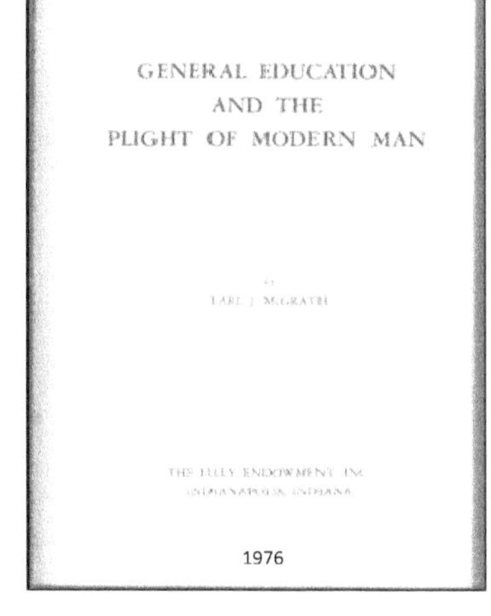

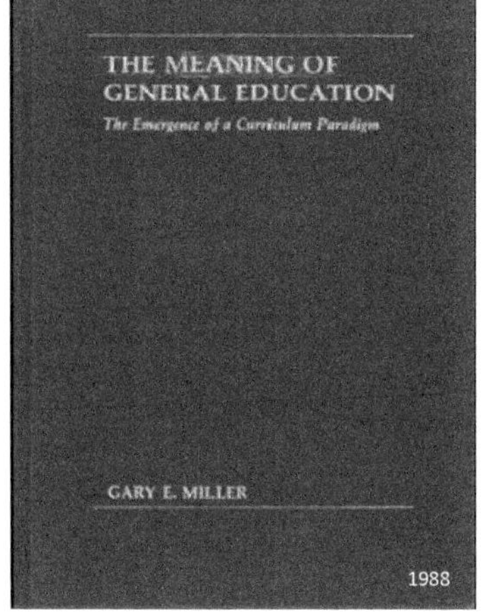

General Education was invented in the United States in the 1930s as an educational reform. In contrast to the "liberal education" in the United States in the 19th century, which is backward-looking (read the books of the ancients) and focused only on liberal arts, general education was forward-looking, with equal emphasis on arts and sciences. General education is born in response to the shortcomings of liberal arts education, in order to cultivate the four qualities of undergraduate students:

1. The scientific spirit of seeking answers to everything.
2. Problem-solving skills.
3. Personal and social values consistent with a democratic society.
4. To acquire knowledge related to these spirits, abilities, and values, and to develop the habit of lifelong learning to become a self-fulfilling individual and a citizen who fully participates in society.

Peking University

In 1998, on the centennial of Peking University, China put forward the vision of creating a world-class university, followed by the introduction of general education in universities at the beginning of this century, including the Yuanpei Program Experimental Class (renamed Yuanpei College 元培学院 in 2007) started by Peking University in 2001, the Fudan College (复旦学院) established by Fudan University in 2005, and the Xinya College (新雅书院) established by Tsinghua University in 2014. These general education have "Chinese characteristics."

In fact, the world's earliest general education did not appear in the United States, but first appeared at Peking University in 1918, more than ten years earlier than the United States. On January 4, 1917, Cai Yuan-Pei became the president of Peking University, and according to his experience of studying in Europe, he believed that arts and science were inseparable [4.1].

In order to integrate the humanities and natural science, he proposed to the meeting of presidents of colleges and universities on October 30, 1918, that in undergraduate education, "remove the boundary between the liberal arts and natural science: Those who study liberal arts must take one of the natural-science courses at the same time (such as history students take a geology class; philosophy students take biology and the like); those who study science should study some kind of liberal arts (such as the history of philosophy, the history of civilization, etc.)." This is exactly the current strategy of general education in American universities, but Peking University has already implemented it a hundred years ago.

United States

Unfortunately, Cai's general-education reform has not been passed down in China. As for the general education in the United States, has it been influenced by Peking University? The answer is unclear, but it cannot be ruled out because the educator John Dewey (1859-1952, 胡适 Hu Shi's doctoral supervisor) visited China from May 1919 to July 1921, and he should have learned about Peking University's educational reforms. And he could report to his peers after returning to the United States, including the president of Harvard University.

Remarks

Regarding the content and role of general education, I would like to add this:

1. General education courses could include a few that are introductory undergrad courses from different departments but the majority should be courses specially designed for general education. The reason is that the purpose of general education is interdisciplinary and its focus is *not* on imparting the expertise of various subjects, but to show the unified perspective behind all the disciplines ranging from the humanities to natural science *and* to teach students on how to seek knowledge. The best time to teach all these is during the undergraduate years.

2. General-education courses should be constantly updated and keep pace with the times, so they should not be judged by the maturity/seriousness of the subject matter but by the *usefulness* to the students. What is useful? For example, what the country needs right now is useful, and lessons that teach innovation fall into this category.

3. Contrary to popular thinking, I think that one of the reasons that there has been no major innovation in China in the past 30 years is that university teaching is too serious and all courses are taught too "solidly." The reason is that innovation requires new thinking, and spending time learning all the existing courses does not help. For example, no matter how well you learn the basic courses of physics, it is not enough or not very effective in innovation. Physics has not made much progress in the past 40 years, and it is obvious that something well beyond the basic courses is needed.

4.3 Research: three-prong advance

Academic research is to understand certain things, to overcome some difficulties, just like fighting a war. There are three levels (or approaches) to the study of any discipline: empirical, phenomenological, and bottom up. These three levels support each other and work together, just like the air, navy and land armies of the military.

1. The *empirical* level is to collect and analyze information, and if you are lucky, you can obtain some laws by summarizing the results. This is equivalent to the Air Force looking at the objects from a high distance.

2. The *phenomenological* level is to write down some formulas/laws without understanding the mechanism. This is equivalent to the medium-range combat of the Navy.

3. The *bottom-up* level is starting from the lowest level, from theory or computer simulation to obtain results. This is equivalent to the short-range combat of the land soldiers.

From the bottom-up level, one can derive the phenomenological level results and know more; from phenomenological theory one can derive empirical results and know more. In other words, bottom-up → phenomenological → empirical.

Physicists are usually aware of these three levels. For example, the gas law can be summarized from the experimental results of several experiments; the phenomenological equations governing the flow of fluids (liquids and gases) can be derived from several basic assumptions (space-time translational invariance and constant density); the above fluid equations can be derived from computer simulations starting with interacting atoms (bottom-up).

Medical people have a very deep understanding of the three levels of research (e.g., starting from the DNA structure and mRNA of the virus to make a vaccine, from the bottom up). And that is why medicine could advance so much in last few hundred years.

Those who are *not* familiar with these three levels of research are scholars in the humanities because their university courses do not teach these (even the teachers generally are not aware of these three levels). However, humanities studies in the West have broken through the empirical level in the past 30 years [4.4] although they have not yet become mainstream. Some Chinese efforts are seen in this area, such as the study of Cao Cao's (曹操) family origin through DNA.

4.4 New humanities

Old Humanities

The study of the humanities (literature, history, "philosophy", etc.), since Plato and Confucius, mainly stays at the empirical level. That is, after collecting and sorting out the information observed and from the predecessors, say a few words on what have been summed up and learned, and call it finished. If it is done well, some personal understanding, conjectures, and "explanations" will be added. Late comers may add more material, make different conjectures or explanations, and that is it. As a result, there are so many conjectures about the same topic, and no one can convince anyone. For example, *Tao Te Ching*'s "The Dao that can be spoken is not the eternal Dao" (道可道，非常道) had three different interpretations before the Northern Song dynasty, and there are more today. This is the *old* humanities.

New Humanities

The *new* humanities is what appeared in the humanities, mostly in the West in the past 30 years, through the combination of disciplines (between humanities or between humanities and natural science), that break through the empirical level and raise the study of humanities to the phenomenological and the bottom-up levels [4.3]. In other words, it raised the *scientificity* (scientific level) of the humanities. For example, in the West, Shakespeare's literary works are studied from statistics analysis, evolutionary theory, and neuroscience; in China, Cao Cao's (曹操) family origin is clarified through DNA tracing. For new aesthetics, see [7.8].

The humanities mainly study individual humans, which is the most complex system in the universe, while social science studies human groups and are easier to approximate, so its scientificity is higher than that of the humanities but still lower than that of natural science (of nonhuman systems).

China

In China, the emergence of the new humanities, in addition to satisfy the basic needs of disciplinary development mentioned above, is more in line with the "needs of the times" initiated by the Ministry of Education in 2018. As I understand it, there are three major signs of the success of The Chinese Dream: all national heads are eager to come and visit, economic power, and *cultural* respect. It seems that the former two have met the standard but the latter is awaited.

Widely respected comes from contemporary cultural *innovations* that are recognized by everyone and have contributed to human civilization, such as Tang poetry in the Tang dynasty and Song lyric in the Song dynasty. The construction of new humanities will be the way for new Chinese culture to be respected.

Remarks

Chinese Educators on New Humanities

- "Declaration on the construction of the new liberal arts." Author: Ministry of Education. Originally published: Ministry of Education website, November 3, 2020.

- "New liberal arts: Needs of the times and construction priorities." Author: Fan Li-Ming (樊丽明), President and Professor of Shandong University, Head of the New Liberal Arts Construction Working Group of the Ministry of Education. Originally published: China University Teaching (中国大学教学), Issue 5, 2020.

Selected Articles on New Humanities

- "Histophysics: A New Discipline." Author: Lui Lam. Originally published: Modern Physics Letters B, vol. 16, pp. 1163-1176, 2002.

- "Studies in human history: An example of Scimat." Author: Lin Lei (Lui Lam). Originally published: *Renke* (人科), China Renmin University Press, 2013.

- "Neuroaesthetics: The secret of art is in the brain." Author: Zhu Rui (朱锐), Distinguished Professor of the Department of Philosophy, Shenzhen University. Originally published: Xinrui Weekly (信睿周报), Issue 18.

Books on New Humanities

Brain science

2009

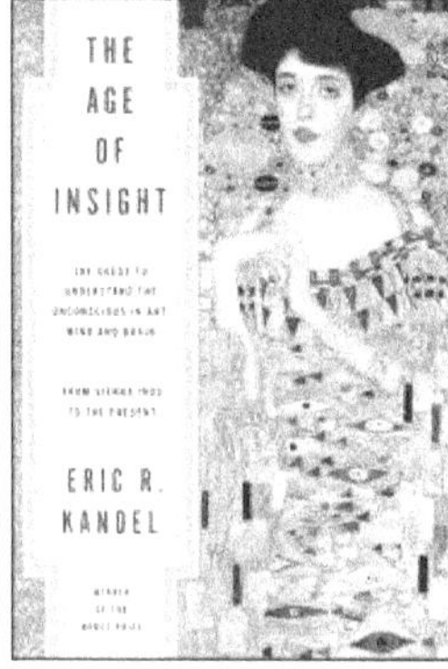

2012

2018

Literary studies

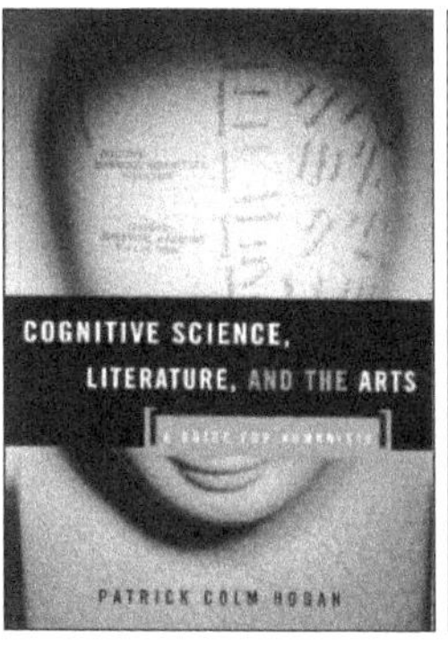

2003

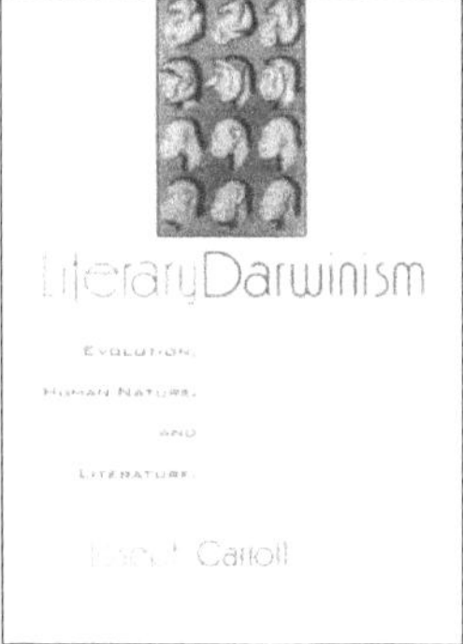

2004

2008

Neurophilosophy

1989

2012

Experimental philosophy

2008

2012

Neuroaesthetics

2007

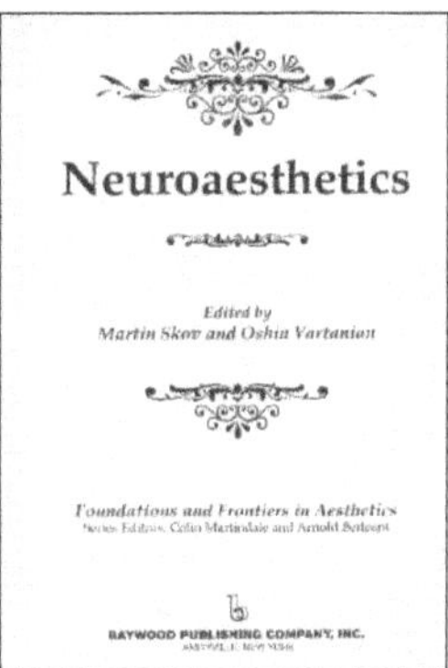

2009

Scimat

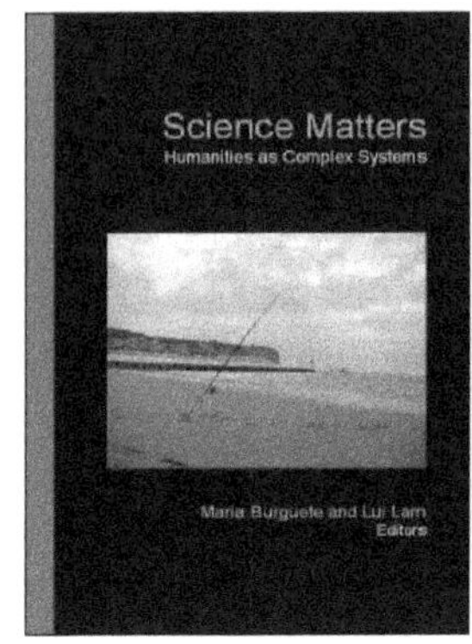

2008

2011

2014

4.5 Sarton: half a mountain

The history of science has existed in ancient times, both in China and abroad, but as a discipline, it is the credit of George Sarton (1884-1956) a hundred years ago. He is a Belgian who studied chemistry, crystallography, and mathematics; his doctoral thesis is "Principles of Newtonian mechanics."

However, Sarton's so-called History of Science is actually "history of science of nonhuman systems + history of medicine."

The word Science used by Sarton is not exactly the 1867 definition of the term, which is: Science is the study of nonhuman (material) systems in nature without introducing God/supernatural considerations. (It is in fact the second definition of the English word Science [8.1].)

Sarton did not point out and emphasize the premise that science deliberately excludes the supernatural, and the inclusion of the history of medicine in the history of science certainly does not fit the scope of science according to the 1867 definition. In fact, beyond the history of medicine, Sarton actually limits the history of science to *nonhuman systems.*

This is a big problem, because humans are recognized as a kind of animal in nature according to Darwin's evolutionary theory, which was established long time ago in 1859. And since science is the study of *everything* in nature, so it is hard to justify the exclusion of studies on human (except medicine) from science. Are the consequences serious?

Yes, very serious. It is as if someone buys a piece of bread and always eats half of it, giving up the other half and throwing it in the trash. Is it because he wants to show off his wealth? Or is it because his stomach is full that he cannot eat the other half? Or, is it because he does not know that the other half is eatable?

Yes, Sarton was the pioneer of the history of science, but he only "opened up" half a mountain.

The consequence is that the history of science is *incomplete* in terms of content, and it also destroys the natural and fundamental relationship between the humanities/social science and natural science. This mistake affects the current effort to synthesize the humanities and natural science, which aims to raise the scientificity of the former [4.1]. And since the humanities include literature, art, history, philosophy, decision-making and other disciplines closely related to people, the development of the humanities directly affect people's happiness, wellbeing and world *peace*, which is of particular concern. In fact, in terms of the survival of humans as a whole, the humanities are far more important than natural science.

4.6 New history of science

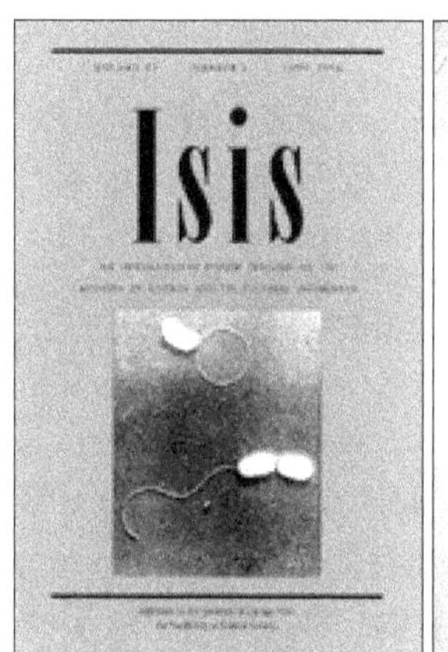

First-class universities are made up of first-class departments, which have one or several masters, and if not, several professors who do first-class research. The main job of a first-class professor is in *research*, not in teaching or translating books (unless it is a literary masterpiece such as *A Dream of the Red Chamber* or Dante's *The Divine Comedy*).

A specialist is one who focusses on a narrow topic. A master or first-rate professor is one whose works will shake up or impact the entire discipline or the whole profession.

What is the motivation to become a master/first-class professor? In addition to the well-known "making contribution to mankind" and "glorifying one's ancestors," in China, there are two *external* reasons:

1. The government hopes to increase its soft power and discourse power.
2. The education department hopes to build "double" first-class universities.

But there are also two *internal* reasons at the individual level:

1. Shake up/influence the entire profession—for fun if nothing else.
2. Dreams may not always come true.

Therefore, set the goal high. If it is not realized you are still high enough: If you cannot become a master, you can become a first-class professor; and if you cannot be a first-class professor, you can keep your job.

In a special research environment (the search engine is not powerful; there are not many foreign books), how can one become an international master or first-class professor? The History of Science discipline is unique in this regard: Unlike science itself, the history of the science as a discipline has only a short history of 100 years, and international competition is not fierce, as can be seen from the fact that the first journal of science history *Isis* was published only four times a year in 108 years (left figure).

If one wants to become an international master of science history, one can consider the following new entry points:

1. Shift the selection of science history topics from simple systems (mathematics, physics, chemistry) and old complex systems (medicine, biology) to other *complex* systems [2.5].
2. Re-examine existing scientific history cases and *rewrite* the article from the perspective of the three research levels: empirical, phenomenological, bottom-up [4.3].
3. Science history is a branch of history. In addition to collecting and sorting out data and writing papers in narrative texts, further quantitative analysis can be carried out to see if some quantitative laws can be found [5.4].
4. Seriously consider extending the history of science to the history of the humanities; after all, the humanities are part of science [3.3]. In this regard, steps has been taken in the West for several years (right figure) but has not become mainstream; it is not too late to catch up and overtake it.

5
History

5.1 History and science

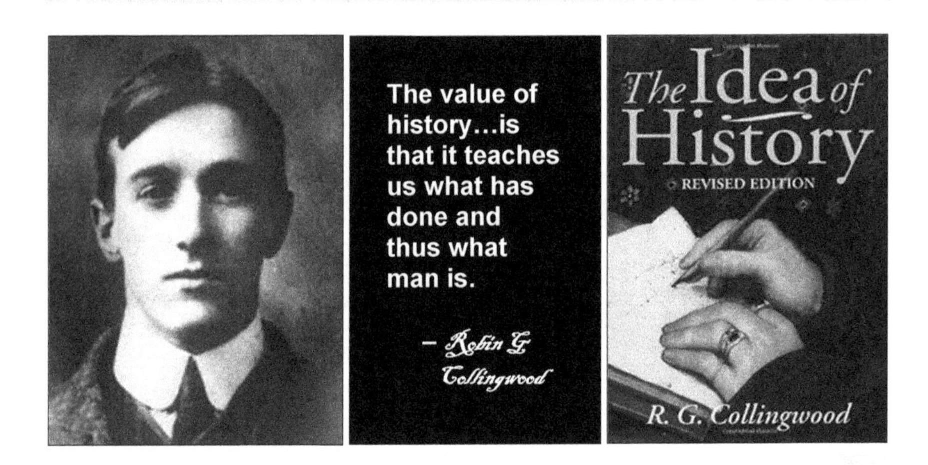

History is about human's past. There is human, there is history. But history as a scholarship begins with the ancient Greek, Herodotus (c.484-425/413 BC). The most discussed relationship between history and science is by the Englishman Robin Collingwood (1889-1943), who believes that history is a science. The conclusion is correct, but his argument is not completely correct.

Collingwood, a graduate of Oxford University and worked there his whole career, studied philosophy, history, and archaeology (his father was engaged in art and archaeology). He excelled in both the humanities and natural science.

His Idea

1. Science is *any* collated knowledge.
2. History *is* a science.
3. History differs from (nonhuman) natural science: The latter deals with the *physical* world while the former is the societal world and all *human* affairs.
4. Natural science studies *real* things while history studies *virtual* things.

5. The research methods of history and natural science are necessarily *different* because the thought processes and historical events of historical figures cannot be directly observed.
6. Historians must put themselves in the position of the historical figures and *reconstruct* every step of their thinking in their brains, based on the information available to the historians.
7. Historians should not ignore information from the person under study that differs from his own knowledge of the world or the account of another historical figure because the difference must have a cause and is valuable information.

My Idea

1. Science is any collated knowledge about nonhuman *and* human systems, without introducing the supernatural. He knew that science includes the human system because the object of history is human, even though he did not mention it explicitly. But he did not mention that science cannot introduce the supernatural. Perhaps because he lived in the early 20th century, more than 50 years after the 1867 definition of science [8.1], that this understanding of science-religion decoupling has become a common knowledge in the academia so there is need to mention it. In other words, he and I use the same definition of science—the 2007 scimat definition [3.3].

2. History is science. (He agrees with me on this.)

3. History differs from (nonhuman) natural science: The latter deals with the *physical* world while the former is the societal world and all *human* affairs. (He agrees with me on this.)
4. What is studied in natural science is physical things (and invisible energy, vacuum, etc.) while history is about virtual things. (He and I *basically* agree on this point.) He is an archaeologist, not a physicist, so he is not familiar with the concept of energy and nonmatter such as the vacuum.

5. Research methods in history are the *same* as in natural science although historical events and the thought processes of historical figures are not directly observed. (He *disagrees* with me on this.) Historical research, like any other discipline, has three levels: empirical, phenomenological, bottom-up [4.3]. History belongs to the humanities, and humanists always think that they must understand people's mind in order to study human behavior and human society. This is wrong.

   Thinking belongs to the "internal states" of a human being. The approach of starting from the internal states of the system to understand the external state and collective behavior of the system, belongs to the bottom-up level in research. And, in fact, historical research also has two other levels: empirical and phenomenological, which can be done without knowing the human-thinking process. For example, the empirical level of history includes statistical analysis and Zipf plot; the phenomenological level includes computer modeling and active walks. All these Collingwood did not know because they appeared after his death. From these two levels, qualitative and even quantitative laws about history can be obtained, and predictions can be made [5.4, 15].

6. It is excellent film actors who can reconstruct the thoughts of historical figures; ordinary historians do not have this ability. (However, to understand why science philosophers are so prone to mistakes, I think it is a good research method to understand their thinking and the details of their life. This method is called *Sciphilogy*—a term I coined [9.7].)

7. Historians should not ignore information they do not like or understand. (He agrees with me on this.) Accordingly, to do any research, we must think calmly and learn to detach from the events under study.

Discussion

1. In historiography, Collingwood and I differ from each other mainly on Point 5—the research methods.
2. Collingwood's account of Bradley from 1935 to 1940 has underwent a process of change. (See Robert Burns [2006]. Collingwood, Bradley, and historical knowledge. https://doi.org/10.1111/j.1468-2303.2006.00356.x.) It is rare for humanists to acknowledge and correct their mistakes; Collingwood is an exception. This may be related to the fact that he is also an archaeologist because archaeology is a discipline that blends humanities and (natural) science, and scientists have a tradition of admitting mistakes [9.7].

Conclusion

Collingwood saw the same thing as I saw: History is part of science (although his arguments are party wrong and mine are completely correct).

Remarks

For Collingwood's views on historiography, see his book *The Idea of History* (1946). Free information: "Robin Collingwood" (Wikipedia) and his article "Are history and science different kinds of knowledge?" (1922).

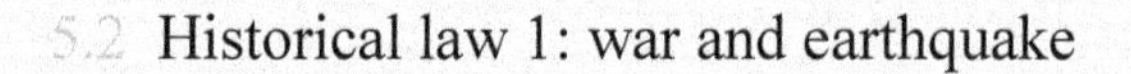

5.2 Historical law 1: war and earthquake

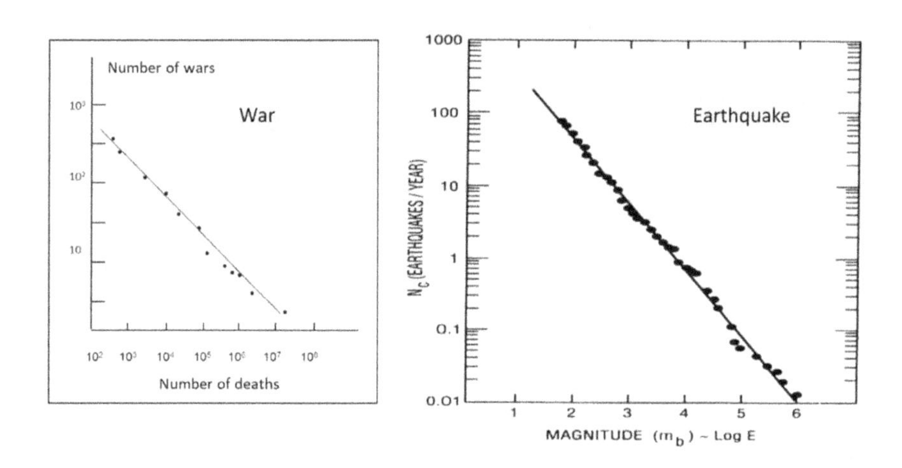

History is the study of what has happened in this many-body system of human beings. It could be about an individual, a family, a village or a province, a country or a continent, or the whole world. It all depends on the researcher's personal interests and academic needs.

Why Study History?

1. People are curious about everything.
2. We gossip and want to know what our parents and ancestors did.
3. People want to know where they came from and how society got to where they are today.
4. Knowing the past may help predict the future. As the Chinese writing *Zhenguan Zhengyao* • *Ren Xian* (贞观正要•任贤) says: With copper as a mirror, you can dress properly. With the ancient as a mirror, you can predict the up and down of history.

Do Historical Laws Exist?

Are there regularities in history? Are there historical *laws*? Many people will agree that history sometimes has regularities (e.g., all Chinese and foreign dynasties have a limited lifespan), but the average person (including many historians) does not believe that there are laws in history. They are wrong.

History has not only laws, but even *quantitative* laws. Example: The relationship of the number of wars (y) and the war casualty (x) obeys a power law. That is, $y = Ax^b$, which is equivalent to log y in a straight line relationship with log x (left figure). This is a bit strange, because in a given war, the number of people killed could depend on the commander's decision (such as crossing or not crossing the river). Yes, a single war is influenced by individuals, but when multiple wars are considered and results independent of individual's decision are found. In other words, the human factor has often been overestimated throughout the history of warfare.

Moreover, the relationship between earthquake frequency and earthquake intensity is also a power law (right figure). It shows that human history and inhuman earthquakes, both complex systems [2.3], have something in common, which is more intriguing and worth investigating.

The power law of war frequency-casualty was discovered by the British physicist and meteorologist Lewis Richardson (1881-1953) in 1941 (*Nature* **148**: 598) [16]. It is the *first* law, and a quantitative law, in human history. The *second* important law concerns Chinese dynasty, also quantitative, was discovered by Lui Lam in 2006 [5.3].

Conclusion

There do exist historical laws, even quantitative ones. To dig deeper, the best way is for historians and physicist to join forces.

5.3 Historical law 2: historical curse

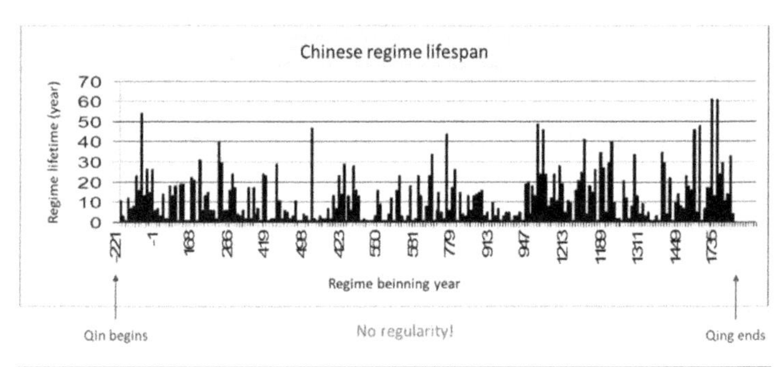

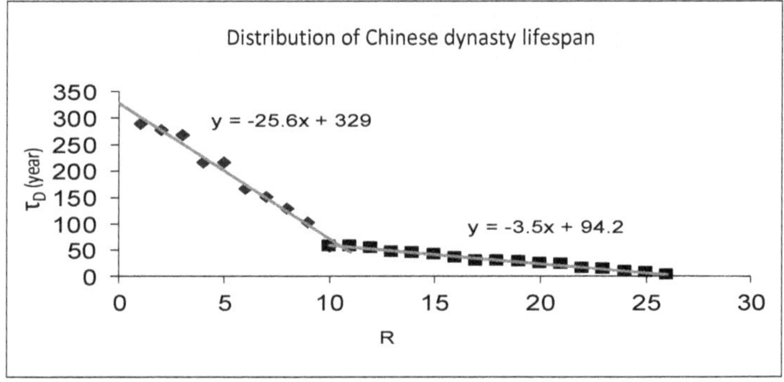

Whether laws of history, like any law of science, exist or not, cannot be predicted. The only way to find out is when they are indeed found. Whether or not one can find the laws, or any law, depends on the person's skills and luck.

Taking the Chinese *regime* lifespan (the number of years an emperor reigns) as an example, there were 231 emperors from the Qin dynasty to the Qing dynasty for a total of 2133 years, with an average reign of 9.2 years (from less than 1 year to 61 years). If the regime lifespan is listed from left to right, it is not clear that there is any regularity (upper figure).

Chinese Dynasties

However, if one adds up all the regime lifespans of the same dynasty to form the *dynasty* lifespan (τ_D) and arrange them from left to right (called the Zipt plot; the longer the lifespan, the more left), you can see the pattern: two cross straight lines (lower figure). It is a quantitative law: Chinese dynasties with a lifespan of less than 57 years may perish every 3.5 years; if they survive 57 years, they may perish every 25.6 years. In other words, dynastic lifespan is not a continuous distribution, but a *discrete* distribution. This is rather strange. Wouldn't you be surprised to be told that human lifespan can only be 3, 6, 9 ... years instead of 1, 2, 3 ··· years (assuming it can only be an integer)? This needs to be explained, but it has not yet been explained.

This result shows that as far as Chinese dynasties are concerned, the so-called *historical curse* does exist.

Prediction

Under the assumption that all dynasties fall in these two straight lines, it can even be *predicted* that *if* there are new dynasties after the Qing dynasty, it will end in one of two possible ways:

1. It has a lifespan of 290 years exactly or less (and obeys the discrete rule).
2. If it exceeds 290 years it will end exactly in its 329th year.

Discussion

There are two reasons that historians, past and present, failed to find historical laws:

1. Historians kept on working at the bottom-up level, one of three levels (empirical, phenomenological, and bottom-up) in research of any discipline [4.3]. That is, they try to understand history from the knowledge of the historical figures [5.1]. This is very difficult and not the place that historical laws been found so far, which are found at the empirical level as shown here and in [5.2].

2. The inadequate science training received by historians. For example, the Chinese dynasty data have been lying there for 93 years; the Zipt plot above could be carried out by hand without computers, and even by high school students. But unless one

knows about power laws and the existence of the Zipf plot, there is no motivation to do so. And these are current topics in the study of complex systems. Ironically, the Zipf plot was first done by George Zipf (1902-1950), a Harvard linguist, with data from the humanities and social science.

Remarks

1. The *Bilinear Effect* of Chinese dynastic lifespan distribution is first published in 2006 [Lui Lam, International Journal of Bifurcation and Chaos **16**(2): 239-268].
2. This effect is also seen in other complex systems (Lam et al., EPL **91**: 68004, 2010). See: www.sjsu.edu/people/lui.lam/histophysics.

5.4 Histophysics

The emergence of new disciplines, like new stars in the sky, is invisible when brewing, and only brightens when it appears. Similarly, when physicists cross the disciplinary boundary to study other areas no one knows, but suddenly, new disciplines are created; examples: Biophysics, Econophysics, etc.

History is concerned with humans' past so it can be combined with all disciplines in the humanities (e.g., sociology and cultural studies), but it can also be combined with those in natural science (e.g., molecular biology and archaeology). The synthesis of history and physics began in 2002 with the introduction of the new discipline called *Histophysics*, meaning the physics of history (see Lui Lam, Modern Physics Letters B

16: 1163-1176, and [15]). Here, physics mainly refers to theoretical physics.

Physics is the study of *all* things in the universe, including humans, of course. There is no certain method of physical research, but it tends to look at problems from the *first principle* (i.e., the basic essence). Yet, doing history with physical methods does not necessarily start from the details, i.e., not necessarily from the interaction between people, although the system studied in history is composed of individuals. The reason is that, in any discipline, physics and history included, there are three levels in research: empirical, phenomenological, bottom-up [4.3]. Examples of these three levels in history studies are listed below, which go beyond the traditional approach of storytelling.

Level	Method	Example
Empirical	Collect data, organize data → Empirical laws Statistical analysis, Zipt plot	War frequency-casualty distribution, Chinese dynasty lifespan distribution
Phenomeno-logical	Computer modeling, active walk	Modeling of economic history, human evolutionary history, human societal history
Bottom-Up	Computer simulation	Simulation of a village's evolution

Need to ask why

A year after the introduction of Histophysics, Peter Turchin proposed *Cliodynamics* in 2003. From the perspective of research methods, Cliodynamics is a branch of Histophysics. It is worth mentioning: Cliodynamics has published several monographs and an electronic journal (published by the University of California).

In 2022, someone called physics of history by the name *Cliophysics*, which is 20 years after the term Histophysics appeared. It is clearly not in line with the academic tradition of "who publishes first will name it" [14].

Remarks

In 1998, my second nonlinear physics book *Nonlinear Physics for Beginners* was published, and I had been doing science (physics, chemistry, and complex systems) for 30 years. The academic reason for shifting to the humanities is that the Active Walk paradigm [13.4], which I invented in 1992, has been widely used in natural science and social science, so I think it should be applied the humanities, too. Influenced by the people around me, I decided to study history after considering studying English literature or law. The details of how it evolved into Histophysics is described in [14].

5.5 Civilization: three-leg table

A society can be supported by *ethics* alone, like a one-leg table. That was a primitive society existing millions of years ago: There must be simple ethics such as "Don't eat me when I am asleep" when they lived together. As society became more complex, simple ethics evolved to something like "Thou shalt not covet thy neighbor's wife."

After *art* emerged one or two millions years ago [7.1], society was supported by two pillars (ethics + art), which was more stable, like a two-leg table.

Later, 2,600 years ago after *science* was invented in ancient Greece by Thales (c.624-c.546 BC) [8.1], society was supported by three pillars (ethics + art + science), especially in modern society, which is more stable, like a three-leg table.

Ethics

Ethics is not only spiritual, but a tool for maintaining social order and living in harmony. When the pillar of ethics collapses, the society "table" lacks a column, which, of course, is less stable or unstable. That is why there is a good reason to restore ethics, the sooner the better. The importance of ethics was recognized by Confucius who devoted himself completely to this.

In the West, the pillar of ethics is mainly provided by religion. In mainland China, is it provided by Confucianism? Theoretically, yes. But Confucianism is incompatible with the scientific spirit, which is the respect of truth, the dare to ask questions and debate everything. This is the main reason why in China there has been no major scientific innovation in the past 30 years.

Relative Importance

Relatively speaking, ethics is the *most* important of the three because ethical destruction leads to social instability.

The *second* most important is art. In addition to "killing time," emergence and development of art has accelerated the breadth and depth of human thinking and imagination, giving human the advantage over other animals in evolution [7.1, 13.6]. More importantly, art has ensured the quality of people's spiritual life and helps to make people willing to live in this seemingly meaningless world.

Science is not just to satisfy human's curiosity; its proper application is good for economics, medicine and world peace. Yet, science is the *least* important because the main function of science is to improve people's material living standards and prolong life. Think about the life before mobile phone appeared: Life was less convenient but more relaxed.

Conclusion

1. Art, science and ethics appeared in human history at different times.
2. In terms of decreasing order of importance, it is ethics, art and science.
3. Yet, art, science and ethics are the three pillars that underpin modern civilization.

5.6 Europe: reason and romance

Don't worry about France. Napoleon stipulated that all high school students should take philosophy and pass the exam before graduation, which still is. Every French person has a philosophical cultivation, sees problems deeply, knows what life is and how to live it.

The Italian *Renaissance* provided the world with good art (Leonardo da Vinci, Michelangelo), followed by modern science (Galileo). Britain provided reason (Magna Carta, Newton, Darwin).

The French *Enlightenment* (1688-1789) opened the search for "science of man" and later provided universal values (freedom, equality, fraternity) and romance (ballet, cinema, modernism in art). Street demonstrations and conflicts are part of the French political tradition, and are effective in solving political matters (Paris Commune, 1968 student demonstrations—sweeping the world, bringing the United States to end the Vietnam War).

The past few hundred years have been a model of Anglo-French cooperation to provide reason and romance for humankind. Don't worry about France. Worry about yourself and get your own things done.

Thanks to Italy, England, France.

Thanks also to China's Tang poetry and Song lyric.

6

Philosophy

6.1 Philosophy shrinks

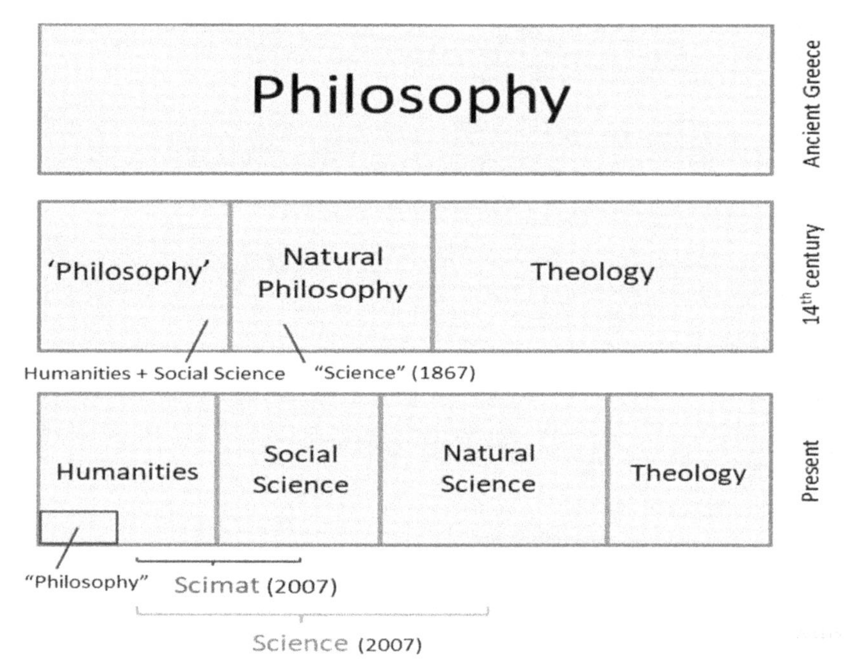

A brief history of disciplines. Blue = No supernatural. Red = Invoke supernatural.

Philosophy, the pursuit of wisdom without invoking the Greek gods, was invented by Thales (c.624-c.546 BC) 2,600 years ago in ancient Greece. While the Greek gods are not mentioned, other supernatural like *soul* could be invoked. Subsequently, in the 14th century, philosophy split into three parts: 'Philosophy' (single quotes), Natural Philosophy, and Theology. Later, the no-God parts of the first two expanded to form the three categories of learning as we know it today: Humanities, Social Science, and Natural Science. The humanities contain the "Philosophy" (double quotes) as we know it today. In other words, philosophy shrinks in its content; there are three different kinds of philosophy in history such that Philosophy > 'Philosophy' > "Philosophy." Discussions without distinguishing the three often lead to confusion and misunderstanding. For clarity, it is strongly urged that people adopt the convention of (single or double) quotes accordingly when they talk or write about philosophy from now on.

Searching History

There are two categories of things:

1. **Man-made** things. Things that are not naturally occurring, invented by people, such as semiconductor and plastic, are given a name (a new term) as soon as they appear. Consequently, there is no need to go back in history to search this item because you cannot find such a thing retrospectively.
2. **Natural** things. After natural elements (such as gold) are named by people, if one goes back a hundred or hundreds of millions of years, gold can be found to exist.

Most of the things in the humanities belong to the second category because the humanities are about people, who have existed for millions of years while the terms came into existence only thousands of years ago after the written word. For example, with the word “human,” humans can be found millions of years back in history, but “modern society” or “lawyer” cannot be found far in the past. For this matter, the origin of the academic “disciplines” needs to be traced backward in history and can be traced backward [2.1].

Disciplines: A Brief History

In ancient Greece 2,600 years ago, Greek mythology explained the world perfectly, but could not predict it, and thus could not satisfy the *curiosity* of Thales (c.624–c.546 BC) and others. Thales then took the crucial step of consciously getting rid of Greek mythology and using “rational” thinking to understand everything in the universe. The learning that comes out of this is called *Philosophy*. Accordingly, Thales is revered posterity as the Father of Philosophy.

The word philosophy stands for “love of wisdom” and was coined by Pythagoras (c.570-c.495 BC). To love wisdom is to love all wisdom, i.e., to care for everything in the universe.

Philosophy was the only discipline at that time that covered everything in the universe. All subsequent disciplines were separated from Philosophy.

Content of Philosophy is divided into *two* parts: one part does not introduce the *supernatural*, and the other part has. For example, Thales' statement that "all things are water" belongs to the former, and his belief that "everything has a *soul*" belongs to the latter. Why? The reason was that Thales, with his limited knowledge of physics at the time, could not explain why stones rolled, and the things he knew could move (actively) were people and other animals. And so, under the assumption that animals moved because they had souls, he naturally believed that stones also had souls.

Note that Thales' so-called "rational" is not the same "rational" as understood presently: The latter does *not* contain any supernatural, and the former does. As a philosopher, Thales did not completely abandon the supernatural. He actually did two things:

1. Change the supernatural to a different form: from the cumbersome Greek gods to a simpler "soul."
2. Some of his studies do not introduce the supernatural. (This part made him the Father of Science [8.1].)

In Europe in the 14th century, the number of people who did philosophy was large, so there was a division of labor or discipline. Philosophy is divided into three: 'Philosophy' (single quotes), Natural Philosophy, Theology. That is, 'Philosophy' is part of the original Philosophy. 'Philosophy' is the study of the human system; Natural Philosophy is the study of nonhuman systems; Theology is the study of everything related to God (then supernatural) under the premise of God's existence. This division of labor was due to the fact that Christianity dominated society at the time and believed that humans are living beings different from other animals: Humans were specially created by God. Note: Both 'Philosophy' and Natural Philosophy contain two parts: no-God and invoke-God.

Subsequently, the no-God part of 'Philosophy' was further divided into Humanities and Social Science, and the no-God part of Natural Philosophy was formally defined as "Science" in 1867 [8.1], which is also commonly called Natural Science. Later, the invoke-God parts in both 'Philosophy' and Natural Philosophy are incorporated into Theology.

At present, the great success of "science" of simple systems [2.5] and its great impact in the economy has not only squeezed the humanities and social science, but also greatly shrunk theology. Note that after several disciplinary splits, the "Philosophy" left in the humanities was "forced" to inherit the hard bones of the original Philosophy: the ultimate problem of human (such as the meaning of life) and the abstract problem of various disciplines (such as the philosophy of biology), becoming the most difficult subject in the university. In fact, "Philosophy" is harder to tackle and make progress than mathematics, say [6.9].

The status of the Medicine discipline is a bit complicated. It belonged to Philosophy in the early days, but in the Middle Ages its status was unclear because it studied the human body and could not be attributed to Natural Philosophy, even though its research methods overlapped with the empirical method used in Natural Philosophy. At the same time, it was excluded from 'Philosophy' because the latter used purely theoretical reasonings. It was not until 2007 when it was grouped into *Scimat* that medicine's classification problem was resolved [3.3].

The history of disciplines over the past 2,600 years has been the division from Philosophy to multiple disciplines, which are now grouped in four categories [2.5]. And it is also the process of expanding the non-supernatural part of learning in Philosophy to cover almost all the area of the *Knowscape* (knowledge landscape [2.3]) at the expense of shrinkage of Theology (see figure).

Three Kinds of Philosophy

Therefore, there are actually three kinds of philosophy: Philosophy, 'Philosophy,' and "Philosophy"—one shrinking more than the other. That is: Philosophy > 'Philosophy' > "Philosophy." The quotation marks cannot be removed, otherwise it will lead to chaos. Here is an example of how it works between two persons A and B:

A: Is Science part of 'Philosophy'?

B: No. Science is part of Philosophy.

Remove the quotes and the two sentences become:

A: Is Science part of Philosophy?

B: No. Science is part of Philosophy.

Here, A asks a different question, and B becomes incoherent. Can the two persons continue to talk? Obviously, it cannot. Without the quotation marks, the conversations will only become more and more confused, and discussion and learning cannot be done.

Conclusion

There are three kinds of philosophy. Philosophy > 'Philosophy' > "Philosophy."

The source of these troubles is that the humanists had not learned from the physicists: one word, one meaning. For example, if 'Philosophy' is renamed Philosophy 2 and "Philosophy" is called Philosophy 3, there is no need for quotation marks, and the academic world will be much simpler.

6.2 Western philosophy: ask everything

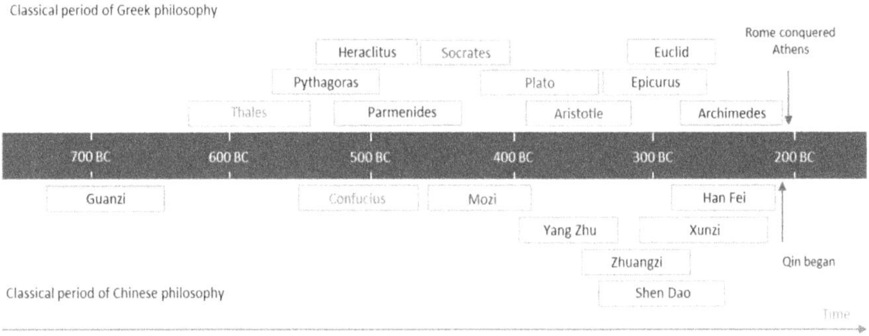

Classical period of Greek philosophy

Rome conquered Athens

Heracltius Socrates Euclid

Pythagoras Plato Epicurus

Thales Parmenides Aristotle Archimedes

700 BC 600 BC 500 BC 400 BC 300 BC 200 BC

Guanzi Confucius Mozi Han Fei

Yang Zhu Xunzi

Zhuangzi Qin began

Classical period of Chinese philosophy

Shen Dao

Time

A distinction of Western philosophy is that the philosophers ask everything, from the universe to things about human, from big questions to small questions. It is argued that the freedom and dare to ask everything and the Socratic method of interrogative debate are the two important techniques that enabled the West to advance knowledge and usher in science.

Ask Everything

Ancient Greek philosophy includes everything in the universe and covered all the basic disciplines of current universities, including arts and sciences. Philosophers at that time were people who enjoyed abundant freedom, and philosophy was not done to pursue freedom, but to satisfy *curiosity* about all things. In addition to asking big questions and small questions, they also care deeply about mathematics. Pythagoras (c.570-c.490 BC) was a famous "number nerd," and although Plato (c.428-347 BC) did not do mathematics, the school he founded in Athens in 387 BC (called the Academy) was engraved on the door: "Let no one ignorant of geometry enter."

Philosophers at that time asked everything. For example: What is everything made of? How are solar eclipses predicted? Should men and women who run government receive the same education? Should civil servants own private property? Should wives and children be shared socially? The only thing that was not asked was: Is slavery good?

Those philosophers also asked big questions: the nature of the universe, man's place in the universe, what is good and what is evil, the nature of god, fate and free will, soul and eternal life, man and state, man and education, man and matter, thought and thinking.

The Socratic Method

The Athenians invented a simple method of divination: oracles. A divine medium is a person who is inspired by the gods to predict the future. You can ask her any questions. For example: Will the woman I want to marry satisfy me? Will I have children? Will I get a good job? Will my next trip be dangerous? To which gods should I sacrifice to stay healthy? Who stole my sheep? Is the child my wife carrying mine? Which god can bless me with a prosperous business?

At one point, someone asked: Is anyone smarter than Socrates? The medium replied: No. Socrates (c.470-399 BC) felt that it was a paradox. To which he said: "I know that I know nothing." After that, he searched everywhere for really smart people, only to find that everyone's ideas were full of loopholes. The interrogative method of research used by Socrates is known as the "Socratic method." Here is an example of how it works:

1. A person A asserts that "courage *is* the endurance of the soul."
2. Socrates persuades A to agree that "courage is a good thing" and "patience based on *ignorance* is not a good thing."
3. Socrates then argues, and A agrees, that these further premises imply a view contrary to the original argument: "Courage is *not* the endurance of the soul."
4. Socrates then claims that he had proved A's argument wrong and that his negation was true.
5. A revises the assertion to "courage is the *knowing* endurance of the soul."

The essence of the Socratic method is this: Through a series of questions, forcing the other person to examine the foundation, essence, and validity of their beliefs, so as to revise or abandon their beliefs and reach a higher level.

Discussion

1. The questioning of slavery is a very recent event in history. Ancient Greek slaves had three fates: servants, prostitutes, and intense laborers. Some servants were treated pretty well. For example, a Greek sent the valet, a male slave, to study accounting, let him manage the family finances, and entrusted him with taking care of his wife and children when he died. On the other hand, those sent to mines generally lived less than eight months. Philosophy (parts of which form science) emerged in ancient Greece in part because slavery made Athenian society quite wealthy and convenient, allowing the Greek intellectuals to have a great deal of leisure time, the prerequisite in asking questions that have no immediate use. This is in contrast to the ancient Chinese intellectuals who had to find a government job to support themselves and their family and so ended up doing political philosophy.

2. If the inside of a circle is regarded as known knowledge, then the circumference represents unknown knowledge. When the known continues to expand, the unknown will expand with it, and the process of "asking everything" will not end.

3. There are three places where high-level intellectuals met in history: Athens in ancient Greece, the current Left Bank of Paris, and Zhongguancun in Beijing. The intellectuals' assembly is conducive to two things: debate and collaboration. In the West, debate is more abundant and important compared to collaboration. It is debate, like in the Socratic method, that advances knowledge and enables innovation. Yet, in China, debate is generally lacking, in the past or present, partly due to the dominance of Confucianism.

Conclusion

Keep asking, and you will reach the frontier of knowledge. If you are lucky you may end up with a discovery or innovation, and even win a Nobel Prize. But more influential than winning a Nobel Prize is to create a new field of study. For example, Einstein's general theory of relativity pioneered Cosmology, but did not win him the Nobel Prize. The one theory that pioneers a new field and wins the theorist Nobel Prize was Heisenberg's quantum mechanics.

"Ask anything" and the "Socratic method" are the two killer techniques for the rise of Western science. And it is also the thing that philosophers/physicists annoy their friends and foes alike, from time to time.

Remarks

History of Western Philosophy by Bertrand Russell (1872–1970) is the first choice for entry to Western philosophy. His mathematical background, clear writing, and clear points of opinion tell you directly where every philosopher is wrong. For more, the first choice is the Stanford Encyclopedia of Philosophy (plato.standford.edu).

6.3 Chinese philosophy: ask nothing

In contrast to Western philosophy that asks everything and advance through debates, Chinese philosophy discourage debates and asking questions. Why is it like that? To answer this question, the case of two ancient Chinese philosophers, Guanzi (c.723-645 BC) and Confucius (551-479 BC), are presented. Guanzi pioneered the model of dual-professions of service and learning, which was followed by other Chinese philosophers that came later. He also proposed the scientific theory that "all things originate from water," about 100 years ahead of Thales. Confucius tried the Guanzi model but failed, and ended up a great educator who established the first school for civilians in the world. Yet, by necessity, the philosophy of Confucius and many like him is confined to political matters. To understand this is to understand the reasons behind the differences between the Western and Chinese philosophies.

The Hunger Games

Classical Chinese philosophers lived during the Spring and Autumn (770-476 BC) and Warring States (475-221 BC) periods. Spring and Autumn period covers the early part of the Eastern Zhou dynasty (770-256 BC); the Warring States period ranges from the latter period of the Eastern Zhou dynasty to the establishment of the Qin dynasty.

Guanzi (管子) and Confucius (孔子) were from the Spring and Autumn period; Mozi (墨子), Yang Zhu (杨朱), Zhuangzi (庄子), Shen Dao (慎到), Xunzi (荀子), and Han Fei (韩非) were Warring States people (Table).

Table. Comparison of Ancient Chinese philosophers. Pink = Spring and Autumn period (770-476 BC). Blue = Warring States period (475-221 BC).

Philosopher	Lifespan (BC)	Class origin	Official position	Proposal
Guanzi	c.723 - 645	Descendant of King Mu of Zhou dynasty; father a *daifu* (大夫) of the Qi state	State Minister of Qi state	People know honor and disgrace when they have enough food and clothing; all things originate from water
Confucius	551 - 479	Nobleman of the Song state; father a daifu of the Lu state	Grand Secretary of the Lu state (acting state minister)	Benevolence, righteousness, courtesy, wisdom, and trust
Mozi	c.476 - c.390	Ancestor Yin Shang royal family; peasant	Song state daifu	Anti-Confucianism; advocate science, universal love, and anti-war
Yang Zhu	c.395 - c.335	(Unknown)	(Unknown)	Love yourself; enjoy life
Zhuangzi	c.369 - c.286	Descendant of the Duke of Song state	Song dynasty lacquer yard official	Enjoyment of untroubled ease, adjustment of controversies, health preservation
Shen Dao	c.350 - c.275	(Unknown)	Qi state daifu	Rule by law; emphasize power
Xunzi	c.313 - 238	Landowner	Qi state Tribute (祭酒)	Human nature evil; educatable
Han Fei	280 - 233	Aristocracy of state of Korea	County Magistrate of the Chu state	Despotism

In the Spring and Autumn period, the king of Eastern Zhou was weak, and the dukes competed for hegemony, with a total of more than 140 duke states. In 242 years, 36 monarchs were killed by subordinates or enemy states; 52 duke states were destroyed; there were more than 480 large and small wars; more than 450 courtship appointments and alliances of dukes.

The fierceness of the competition among the duke states is only comparable to the high-tech business competition in California's Silicon Valley. At present, Silicon Valley has thousands of emerging companies. The demise and listing of companies, and mergers and acquisitions between companies occur almost every day. No company has an absolute sense of security.

The difference is that the Spring and Autumn competition involve human lives, while the Silicon Valley only involves money. So, the drama played out between the duke states in the Spring and Autumn era was more like that in *The Hunger Games*, the 2013 movie.

The Guanzi Model

Athens in ancient Greece was a city-state, with constant wars with the surrounding city-states. Similarly, in the Spring and Autumn period of ancient China, the "central plains" was also a place with continuous wars between the duke states. Apart from this similarity, there are various obvious differences between the two. For example, weather, population, and area.

However, it is not these superficial differences that determine the direction of ancient Chinese and ancient Greek philosophies. It is that Athens had a relatively stable and rich society with a group of relatively leisure people while the central plains did not [6.2].

The Athenian philosophers' methods of subsistence included private tutoring (for a fee or receiving donations), finding wealthy patrons, and being wealthy themselves. The philosophers of the central plains also had these paths, but Guanzi, Confucius, and Mozi were all poor and had to find jobs to support their families, and the rich patrons at that time were the governments of various duke states. This meant that the Chinese philosophers' first choice of job was to be civil servants. Of course, the higher the civil servant, the better: high salary and high social status. And the basic skill of senior civil servants is to govern the country well, resulting in the vast majority of philosophers in ancient China engaged in *political* philosophy.

Yet, these philosophers who successfully became officials were all scholars, and they did not forget to be academic. What vividly interprets this dual-professions is the *Guanzi Model*: to cultivate learning supported by an official's salary and apply what they have learned to governance in their jobs. Guanzi stood high in the state of Qi—under only one person, the king. He not only governed the country splendidly but also put forward a breakthrough philosophical proposition: "All things originate from water." This was about 100 years ahead of the ancient Greek Thales' "all things are water" (see Discussion). In essence, the Guanzi model is "grasp service and learning with two hands."

Confucius' Proposal

Confucius' mother was concubine No. 2 who gave birth to only one child—Confucius. Confucius lost his father at the age of 3, and the mother and son were expelled from the house by his wife. Subsequently, Confucius was raised by a single mother and studied hard. At the age of 20 he had a wife and a son. He tried to be a civil servant but failed—the Guanzi model did not work for him. Before that, he began to tutor students at the age of 23 and, after a break of several years, expanded the class to a school at the age of 30, which was the world's first school for civilians [6.4].

Confucius' way of governing the country: each in his place, obedient at each layer. Basically, it is a theory of stability maintenance: starting from ethics, starting from the family, and building a super-stable society. In practice, he added to it the doctrine of "father for the son, son for the father." That is, for example, when the son commits a crime, like stealing the neighbor's sheep, the father should not report it to the police, vice versa. In other words, create a "father-son alliance" through mutual secrecy, which shows clearly that Confucius did not care about "seeking the truth," not to mention rule by law.

Asking Why Forbidden

How did Confucius argue? Roughly, it goes like this. The child asked dad why he had to be obedient, and dad replied: "If you are disobedient, you don't have food to eat." Not enough? Dad added: "Your grandfather is obedient and became extremely successful." That is, there are punishments and rewards as well as thankfulness. This is different from the ancient Greek approach, which is through reasoning.

Confucius discouraged asking why because many of the questions he was not necessarily able to answer or willing to answer. Most of the Confucius students ask the teacher as a formality so the master can go on to elaborate, not to discuss with the teacher, let alone debate. Debates with teacher are called "rebuttals," which are against your superior and are straightly not allowed. This is opposite to the ancient Greeks' way of education, in which debates are daily exercise and encouraged, as shown in Plato's Academy, the first university-like institution in the world [6.4]. Confucius did not encourage people to dig deep in any topic, nor think thoroughly and independently, not only about people or society, but also about all things, including natural phenomena, which, of course, was not conducive to scientific development [9.5].

Why Not Marketable

Compared with his peers, Confucius' philosophy of governing the country is not bad, why can't it be sold? Reason: In the hunger games of fierce competition between the dukes of the Spring and Autumn period, Confucius' proposal is all about maintaining stability, which is slow to take effect. It is only suitable for a peaceful society after all the states are unified so that there is a stability needing sustention. Some years after Confucius' death, the Qin dynasty unified the country and ruled for only 15 years with two emperors in succession. Qin is thus too short to have the chance to use Confucianism as a governing tool. It was not until Emperor Wu of the Han dynasty (156-87 BC) that succeeded Qin, that Confucianism was revered exclusively. In other words, Confucius' proposal, like Van Gogh's paintings, had no buyers at the time of their appearance because they were ahead of their time.

Confucius' career was not smooth but he was successful in establishing a school for civilians, out of financial necessity because being a senior civil servant, not education, was his priority in life. In his later years he studied the Six Classics: *Poetry*, *Book*, *Rites*, *Music*, *Yi*, *Spring and*

Autumn, and became a master of the generation. After his death, his disciples compiled Confucius' words, deeds, and thoughts into the *Analects*, which became a classic of Confucianism.

Discussion

1. **Guanzi and Thales**. Guanzi's proposal that "all things originate from water" is ahead of Thales' "all things are water" by about 100 years [9.1]. Both are right then but are wrong later. But both are theory to explain nature without supernatural. And so, by science's 1867 definition, both are science [8.1]. Despite this, Thales is called the Father of Science.

2. The **Guanzi Model** means that Guanzi is a full-time civil servant but a *part-time* scholar. The advantage is that this solves the financial-support problem of doing scholarship in the ancient times when modern university was not yet invented (which appeared in 1088 in Italy). The disadvantages are equally obvious and important: (1) Doing scholarship part time has the consequence that many topics are not investigated thoroughly, unlike the case in ancient Greece that the scholars were doing it full time. (2) Part-time scholars do not teach and have no students, resulting in the inefficient passing-down or extension of their thinking. In contrast, Thales' theory was revised by his disciple Anaximander and subsequently Anaximenes, the disciple of Anaximander. Likewise, Socrates had student Plato who was the teacher of Aristotle, and their philosophical thinking kept on being revised or replaced. In this regard, Confucious' failure to be employed as a civil servant except for a few brief years should be counted as a blessing to him as a scholar, meaning that he has time to teach and do scholarship full time and thus ends up more influential in later cultural history.

3. *Analects, Yan Yuan*: "Sovereign and courtier, father and son (君君臣臣父父子子)." This example exemplifies the characteristics of Chinese philosophical writings: not many words, no details, no arguments (or very short arguments). In contrast to the Greek philosophy, which is analytical by keep on asking why, this style of philosophy is called "**fuzzy philosophy**." There are several possible reasons for the

ambiguity: (1) If it is not vague, it will be finished in two minutes. (2) It cannot be explained clearly because the author has not thought it through. (3) It is inconvenient, for political reasons or otherwise, to explain clearly.

4. The ancient Chinese books are **brief** with few words, probably because the enactment of writing is not an easy task (using bamboo slips or silk; the former is heavy and the latter expensive), but more likely it is intentional (see discussion #3 above). In this regard, it is interesting to note that there are rich people among Confucius' disciples. So, there should be no financial difficulties in producing the *Analects* with more words. By the way, paper in China was invented by Cai Lun (蔡伦) in 105 AD during the Eastern Han dynasty, way after the *Analects* was published.

5. The ancients could not be smarter than modern people because the two are merely about two thousand years apart, and human evolution is extremely slow. So, the **IQ** of the two should be almost the same. Besides, we know more than the ancients, much more. I think the reason that some people find the ancient Chinese books broad and profound is because they are written vaguely, lacking details like using broad strokes in a painting. As an example, the left picture below is definitely broader in content and more profound in meaning than the one on the right. In the left picture, you can imagine it to be anything that comes to mind; you can never guess precisely what the painter is trying to say. As for the right picture, it just tells us that there is a woman smiling—no big deal.

6. Doing **physics** requires analytical thinking, and not that many philosophers do physics at the same time that they do philosophy. Thales (c.624–c.546 BC) and Aristotle (384-322 BC) in ancient Greece both did physics; Plato did not. Plato did have mathematical training though so he could probably do physics if he wanted to. Among the ancient Chinese philosophers, only Mozi did physics. Can Confucius do physics? No. Can Confucius' disciples do physics? No. They could not even express themselves clearly, and did not dare to ask questions.

7. There are three similarities between the **history** of ancient China and ancient Greece, which occurred at about the same time [1.4]:

A. The classical period of philosophy in China ended when the Qin dynasty unified the warring states in 221 BC. That in Europe ended when Rome destroyed the Athenian city-state in 146 BC.

B. New ideas/ideologies stopped coming when a country is unified. That happened in China in 221 BC with the beginning of the Qin dynasty, and in Europe, in 117 AD when Rome conquered a large part of it.

C. Confucianism was worshiped exclusively in 134 AD during the Han dynasty. Christianity was recognized by Rome in Europe in 313 AD and established as the state religion ten years later.

8. Historically, there were three types of **societies**:

A. Everyone is comfortable and happy.

B. Some people are comfortable and happy.

C. Everyone is uncomfortable and unhappy.

The top ruler usually goes towards A or B. But the ideal and reality do not necessarily overlap, and a misstep or misfortune the society will end up as C. All ideals are constrained by tradition and human nature.

Conclusion

Generally speaking, Chinese philosophy is to stabilize society and Western philosophy is to understand the world.

Between the Chinese philosophy and Western philosophy, the purpose, content, and approach are different. Like coffee and tea, each has its own followers. It is not about good or bad. Good or bad is related to geological location, aim of life, and personal worldview. These are the topics of futurology.

Two Fables

When Confucius Meets Plato

Confucius and Plato, both political philosophers, meet unexpectedly in Heaven's Tea & Coffee House. Each took out the mobile phone. The former used Huawei; the latter used Apple. Both have voice translation; conversation is not a problem. Confucius ordered the West Lake Longjing tea, and Plato ordered a Frappuccino. The dialogue is as follows:

Confucius: Only women and villains are difficult to support.

Plato: Why?

Confucius: If you are near, you will be disrespectful; if you are far, you will be resentful.

Plato: Why?

Confucius: Knowing is knowing, not knowing is not knowing, is knowing.

Plato: Why?

Confucius: Is climate warming real?

Plato: This question you have to ask my student Aristotle.

Old Shop New Business

It is said that there is a century-old shop that produces and sells all foods related to noodles: noodles, dough, gluten, etc., and its business is prosperous. Although the shop has been maliciously acquired by the barbarians at the door twice the new owner never changed the name of the shop and thus the shop, on paper at least, never went bankrupt.

Since then, the store has been robbed several times but the shop survived. In order to restore the glory of the ancestral business, the old shop has a new idea: In addition to noodle-related food, the shop also wanted to enter haute couture, on the grounds that it was profitable and could be sold far away.

New ideas bring in a new question: Can the same thinking and practice that underpin the noodle business be used for fashion designs that require creativity? Yes, excellent. Can't, what should be done?

6.4 Confucius and Plato's schools: a comparison

When physics professors in their school days studying for a PhD, they know that they like to do physics, not teaching physics. As professors they are forced to teach. Confucius, on the other hand, was "forced" to teach in order to raise his family after failing to follow the Guanzi model of being a scholar while working as a civil servant [6.3].

However, evaluation of historical merits is based on the effect, not the initial motivation, which is not the most important. Accordingly, Confucius, as the first educator in the world to open a civilian private school, is a pioneer in education.

School Experience

Confucius, born in Zoyi (陬邑), Lu state (now Qufu 曲阜, Shandong Province), in 551 BC; died in 479 BC at the age of 73 and was buried in Sishui (泗水), north of the Lu city (鲁城). His brief biography is given in Table 1.

Table 1. Biography of Confucius. Green = government employment, Pink = teaching.

Age (year)	Experience	Timespan (year)
0+	Born in Zoyi, Lu state (鲁国)	
3-20	Lost father at age 3, mother at 17; married at 19; had a son at 20	17
19-22	At age 19, became a petty official as keeper of granaries; at 20, same, in charge of livestock	3
23	In hometown: began teaching private students	1
24-30	At age 27, learned ancient official system and musical instrument playing	6
30-35	In hometown: around age 30, started a school and met Duke Qi Jing-Gong (齐景公) who was visiting Lu state	5
35-37	In Qi state (齐国): at age 35, left Lu state and went to a neighboring state in the north, becoming a vassal of the Qi state; that year, Duke Qi asked him about politics, and Confucius replied with hierarchical absolute obedience; at 36, learn Shao music (韶乐); at 37, returned to Lu state for fear of being murdered by colleagues	2
38-50	Teaching; at age 47, edited 5 classics: *Poetry*, *Book*, *Rites*, *Yi*, *Music*, and took disciples to travel to various states	12
51-54	Starting his official career: at age 51, worked as minister of works and at 53 as minister of crime, doubling as prime minister, in Lu state; disappointed by the Lu ruler at 54, left Lu state and went to Wei state	3
54-68	At age 54, began to visit many states to seek official posts (taking disciples along but not family members); at 56, Wei state provided the same salary as Lu state did but did not let him participate in politics as an official, and so left after 10 months; failing to obtain a government job along the way, returned to Lu state at 68	14
68-73	In hometown: After returning to Lu state at age 68, found that his wife had died a year ago: at 69, his only son died; completed *The Spring and Autumn Annals* at 71; died at 73	5

Confucius was poor, and he worked as a petty official at the age of 19 and 20. Since his 20s, he wanted to follow the Guanzi model in his career; i.e., work in the government while remaining a scholar [6.3].

At the age of 23 in hometown, Confucius began to take in apprentices to lecture. His students included Yan You (颜由, father of Yan Hui 颜回), Zeng Dian (曾点, father of Zeng Shen 曾参), Ran Geng (冉耕). Around the age of 30, some of the disciples returned and he expanded the enrollment to form a school. Since then, Confucius has been engaged in education. Confucius spent a total of 23 years in education. But he did it part time for 12 years while traveling, and only 11 years full time when he was staying in his hometown.

Confucius worked as an official in his own Lu state for three years between the ages of 51 and 54, reaching the rank of minister of crime (大司寇) and doing the work of the prime minister (i.e., exploited by his employer). During this period, he showed his harsh face: Sentenced Shao Zheng-Mao (少正卯) to death and exposed the corpse for three days; tried to demolish the castle built by three high officials (but only succeeded in two). The record was mixed.

Confucius resigned from the government at the age of 54 and traveled to various states with his disciples—not a study tour, but a teacher's journey to seek government posts. After 14 years of unsuccessful job seeking, he finally returned to Lu state at the age of 68 to concentrate on teaching and writing, and died five years later.

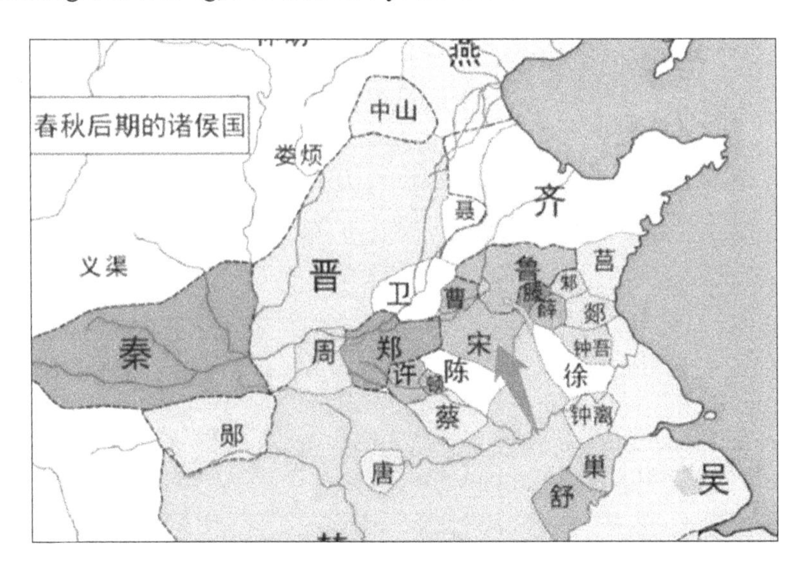

Before Confucius opened his school for common people there were government schools that only accepted aristocratic children. The increasing number of declining aristocrats like Confucius, together with the poor, have no school to go. Confucius learned through self-study. His running of an affordable private school (tuition fee 10 strips of dried meat or more), though forced by making a living after failing to find a state job, did meet the market demand and has the effect of breaking the government's monopoly of education. In this sense, he was a pioneer. There were as many as 3,000 disciples over the time, 72 of whom were outstanding. Many of the 72 later became high-ranking officials in various states and continued Confucianism, but none of them became great scholars.

Two Educational Models

Confucius began his private teaching in 529 BC and began running a school in 522 BC, 135 years *before* Plato. Both ran *civilian* private schools, but they followed two different paths and created two educational models with far-reaching impact (Table 2).

1. Confucius' school was not full-time for more than half of the 43 years from its opening to its closure.
2. Plato's school was run full-time for a total of 304 years, with first 40 years under his leadership and then took over by successors after his death.
3. Confucius' school emphasized inheritance, and Plato's school emphasized innovation.

Discussion

1. In the Spring and Autumn period, people's life expectancy was not as long as it is now; e.g., Confucius' son lived only 49 years. When Confucius finished his job as a senior official in his own state of Lu for three years, he was already 54 years old—an "old" man. He could be near the end of his life and had no idea how many years left of him. He could choose to return to his hometown, concentrate on running his school and writing books (*The Spring and Autumn Annals* did not begin or finish, which was only completed at the age of 71). But instead, he chose to travel to other states to seek high government jobs, and there were thrills and dangers for 14 years along the journey. This obviously is not a mid-life crisis, but an "end-life crisis." This choice shows that even at that time, he valued a political career as more important than running a

school or doing scholarship. Is it because he felt that the three years of unfulfilling works he did at Lu state was due to bad luck only, and so wanted to prove himself in other places? Or is it something else?

Table 2. Two models of education: the Confucius model and the Plato model.

Content	Confucius' school	Plato's school
Years	Classes only in 529 BC; school established in 522 BC; school closed when Confucius died; school lifespan is less than 43 years	School founded around 387 BC; Plato died 40 years later; school lasted until 83 BC; school lifespan is 304 years
Name	(unknown)	The Academy
Place	Zoyi, Lu state	Athens, Greece
Type	Grade school to university	Comprehensive university
Entrance exam	None	Knowledge of geometry
Fee	10 strips of dried meat or more	Free (at least in the first 40 years)
Scholarship	None	(no need)
No. of students	3,000 total (no female students)	Tens of thousands total (at least 2 female students in the first 40 years)
No. of teachers	One	Numerous
Ranking	No. 1 in the world	No. 1 in the world
Teaching aims	How to behave and do things; pass on the teacher's teachings to all generations	Seek the truth; understand the society/world; produce new knowledge
Teaching policy	Teaching without discrimination; teaching according to one's aptitude; integrating knowledge and action; applying what is learned	Everyone is equal
Teaching content	Confucius' doctrine	Plato's doctrine; mathematics, physics, astronomy, dialectic, philosophy, politics
Teaching methods	Indoctrination; teacher's inability to answer student's question is considered undermining dignity of the teacher	open-ended (lectures and seminars); if student's question cannot be answered, the whole class will discuss it together
Discussion mode	A simple sentence or two	Interrogation (Socratic method)
Study travel	Traveled various states for 14 years	None
Famous alumni/ achievements	72 people (Yan Hui, Zilu, Zigong...)/editor of the *Analects*	Aristotle (studied for 20 years, 367-347 BC, from age 17 until Plato's death)/flipped teacher's theory/ excelled in both arts and science
Essence	Study and pass on Confucius' teachings	Truth-seeking; study both arts and science; innovation

2. About 30 years after the death of Confucius, the Jixia Academy (稷下学宫) was opened in the Qi state during the Warring States period. It was the earliest government-run institution of higher learning and think tank. Many schools of thought, including Confucianism, Daoism, Legalism, Famous Scholars, Soldiers, Peasants, and Yin-Yang, were allowed and gathered there; freely lecturing, writing books and arguing were allowed and encouraged. (In later generations in China, only Peking University under Cai Yuan-Pei was comparable.) This Jixia Academy was taking the exact opposite path to the Confucius School. Was it a reaction to Confucius' school?

3. For more than 2,000 years, Western education has followed the Plato model and, as a result, the development of science is promoted. Since Confucius, except for the few decades of the Republic of China, Chinese education has used the Confucius model, including the purpose/method of teaching and the concept/method of accepting apprentices (such as the admission of graduate students to the universities). The lack of major scientific innovations in the last 30 years is related to this. Whether independent innovation can be made in the next 30 years is related to whether the education model can change course.

4. Confucius opened a school because his career was not smooth and became a Master in education. Sima Qian (司马迁, c.145-c.87 BC) was imprisoned for a crime and wrote the *Historical Records* (史记) and became the father of (Chinese) history. These two cases are examples of "a blessing in disguise," just as some people did not get into a prestigious school but thus achieved a great career; there are many such examples.

5. Mozi (c.476-c.390 BC) was the first person in history who excelled in both arts and science [6.6]; Aristotle (384-322 BC) was the first such person in the West.

Conclusion

1. Chinese education focuses on "defense." Western education focuses on "attack."
2. The essence of innovation is to attack, and the educational model should follow accordingly.

6.5 Confucius' dad: must have son

The self-evident saying that "women can hold up half the sky" has been less than a century old in modern China. But in older times, no one has believed it.

In ancient China, it was not only "preference of boys to girls" but "women do *not* count," except for such things as cooking rice and giving birth. Where did this concept come from?

Mencius (c.372-289 BC), more than a hundred years after the death of Confucius, was a representative figure of Confucianism during the Warring States period. When Mencius said, "There are three ways to be unfilial, and having no offspring is the greatest," he only singled out the phrase "women do not count" but it was not his invention. (Here, according to the general interpretation, "no offspring" means that there is no son.) Confucius' father was more demanding than him.

Institution of Marriage

More than 10,000 years ago, during the pre-Neolithic period, families consisted of loose organizations of up to 30 people which consisted of several male leaders, sharing multiple women and children. As hunter-gatherers settled into agricultural civilization 10,000 years ago, society needed more stable arrangements, and *marriage* was established.

The earliest marriages date back thousands of years, and one of them was *monogamous*. The earliest documented monogamous wedding took place in 2,350 BC in Mesopotamia in the Middle East. However, for thousands of years before 100 years ago, the essence of marriage in China and beyond was that women became the private property of men, and had nothing to do with love.

In the animal kingdom, only about 3-5% of the more than 4,000 mammals on Earth practice monogamy even though about 90% of birds are monogamous. Monogamous animals include penguins, dune cranes, seahorses, gray wolves, barn owls, wall-backed stone dragons, bald eagles, gibbons, black eagles, beavers, and swans. Therefore, monogamy is not natural, but can only be enforced by societal norms through culture or laws. On the other hand, from the post-mortem analysis of evolutionary theory, monogamy is beneficial, which is why it appeared in history. For example, monogamy avoids favoring men with superior conditions over poor men and ensures that almost all men have the opportunity to marry a wife, otherwise it will cause social problems. Another advantage is to ensure that the child is raised by parents together, which is very beneficial to the child's healthy growth.

In China, *polygamy* was abolished in 1912 by law, and earlier in Hong Kong which was governed by the British. In mainland China, the actual practice of monogamy did not take place until 1949.

Marriage System in the Spring and Autumn Period

In the Spring and Autumn period, wife and concubine were legal terms, and duties and rights were stipulated according to law. There are two levels of concubines: *yingshi* (媵侍) and *qieshi* (妾侍). Yingshi refers to the woman brought along by the bride, but qieshi is not. At that time, when the royal family or nobles took a wife, the wife had to bring along one or more yingshi; i.e., to marry a wife will end up with a wife and several accompanying secondary "wives." There are three types of yingshi: high-level ones are sisters of the bride, medium-level ones are women of the same clan, and low-level ones are handmaids.

The wife should come from an appropriate family that matches that of the groom but the concubine does not have to, as long as the man likes her. The gift given to the bride's family from the groom's parents is called "dowry," and the gift given when taking a concubine is called "the capital to buy a concubine." The wife listens to the husband, and the concubine listens to the husband and the wife. Also, the yingshi has a higher status than the qieshi and can attend formal banquets. The son born to a wife is called the "legitimate son" (嫡子) while the son born to a concubine is called the "commoner son" (庶子). Only the eldest legitimate son has the right to inherit the estate.

In short, the marriage system in the Spring and Autumn period was "one husband, one wife and several concubines."

Confucious' Dad Fouled

Confucius' father was a famous warrior in the Lu state, with excellent health. With two battle merits, he was appointed a middle-level official in his hometown. Because he was not a noble, his wife did not bring any yingshi concubines when she passed the door. The wife gave birth to nine daughters but no son. He also married a qieshi concubine who, unfortunately, gave birth to a son with a disabled foot. Since neither his daughters nor the *crippled* son are qualified to perform ceremonies to worship ancestors, the 72-year-old father of Confucius begged a good friend of three daughters to marry one of them to him. The friend asked his daughters who would like to marry, and the two eldest were silent. The youngest was only 18 years old, and knew that fate had been arranged, so she said: "The righteousness of a woman at home is to follow her father's order. Why ask?"

The 72-year-old man took an 18-year-old concubine (the age difference is the same as 82 to 28), which was a beautiful thing (from the man's point of view), but it was not in line with the regulations at the time. What regulations? This article:

> A male starts to have teeth when eight months old, which become destroyed at the age of eight. His sperms start at the age of sixteen and stops at sixty four. A female starts to have teeth when seven months old, which become destroyed at the age of seven. Her ability to bear children starts at the age of fourteen and stops at forty nine. Marriages beyond these limits are all "wild unions" (野合).

That is, the age of marriage is limited to 16-64 years for men and 14-49 years for women. Confucius' dad, aged 72, had exceeded the *upper* age limit for marriage of 64, and this marriage was not in accordance with etiquette. Confucius' dad *violated* the rules. However, it was precisely because Confucius' dad dared to foul the rules that there was a great commoner educator in the world, whose name was Confucius.

Unfortunately, the wife of Confucious' dad was not a kind person. When Confucius' dad was still alive, concubine No. 1 (the disabled son's mom) was repressed by her and died. And when Confucius' dad died, Confucius was three years old. Confucius and his mom (concubine No. 2) were forced to leave the house. Confucius' mom was a kind person: When she left with the three-year old Confucious, she took the crippled son of concubine No. 1 with her.

Conclusion

The story of Confucius father's three marriages shows that at that time, not only "women do not count," but also "disabled men do not count." The economic foundation behind gender inequality and inheritance bias against women is due to the lack of a societal safety net (adequate pension, full medical insurance), and so the elderly need the son's support to survive. This matter has been proven in history.

Confucius dad's violation of the rules adds one more Master to the world, showing that human problems are complex enough and not all rules are 100% sensible. Any simplistic regulations have inadequacies. Individuals and society should learn to face them and have a way to remedy them.

6.6 Mozi's governance

Mozi (c.476-c.390 BC) was the only peasant among the ancient Chinese philosophers. He was also an innovator in science and engineering. He led a group of volunteers to help ordinary people and small states to defend themselves. His proposal of how to govern a country looks very familiar to what is seen today. Here, his life, career and achievements are outlined.

Mozi's Life

Warring States (475–221 BC) was a period of chaotic wars between many duke states before the Qin dynasty unified the central plains. A year before the beginning of the Warring States and three years after the death of Confucius, Mozi was born. Mozi, descendant of patriarch of the Yin Shang royal family, was a peasant and a *daifu* (大夫) official of the Song state. He promoted anti-Confucianism and advocated science, love others, non-offensive, respect sages, frugality on expenses and burial, and "not happy, not life." Mozi is a thinker, educator, scientist, engineer, and military strategist. Mozi had established a school to train people to serve in the governments.

A Romantic

Hundreds of Mozi's close disciples, including a large number of handicraftsmen and lower-class scholars, organized a Mozi guild (civil armed group). All the members wore short cloths and straw shoes, abled to work and endure hardships, stressed discipline, and were good at fighting. The guild acted chivalrous and did not seek rewards.

After Mozi's death, his disciples scattered to various states; some became scholars, others became chivalrous.

Science and Technology Pioneer

Because of his upbringing as a peasant and work as a carpenter, Mozi thought and acted practically. Unique among the philosophers, he assumed the triple roles of scholar, scientist, and action man. He constantly asked "what happened" and "why," and had done first-class research in mathematics, physics, and mechanics.

He was a pioneer of optics, developed the principles of logic, and invented very creative things: mechanical birds, transport carts, ladders for sieges (see *Mohism*).

Keen and creative, Mozi is a successful innovator in the field of science and technology.

Mozi's Governance

Mozi is not an anti-system rebel. Like Confucius, his goal was to improve and stabilize society.

He suggested that officials at all levels collect people's thoughts, report them one layer up, layer by layer, and finally reach the highest level, so that the king can make the best decisions based on big data. This, of course, assumes the king is super smart and can digest big data without the help of a quantum computer.

Mozi's rule of the state is a completely different approach compared to Guanzi or Confucius' [6.3]. Mohism was a powerful rival to Confucianism. It was listed as "outstanding learning" alongside

Confucianism, with the saying that "if not Confucianism, it is Mohism." Unfortunately, Mohism was suppressed in the Qin dynasty, and nearly died out in the Han dynasty after Confucianism was revered exclusively.

Conclusion

As a scholar Mozi's problem is that he had too many interests, more than Thales, so none of them was done thoroughly and deep enough, unlike Plato and Aristotle, say. But he is unique: can write and fight, and excel in both theory and action.

Unfortunately, Mohism's impact on China's later generations is incognizant when compared to Confucianism—not his fault, though. The problem is that in a one-dimensional society, the preferences of the top decision maker determine everything. Imagine if the Han dynasty revered Mohism instead of Confucianism, would China be the same today?

Mozi is different from other philosophers in that his attitude towards the world is rational, empirical and pragmatic—very modern and close to the present.

Mozi excelled in both the humanities and science, more than a hundred years before Aristotle.

Mozi, a *rational romantic*, an excellent humanities-science person.

Remarks

A rational romantic is a person who analyzes all the possible options but not always picks the one best for himself/herself.

6.7 Kant's mistakes

Any system of philosophical thought is based on established scientific knowledge at the time. Due to the continuous emergence of new scientific knowledge, especially in the past hundred years or so—evolutionary theory, electromagnetic theory, quantum mechanics, relativity, cosmology, big bang, chaos, emergent properties, etc.—the various philosophical theories proposed more than 150 years ago are largely outdated because they no longer conform to current scientific understanding. Moreover, there are many hidden assumptions embedded in any philosophical theory that the author may not even be aware of, and that render their theory wrong or useless. The case of Immanuel Kant (1724-1804) is presented here as an example.

Kant's Mistakes

Here, philosophy = "philosophy" (double quotes), the philosophy of recent centuries [6.1].

In Kant's times, science was Newton's science of deterministic systems: Results are certain, given initial conditions. Thus,

1. Kant first *mistakenly* believed that science belongs to the realm of certainty, the kingdom of *necessity*.
2. Following this, he *correctly* realized that morality is uncertain and belongs to the kingdom of *freedom*; i.e., human has the freedom to choose in moral matters.
3. He then asked: If everyone has the freedom to choose what they want or do, how can we ensure that the world is *rational* and *meaningful*?
4. To solve this problem, he directly introduced the *religious* kingdom that governs morality. He declared: We need a "kingdom of heaven" (or a higher principle, such as "one should not impose on others what he himself does not desire"), which will govern moral matters and bring meaning to our lives. (See *General Education Lectures on Humanities: Philosophy*, Vol. 1, pp 6-9, Cultural Arts Publishing House, Beijing, 2007.)

From what we know today, this set of arguments is no longer valid. Because we now know:

1. Science is not only about deterministic systems, but also about probabilistic systems [2.5]. In addition,
2. Although inconclusive, morality may be scientifically explained by evolutionary natural selection, or it may even be innate and already wired in the brain (see *The Altruistic Brain: How We Are Naturally Good*, 2014). Thus,
3. Moral issues need not involve the supernatural such as heaven. Finally,
4. "Life must have meaning" is just one of Kant's hidden hypotheses. Who said that human life must have a priori meaning?

Kant's argument is full of metaphysical assumptions and logical holes. Like many other philosophers, Kant's philosophy is outdated.

Peer Review

Historians and philosopher Robin Collingwood (1889-1943) have the following to say about Kant's work:

> So long as he confines himself to drawing the distinction between philosophical method and mathematical, his touch is that of a master; every point is firm, every line conclusive. But when he turns to give a positive account of what philosophy is, his own distinction between a critical propaedeutic and a substantive metaphysics, hardened into a separation between two bodies of thought, becomes a rock on which his arguments splits. (*An Essay on Philosophical Method*, Oxford U. P., 2008, p 25.)

As for the mathematician and philosopher Bertrand Russell (1872-1970), he said:

> Immanuel Kant (1724-1804) is generally considered the greatest of modern philosophers. I cannot agree with this estimate, but it would be foolish not to recognize his great importance. (*History of Western Philosophy*, Simon & Schuster, 1945/1972, p 704.)

In other words, Collingwood believed that Kant's writings on philosophy were mostly *wrong*, and Russell believed that Kant was *overrated.*

It is worth mentioning that Collingwood and Russell are not ordinary philosophers. Collingwood affirmed that history is science nearly a hundred years ago [5.1]. Russell was solid in mathematics and had long been engaged in philosophical logic. And although his philosophical approach was buried by the two incompleteness theorems (1931) of Kurt Gödel (1906-1978) [11.3], it was not entirely his fault. It was bad luck and he lost the bet.

Discussion

1. Kant, the founder of German classical philosophy, was born in the age of the Enlightenment. At the age of 16, he entered the University of Königsberg and studied arts and sciences. At the age of 31, he began teaching mathematics, physics, logic and metaphysics at his alma mater. He is the author of books such as *Universal Natural History and Theory*

of Heaven and *What Is Enlightenment?* At the age of 57, he began to establish his critical philosophical system with three thick books: *Critique of Pure Reason*, *Critique of Practical Reason*, and *Critique of Judgment*. Note that Kant's so-called "reason" is the possibility of introducing supernatural, which is different from the current term of rational reasoning.

2. "Outdated" means it was right at the time and not right later, which is quite common in science. For example, Aristotle's physics was supported by those in power and was popular for two thousand years, but when Galileo's (and Newton's) physics came out, Aristotle's became obsolete. Similarly, as soon as the BCS theory of superconductivity came out in 1957, all but one of the theories published in the previous 46 years immediately became obsolete. Philosophy, as part of science, is no exception. The difference is that most physical theories have experimental tests, making them easier to know if they are wrong and can be modified or abandoned early, while philosophical theories related to human belong to complex system theory, which is not easy to confirm, and can be wrong for hundreds of years. Thus, it is understandable that Kant could become the most overrated and misestimated philosopher in the history of philosophy.

3. The reason that a philosophical theory or philosopher is still receiving attention could be because it is worth studying from the point of view of the history of philosophy. It is like the case that there are still people studying Aristotle's physics, which does not mean that his physics is correct. Of course, there will always be scholars who continue to think that this or that philosopher's stuff still makes sense, and it is inevitable. If only individuals are involved, there is no harm. But if a whole profession is involved, one can only wait for time to correct it.

4. If life has no *a priori* meaning, it does not mean that life has no meaning. Instead, it means human is free. For example, individuals may choose to help others or watch the sunset to define the meaning of their lives. It is a world full of wonders waiting for everyone to discover.

5. Even though Kant had studied and taught physics he apparently was not using debate as a tool in clearing up his thoughts—the Socratic method, especially when he was reclusively doing his *Critique* series, and ended up building his theories on sand—quicksand.

Conclusion

To find something useful from outdated philosophical theories to help construct one's own theory is usually very difficult. It is like finding useful parts that are lying at the scene of a plane crash. It is not completely impossible, but the success rate is too low and the cost is too high. It is better to start anew and construct something unique yourself.

Is it too hard to make a completely new philosophical system? Yes, but it is alright and is worth it. In physics, only three people (Newton, Heisenberg, Einstein) have successfully built a completely new system. On the other hand, there are simpler ways to do new philosophy: territorial expansion, interdisciplinarity, humanities-science synthesis, and cross-disciplinary cooperation [6.9].

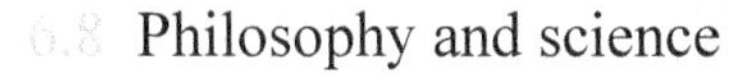

6.8 Philosophy and science

In ancient Greece, philosophy was to get rid of Greek mythology (but to bring in other supernatural such as soul) to understand everything. The current "philosophy" (double quotes), a small part of Greek philosophy [6.1], is to ask the nature of things: What does this mean (but not necessarily why)? If why is asked, the "philosopher" has stepped into the door of *science*. Asking why *and* pursuing the answer (without introducing any supernatural) is two feet past the door of science; i.e., the "philosopher" becomes a scientist.

So, by definition, all scientists are "philosophers," not vice versa. The point is that scientists need to ask why, and "philosophers" don't have to be.

Note that it is the second (1867) or third (2007) definition of the word Science that is adopted here [8.1]. The former only studies nonhuman systems while the latter studies all systems of nature, both human and nonhuman systems.

Physics and "Philosophy"

It is a *nonsense* that "philosophy" can influence the development of physics, and Feynman has said it many times. Bohr, Einstein and others know a little about "philosophy," but in general, "philosophy" has no *decisive* influence on physics.

At most it serves as a starting point for inspiration because those "philosophies" are the conjectures about the world/universe put forward by "philosophers" based on what they know about the limited scientific knowledge at the time [6.7]. In fact, when the term Science in the modern sense was defined in England in 1867, philosophy in the form of metaphysics, was explicitly excluded from science along with the supernatural [8.1].

It is the experiments that determine the progress of physics. "Physics starts with experiments and ends with experiments," which I heard and learned in the hallway when I was a graduate student at Columbia University.

6.9 Philosophy's future

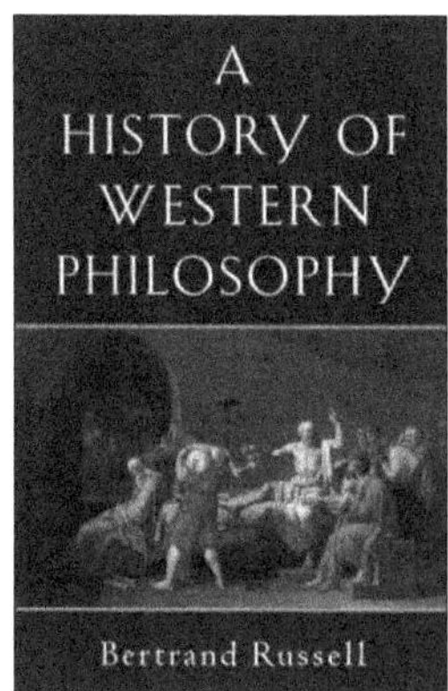

The philosophers of ancient Greece were concerned with all things, including the problems of everyday life. They asked *questions*, sought *answers*, and tried to influence the public with their answers. For example, Crates broke into people's homes and told them why they were morally lacking; Xenophanes wrote an article on how to throw a successful party; Socrates kept arguing in the streets. Today, the common impression that doing philosophy is just to contemplate quietly probably comes from philosophers like Plato.

Concepts Only

For more than 2,000 years, philosophers asked questions and sought answers. Only in the last hundred years or so did some philosophers believe that the sole purpose of contemporary "philosophy" (double quotes [6.1]) is to talk about *concepts*, not to seek answers. Why? It is a bit strange because no other discipline does so. For example, if a physics professor says that he only studies physical concepts and does not do any physical theories or experiments, the department chair may ask the person to transfer to the philosophy department.

Philosophers who believe that "philosophy" should be only about concepts include Bertrand Russell (1872-1970), the author of *A History of Western Philosophy* and a Nobel laureate in literature. In *The Problems of Philosophy*, he said:

> Philosophy is to be studied, not for the sake of any definite answers to its questions, since no definite answers can, as a rule, be known to be true, but rather for the sake of the questions themselves.

What does this mean? Imagine a boy who looks at the transformer through a toy store's window and wants very much to own it but lacks the money to buy it. A classmate walks by and encourages him to go in and buy. The child replies: Transformer is for watching, not for playing; I don't want it. That is: To cover up one's "can't" with "won't." That is it.

Talking only about concepts is definitely not the attitude of the ancient Greek philosophers or latecomers such as Kant, although this may be the attitude of many contemporary philosophers when it comes to difficult problems. No one knows if the ultimate answers exist or can be reached, but that has not stopped natural scientists from looking for answers, even if only temporary answers. There is no reason to expect less from the philosophers, although philosophy is indeed more difficult to do than physics.

It is not that philosophy can't deliver; it is that philosophers have not kept pace (see Remarks).

Way Out

The way out is for "philosophy" is to return to the down-to-earth tradition of the ancient Greek philosophy, through the following four methods.

Territorial expansion. Do philosophy of new disciplines, such as philosophy of artificial intelligence. This is the classic way out by avoiding the old, difficult "philosophical" problems. It is relatively easy to do.

Interdisciplinarity [4.4]. *Experimental philosophy*, which began about 20 years ago, attempts to merge philosophy with rigorous experimental psychology. For example, use controlled and systematic experiments to explore people's intuitive ideas (e.g., "human are born good") and conceptual usage. The major figure in this kind of work is Joshua Knobe of Yale University. Also, Timothy Bayne of Monash University in Australia leads the "Measuring the Mind" project, which attempts to develop a new framework for measuring consciousness.

Humanities-science synthesis [4.1]. Patricia Churchland of the University of California, San Diego, published *Neurophilosophy* (MIT Press) in 1989, an attempt to use neuroscience to answer the mind-brain dualism problem in philosophy.

Cross-disciplinary cooperation. Daniel Dennett of Tufts University collaborates with brain scientists on tackling the consciousness problem. Jennifer Baker of Charleston College works with marine biologists to explore how and whether coral reefs should be protected.

Conclusion

Two things worth advocating:

1. Improve the level of science education of philosophy students.
2. Encourage teachers and students of the Department of Philosophy to engage in interdisciplinarity, humanities-science synthesis, and cross-disciplinary cooperation.

Remarks

1. The big questions that early philosophy focused on, such as the origin of the universe and the nature of space-time, have been included in science but not in "philosophy." Yet, they can be studied in the history of philosophy, even though philosophers a hundred years ago still talked about cosmology. But that was before the development of modern cosmology based on general relativity.

2. The Western philosophical tradition is divided into two schools: Plato and Aristotle. The latter is close to the spirit of modern science; the former believes that meditation and logical reasoning alone can obtain true knowledge—a path that has been disproved.

3. The study of *truth* is an important topic in early philosophy, and the current consensus is that truth cannot be defined or studied. But the truth of what happened is a subject that can be seriously studied.

7
Art

7.1 Origin of art

The *origin* of art and the *nature* of art are the two most basic, important, difficult and fascinating topics in art studies.

Regarding the origin of art, there have been two schools of thought: the evolutionary origin (innatism) and the cultural origin (acquired). Innatism emphasizes that in the process of evolution, art is beneficial to the continuation of humans and is therefore passed on from generation to generation through genes. Acquired theory holds that after the emergence of art, it can be maintained as part of culture, and the role of genes is not necessary. There is not necessarily a direct contradiction between these two hypotheses, they can coexist and work at the same time. The problem is that either the innatism or acquired theory only says why art can continue after it appears, does not say how art began, does not talk about the raison d'être of pure art, and cannot prove it right. In this context, a new point of view emerged ten years ago to discuss the origin of art from the perspective of *scimat* [3.3].

A Brief History of Human

The view of scimat is that human is a kind of animal whose behavior is constrained by human nature and its evolutionary history. Therefore, to understand deep questions related to human, it is often (not every time) necessary to go back in history, sometimes to the human history millions of years ago [1.1, 1.2]. The origin of art falls into this category. To understand the origin of art, it is necessary to first understand the origin of *human* (Table).

From the perspective of painting, the cave art of 35,000 years ago is not simple but quite mature, and it is not something that we ordinary people can easily draw (upper figure). The not-so-simple invention of using fire (1.6 million years ago) and cooking food (400,000 years ago) predate the advent of *Homo sapiens* (195,000 years ago; lower figures). All of this points to the fact that art may have originated one *million* years ago or earlier.

Kill Time

No one can be sure how art originated because it is too old for any record. But because humans evolved so slowly, the instincts of our ancestors were *not* too different from our own. Therefore, it is reasonable to guess that art might have started like this:

Let us imagine that one million years ago or earlier, somewhere in Africa, there was a group of more than a dozen people living together in caves. What would they do when a heavy rain lasted for three days and three nights? Let us say they had just had a harvesty hunt and there was plenty of food so there was no rush to prepare for the next hunt. Sexual activity did not take up too much time; moreover, sex was so frequent that most women were probably pregnant by this time. So, some of them may start doing "useless" things just to *kill time* [13.6]. One person may use a branch to outline another person's shadow on the ground on the dirt floor—early *drawing*; or bang a tree trunk with a branch to imitate the sound of raindrops outside the cave—early *music*; or dance or use his hands to "tell" others about his hunting experiences—*pantomime*; or balance on a tree trunk resting on a stone—early *performing art*. All these activities appeared as leisure at this time. Performing these activities requires not high intelligence but *mimesis*—mime, imitate, gesture and rehearsal of skill—which were already available two million years ago (Table).

Years ago	Evolution	Migration	Lifestyle	Art related
6 million	Human and chimpanzee lineages split			
3.5-1.8 million			First hominids move from forest to savannah; meat eating begins	
2.5 million	*Homo habilis* appears			
2 million	*Homo erectus* appears; brain enlarged; mimesis capabilities			
1.8 million		First wave of migration out of Africa begins		
1.6 million			First use of fire; more complex stone tools created; art could begin	
400,000			Earliest evidence of cooking	
195,000	*Homo sapiens* (early modern humans) appears			
150,000				Language begins
120,000				Pigment use gives first evidence of symbolic culture
72,000				Clothing invented and earliest evidence of jewelry
60,000		Second wave of migration out of Africa		
50,000				Cultural revolution: ritualistic burials, clothes-making, invention of complex hunting techniques
35,000				Cave art (in France, Spain)
10,000			Agriculture begins; first villages appear	
5,500				Bronze Age begins
5,000				Earliest known writing

After the rain stopped, everyone had to go out hunting again, and at this time, because the women who were pregnant or had to take care of the children had to stay, they asked someone who had performed well before to accompany them and kill time. This person was the first professional *artist* in human history, entertaining others in exchange for food. He or she stayed at the residence without having to go out hunting. In other words,

1. The first artist can be male or female.
2. Artists were the first *safe* profession in human history. (There was only one other profession before—hunting, which was dangerous work.)

Safe Job

This first artist would have appeared, if not in the cave, a million years ago because there was so much *leisure time* for humans at that time, and all kinds of modern methods of killing time (newspapers, television, etc.) were not there. Finally, this would happen even more when the population reaches a certain number, because the group is large enough to support such an artist. Equally important:

1. Because the artist is a safe and good job, there has been fierce *competition* in the profession from the beginning. And the effective way to compete is through *innovation*, so innovation has existed in the art profession from the beginning, too.
2. Apart from innovation, the way to keep this good job is to reduce competition. So, early artists kept their skills strictly *secret*, passing them on only to their sons, or adding mystical elements to their activities—the origin of *sorcerer*.
3. Market demand and *positive* feedback effects ensure that art, as a profession, does not disappear after it is established.

With the passage of time and the development of humankind, more types of art appeared. For example, the invention of pigments gave color to painting, singing with language, and literature with words. Only when humans have a lot of leisure time and sufficient market demand will *pure* art finally appear, and before that it was only *applied* art [7.2]. Of course, since then, applied art and pure art have existed at the same time.

Conclusion

That art could appear millions of years ago is based on three preconditions:

1. Humans already have the mimesis capabilities.
2. Ancient humans have a lot of leisure time.
3. Art does not necessarily need language and writing (that was available only after the advent of *Homo sapiens*).

What differentiate humans from other animals in the process of evolution is the emergence of upright walking and enlarged brains in terms of *hardware* and the emergence of art in terms of *software*. The emergence and development of art has accelerated the breadth and depth of human thinking and imagination. Software is just as important as hardware.

Even today, what differentiates different kinds of animals is how they spend their leisure time. The same holds true in differentiating different members of the humankind [13.6].

7.2 Nature of art

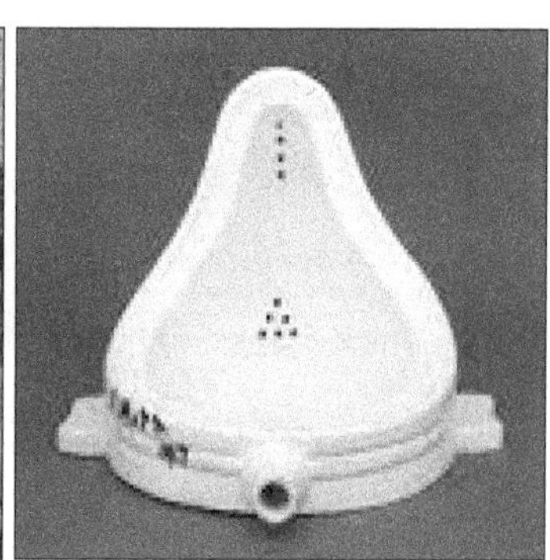

Left: Johannes Vermeer's *Girl with a Pearl Earring* (c. 1665). *Middle*: Edvard Munch's *The Scream* (1893-1910). *Right*: Marcel Duchamp's *Fountain* (1917).

The confusion about the nature of art lasted for 2,400 years since Plato and entered a state of crisis after 1917, until the emergence of a new theory in 2010, which is the interpretation of art from the perspective of Scimat proposed by the author [3.3, 16].

Art is a type of human *creative* activity whose purpose is to stimulate the receiver's neurons (which may or may not cause considerable consequences) through the senses of the *receiver*. In fact, good works in other fields such as physics, require creativity too, so creativity is not specific to art. But art is directed at the *neurons* of the receiver and this is what makes art special. For example, pure science aims to understand how nature works, and it does not target anyone's neurons, or even require anyone (except the researcher herself) to receive the results.

A work of art reflects the artist's *perception* or *worldview* of the real world. Art is not always about beauty (left figure, above), it can be about all kinds of *emotions* (middle figure, above), or it can even be about nothing specific. In fact, art can be about the expression, description, and articulation of *anything* in nature—both nonliving and living systems—such as human feelings and relationships, just like physics. As a reflection or interpretation of all that nature has, art is conditioned by the *human nature* of the artist and the receiver. In other words, art is *not* completely free, and artists are not completely free, otherwise a work of art can be completed in minutes or hours. Of course, some important works of art are indeed completed in minutes, but this does not count towards the time the artist spends in the ideation phase.

Applied Art and Pure Art

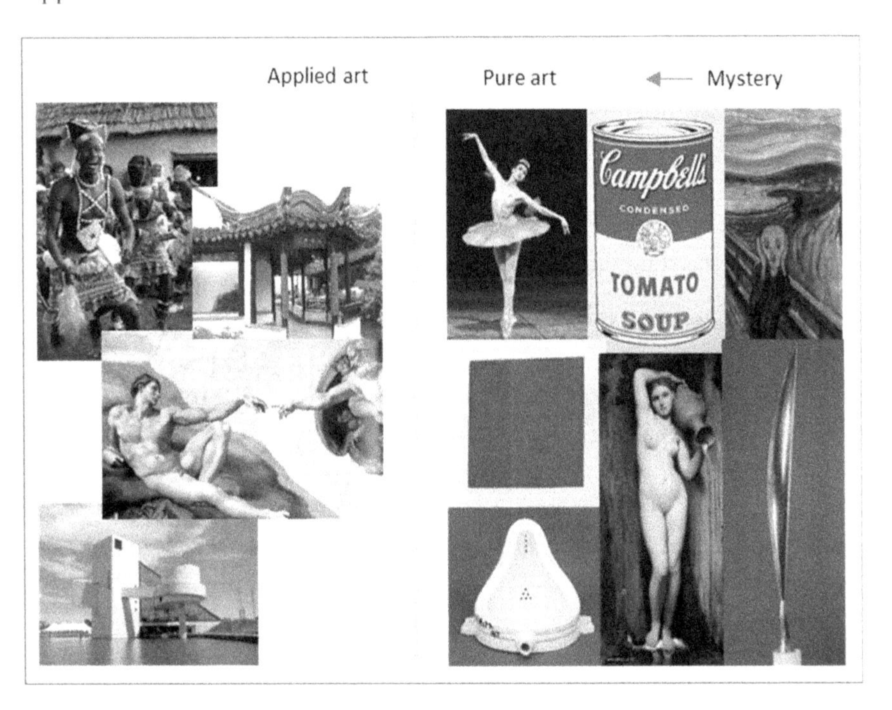

Applied arts have some practical considerations in the creative process. For example, decorating a vase beautifully, in addition to placing it in the living room, can increase its aesthetic value—as a commodity, it can increase sales; a skillful political novel may change the reader's worldview, transforming her into a warrior or a revolutionary; group dances in tribal ceremonies are meant to strengthen the cohesion of the group. Obviously, architecture belongs to the applied arts.

Pure art is all art except applied arts. Pure art falls into *two* categories: Class A (purposeful) and Class B (purpose unclear). Thus, Class A is (relatively) easy to understand and Class B is (almost) incomprehensible.

For example, ballet or Leonardo da Vinci's *Mona Lisa* belongs to Class A; Kazimir Malevich's *Red Square* (1915), Class B. Class B of pure art is the mystery. Why does it exist? What is the use? To answer these two questions, we must first look back at the history of art.

A Brief History of Art

Although art first appeared millions of years ago to simply *kill time*, over time, due to the *innate* competitiveness of the artist profession [7.1] and the complexity of human life (such as the belief in gods and ghosts, language, and writing), the content and form of art have become increasingly complex. For example, there are landscape paintings, figure paintings, religious paintings, and sculptures such as bronze sculptures, stone sculptures, and jade sculptures.

In Europe, however, the *freedom* of artists to choose their subjects was only after the advent of the free art market in the last 200 years. Before that, the subject of the artwork was determined by the employer (such as figure paintings or religious content on the walls). It was the free choice of topics that allowed artists to do art for art's sake, which led to the emergence of pure art's Class B.[1] Specifically, the advent of photographic technology more than 100 years ago forced the *Impressionist* style of 1874, and the modern art pioneered by Paul **Cézanne** (1839-1906) led to *abstract* art, resulting in visual art (painting, sculpture) moving away from realism (top right figure, below).

As the perception of the nature of art is concerned, the turning point in the change occurred in 1917 (the year of the establishment of the Russian Soviet Federative Socialist Republic) as the *Fountain* (right figure, beginning of article) by Marcel **Duchamp** (1887-1968) was accepted as a work of art to be exhibited in New York. The work is a ready-made male urinal bought from the store, and the author simply signs the (fake) name and writes the year on it *and* decides how to place it for display. The appearance of *Fountain* abolished all previous theories of art, especially the theory that art must be related to aesthetics, and made art almost *free* of the creator—the artist (see below). It is this kind of nonrealistic or "nonsensical" Class B of pure art that is particularly puzzling. What is their essence?

Class B of Pure Art

To answer this question, let us compare the kill-time features of various types of artwork. While all arts can kill time, for the applied arts, their time-killing function has taken a back seat. For example, adding a beautiful pattern to a coffee cup is just to make the cup look better and sell for a higher price, not to prolong your coffee-drinking time. For pure art's Class A works such as *Mona Lisa*, in addition to meeting the basic requirements of the employer (painting portraits of designated people), the artist uses all his skills to make the smile of the character having multiple interpretations, increasing the viewer's return rate—killing more of his time.

Pure art's Class B goes further. Although the picture may be simpler or super simple on the surface it can arouse various associations of the viewer, so it can kill even more time. In other words, the essence of pure art is to return to the function that art began millions of years ago: *kill time*.[2]

Five Criteria of Pure Art

Here, we give five criteria for *good* (long-standing) pure art:

1. Aim at receiver's neurons
2. Kill a lot of the receiver's time
3. Kill time gently and harmlessly
4. Passivity
5. Human creation or intervention

It needs to be explained point by point. 1. This is the basic characteristic of all art. The receiver is a viewer/listener who may like or dislike the artwork (e.g. an audience in an art museum).

2. An important function of pure art is to *kill time*—the receiver's time. And good pure art can kill a lot of time. In fact, an important artwork will make the receiver spend a lot of time thinking about it when they

first appreciate it, and later. Leonardo da Vinci's *Mona Lisa* and Duchamp's *Fountain* are examples. The same applies to an art film or a good play although the viewing time is only two hours. In this regard, ambiguity with multiple interpretations is a trick that works. Another trick is give the artwork a title that does not seem to have anything to do with the content on first encounter.

3. However, just being able to kill time cannot be called pure art. Entertainment and drugs can also kill time. The difference is that pure art can kill time *gently* and *harmlessly* while entertainment—like a great World Cup soccer match—may stimulate your neurons every 10 or 15 minutes, and the stimulation cannot last. Similarly, the effects of drugs are usually not gentle and drugs can send you to jail. In other words, pure art allows us to kill time with good feelings. They do not stimulate our neurons violently and therefore make us willing to revisit them frequently.

4. People do not want to be actively involved in everything they do. At the end of the day's hard work, many people want to *passively* relax by watching TV and those who love art relax by listening to classical music or through other passive methods. On weekends, they might read a book or visit an art museum to passively appreciate art. In fact, from painting to literature to the performing arts, passivity is the hallmark of all great arts as far as the receiver is concerned. That is why interactive art has not caught on and will not be so in the future. Too much interaction is bad for pure art. (Passivity carries no negative meaning here.)

5. By definition, art must be created or intervened by human.[3] This is not to say that artists cannot work with (natural or artificial) materials or with the help of machines or computers. Of course, they do and have been doing it all the time. According to this criterion, a stone lying on the roadside, no matter how beautiful, is not a work of art. However, if you take a picture of the stone the photograph can become a work of art—*photographic* art, because of your intervention.[4] You can also take this stone home and put it behind a frame (bottom right figure, below) and you immediately become an artist, because through that frame, you tell the receiver (viewer) that you want her to see the stone from that angle—

you intervened. Of course, this does not guarantee that this *geological* art is a good art.

Conclusion

To sum up, *pure* art is *created* by humans or with human *intervention*, to kill time *gently* and *harmlessly*, and let the receiver to experience it *passively*. From this understanding, it is clear that the *content* or *form* of pure art is *secondary*;[5] they are there to serve Criteria 2 to 4.

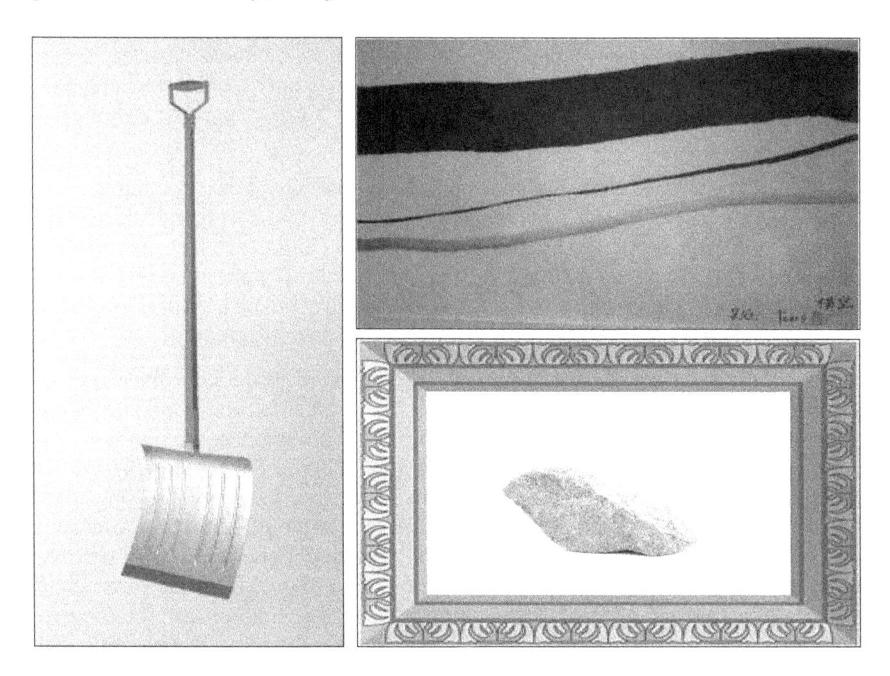

Notes

1. Early European artists needed employers (or patrons), partly because of the considerable cost of oil painting materials and stone. In China, painting and *calligraphy* only use pen, ink and paper; the cost is not high, and painters can freely sell paintings and calligraphies. But the selection of art topics is not wide, and there is no pure art's Class B artworks. The reason is related to the fact that traditional Chinese culture does not encourage personality development and innovative spirit.

2. This interpretation comes from the perspective of bystanders other than the creator and receiver. The creator and receiver are less likely to think of killing

time when creating and watching. The creator may just want to express her thoughts and emotions about something while the receiver may get other ideas or enjoyment from the work. The direct participants in anything usually do not think about or realize the nature of what they are doing, just as the driver and passengers in a car do not care or are aware of what a car is about: A vehicle that can move people or goods from one place to another, move on four or more wheels in touch with the ground, and use non-animals (including human) to provide power.

3. Accordingly, a painting created by a nonhuman (e.g., chimpanzee, horse, fully autonomous robot), no matter how good, is of little value as long as the nonhuman identity of its creator is disclosed. If desired, they can be called *chimpanzee* art and so on, and art, by definition, must be *human* art. Sports are similar: People are particularly concerned about how many seconds a person can run a hundred meters fast—the speed of a person, not the speed of a dog even though a dog can run really fast.

4. In the same way, the clouds and rainbows in the sky, sunrises, sunsets, and natural scenery, no matter how beautiful, are the "masterpieces" of nature, without human intervention, so they cannot be called art. But if someone photographs it, it may become art. By analogy, beautiful patterns seen in the lab (perhaps through a microscope or other tools) are not art. But if someone photographs them, frame them, sign them, that will be a different matter.

5. It is not that content is not important. In Duchamp's *Fountain*, for example, the bought urinal is a not-so-simple object that evokes all kinds of interesting associations and fits Criterion 2. For example, the direction in which the urinal is placed, the hole, can even evoke sexual associations. If he had replaced the urinal with a simple rice bowl, the artwork would not have been so powerful. Or, if he replaced the clean urinal with a dirty one it would not work because it will make the receiver feel uncomfortable, violating Criterion 3. In fact, two years before the *Fountain*, Duchamp bought a snow shovel from the store, hung it from the ceiling of his studio, and titled it *In Advance of the Broken Arm/(from) Marcel Duchamp* (left figure, immediately above), which did not cause any sensation, even if he called it art. Reason: The shape of this snow shovel is too simple; it cannot attract any complex associations; and it will not kill much time—not meeting Criterion 2. It *is* art, but *not* a good art.

7.3 Art and science

目 录

第一部分 科学与艺术的关系

第二部分 设计与建筑领域的科学与艺术

第三部分 新媒体领域的科学与艺术

Science is human's effort to understand *all* things in nature (including human and all nonhuman systems) without introducing any supernatural considerations. Science consists of the process and results of this effort, which includes all the disciplines in natural science, social science, the humanities, and medicine [8.1]. The meaning of the word *Art* includes the thinking and process of artistic creation, artwork, and art studies. Art *studies* or research is part of science, and the thinking and process of artistic *creation*, if sorted out, can also be part of science, but *artwork* is not.

Relationship between Art and Science

When people talk about "art and science," they generally mean the relationship between art and *natural* science, and here too. There are many books on this, such as Eliane Strosberg's *Art and Science* (2015), David Edwards' *Artscience* (2008), and Sian Ede's *Art and Science* (2005). Unfortunately, all these authors have misunderstood this question, mainly because they have a wrong understanding of the nature and scope of science as well as the nature and origin of art. The views of *Scimat* [3.3] are shown in the Table.

Characteristics	Art	Science
Aims different	Aim at receiver's neurons	Aim to understand how nature works
Receiver yes/no	Need a receiver to appreciate the artwork	Need no receiver (but has to compare with nature, the ultimate judge; not even have to publish)
Content same	Everything in universe (especially human system)	Everything in universe (especially nonhuman systems)
Methods different	Express feeling/thoughts/ideas; don't ask why	Ignore feeling; always ask why
History different	Started at least 35,000 (and could be a million) years ago	Started about 2,600 years ago since Thales (c. 624-c. 546 BC), after the invention of language and writing
Relationship between art and science	• Both involve creative process (for different reasons)—but same in many other human activities • Art is humans' creation, reflecting on the world of human and nonhuman systems; the principles governing this world are the same principles (e.g., symmetry, spontaneous symmetry breaking, fractal, chaos, active walk) studied by scientists • Progress in science (and related technology) advances the development of art; e.g., pigments → color painting, film/camera → photographic art, electricity → cinema, laser → photon art, computer → digital art	

In particular:

1. Art and science are concerned with the *same* things—everything in the universe, but with different emphases. So far, art has been primarily concerned with the human system, a complex system, while science has been primarily concerned with nonhuman systems, particularly simple systems. Therefore, it is more difficult to do art well than science, which is why there are fewer good artists than good scientists.

2. *Innovation* in science and technology often leads to innovation in art, but *not* vice versa. But there are examples of mutual inspiration between artists and scientists.

Importance of Art and Science

Many people have talked about the importance of art and science. In China, Cai Yuan-Pei (蔡元培) pointed out a hundred years ago: "Those who value morality all rely on art and science, which are like the two wheels of a cart or the two wings of a bird." Later, Nobel laureate Tsung-Dao Lee (李政道) put it another way, comparing art and science to two sides of the same coin.

Unfortunately, these tropes are problematic because:

1. Art and science are not like the wings of a bird, the wheels of a cart, or the two sides of a coin because art and science did not appear at the same time in history.
2. Nor are art and science equally important in the history of human development.

In fact, several millions of years ago, primitive humans must have had *ethics* when they lived together. Then, *art* appeared one to two millions of years ago [7.1], followed by *science* 2,600 years ago [8.1].

In terms of human survival, ethics is more important than art, and art is more important than science. Instead of saying art and science are equally important, a better description would be: Art, science and ethics are the three pillars that underpin modern civilization [5.5].

Artistic Level and Scientific Level

The level of art is not necessarily related to the level of science.

In the West, the artistic level of the European cave paintings was already high 35,000 years ago well before science was invented in Greece 2,600 years ago. After that, in Europe, art and science developed side by side and promoted each other (e.g., perspective was first adopted in Leonardo da Vinci's *Mona Lisa*, and the invention of the camera forced the emergence of Impressionism), giving rise to the *illusion* of "artistic level = scientific level."

In China, the level of ancient art is extremely high. For example, the bronze and jade sculptures of the Shang dynasty 3,300 years ago is amazing; the Tang poetry and Song lyric are excellent. The decline of artistic level is a later event. Compared with the level of art, the level of ancient Chinese science is much worse, mainly supported by traditional Chinese medicine [9.4]. Modern science has been introduced since the end of the Ming dynasty, and flourished from the beginning of the Republic of China to 1937. But science has not really developed until the past few decades; some of them are even world-class [13.1, 13.3]. However, the level of art has yet to recover since its decline.

The revival of China should include artistic revival: not to restore old art, but to have its own *new* art.

Conclusion

1. Both art and science are concerned with everything in the universe, but the emphasis is different.
2. Many artistic processes use scientific knowledge.
3. Scientific and technological innovation promotes artistic innovation, not vice versa.
4. Art makes life worth living, and science makes life more comfortable. Art is more important than science!
5. Art, science and ethics are the three pillars that underpin modern civilization.
6. The level of art is not necessarily related to the level of science.

7.4 Modernism: Su Dong-Po and Cézanne

The idea of modernism in art is to portray the essence of the subject, not their appearance or the artist's impression. This idea of artistic modernism was proposed and practiced by Su Shi in China, about 800 years ahead of Paul Cézanne in France.

Su Shi (苏轼, 1037-1104), also known as Su Dong-Po (苏东坡), a writer and calligrapher of the Northern Song dynasty, was known for his poetry and painting as well as the Dong-Po Meat.

Su left a lot of calligraphies but only two paintings remained: *Dead Wood and Strange Stone Drawing* (枯木怪石图, entered Japan during the Sino-Japanese War) and *Xiaoxiang Bamboo Stone Scroll* (潇湘竹石图卷, in the National Art Museum of China in Beijing). He mainly painted ink bamboo and strange stones. He advocated that painting bamboo should not be realistic, but should draw the *essence* of bamboo, the so-called "gain outside the image" (得之象外).

He believed that the essence of bamboo is *ethereal*, so the bamboo he painted has no bamboo knots and the leaves can *detach* from the bamboo branches and suspend in the air—just to show the ethereal (left figure). In contrast, in his contemporaries' bamboo paintings each bamboo leaf is connected to the branch, looking very dense and annoying, and quite tacky.

Paul Cézanne (1839-1906), called father of modernism in the West [7.5], also advocated painting the essence of things, not their images. A year before his death, he painted the essence of a mountain near his home—the germ of abstract painting (right figure). The essence of modernism is *elegancy.*

Su Shi's claim that "painting is showing the essence of things" is exactly the same as Cézanne's claim. But Su was 800 years before Cézanne and was really ahead of his time [9.2]. The fact that Su failed to become the pioneer of modernism is because there was a lack of successors to carry on his torch. In contrast, Cézanne had a lot of fans and converts such as the two masters, Picasso and Matisse.

Lesson: Be culturally confident and tell well your own story.

7.5 Cézanne's self-confidence

Paul Cézanne (1839-1906), pioneer of modern art. Picasso called him "my one and only master." Picasso and Matisse called him "the father of us all."

Not only modern art but Cézanne was also the pioneer of *modernism*—he pioneered a new style. The same is true for doing research in any discipline: The goal is to be a pioneer in a field. It is not just to publish papers and it is not that difficult to start a new field. Trying is important even if it may not be successful in the end. The secret lies in the art of picking research topics. Like they say in physics: Picking the right topic determines everything.

Doing art and physics requires self-confidence. Cézanne had a great deal of self-confidence. He once said: If there is only one painter in the world, that painter is me. Albert Einstein (1879-1955) also had super-confidence. When the results of an experiment did not match his theory, he said: The experiment is wrong.

Therefore, to promote innovation, it is necessary to cultivate students' self-confidence from kindergarten on. If you do the opposite, no innovation will come.

7.6 Da Vince, the folk scientist

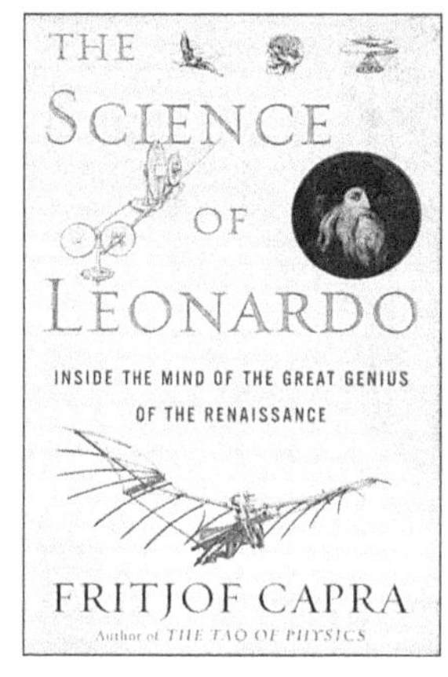

Folk scientist refers to those who have not received systematic scientific training. According to this, China's Mozi was not a folk scientist since there was no science education at that time, but a scientist [6.6]; Italy's Da Vinci, who came much later, is a folk scientist.

Florence

In Italy in the early *Renaissance* (15th-16th centuries), the Holy Roman Empire and Catholic governance slowly failed; local literature and the study of classic works rose. Florence appeared in the form of a city-state, politically speaking, and became a cultural center in the whole of Europe at that time, manifested in literature, painting, sculpture, and architecture. By the end of the 15th century, three great art masters had emerged at the same time: Leonardo da Vinci (1452-1519), Michelangelo (1475-1564), and Raphael (1483-1520).

Leonardo

Leonardo da Vince, the Leonardo from Vince, was an illegitimate child. His father was a notary by profession and his mother was a peasant girl. Dad did not marry mom but married someone else. Leonardo lived with his mother, grandmother and a farmer (mom's future husband) until he was five years old—a wild child who ran in the wild all day and loved nature.

When he was five years old, his father's wife did not give birth, so his father turned back to recognize his son and took him home, where Leonardo lived for ten years. Dad did not give him the family title, did not send him to school, only let him take some lessons at home. He did not learn Latin or Greek and so could not read the classic books—a blessing in disguise—which made him free from the mainstream thoughts.

Leonardo showed some talent for art and music, and so, at the age of 15 he was sent to Florence as an apprentice with a master artist, where he learned a knack for grinding rare-earth elements to create colors, making brushes, casting bronzes, perspective and compositional principles, using chiaroscuro in painting, mixing chalk strokes to create seamless smoky shadows (invented by his mentor), and understanding the importance of anatomy.

Throughout his life, however, Leonardo was not content to use known artistic elements, but constantly experimented with creating new artistic techniques. His dedication to exploration as a folk scientist is the basic difference that distinguishes him from other artists, and ultimately enables him to become a master artist.

On a certain occasion, the Master received a commission to paint the baptism of Christ in a church. He asked his apprentice to draw one of the two kneeling angels. After Leonardo finished, the Master took a look and was so impressed, knowing that he could not surpass his apprentice, and sealed his brush from then on. Awesome, right?

Such a talented Leonardo was not all smooth sailing. At that time, there was no free market for art and no auction house, which was only in the last two hundred years. Painters all work on orders, and the employer was either the church or the aristocracy, so the subject matter was either related to religion or portraits of aristocratic relatives. Leonardo sometimes took an order, had many ideas, had distractions, did not finish for three years. As a result, he not only did not get paid but also ruined his professional reputation, and had no job.

Beginning in 1482, at the age of 30, Leonardo stayed in Milan for 17 years, working as a court *engineer*, designing buildings, sewer systems, movable bridges, weapons of all kinds, and painting in his spare time. Here, at the age of 40-46, he completed *The Last Supper* and drew a lot of advanced designs in his notebook.

Leonardo da Vinci did not love women and painted only three non-religious portraits of women in his lifetime, each 15 years apart, the third being the *Mona Lisa*. He began painting *Mona Lisa* at the age of 51, constantly revised. At the age of 64 he was invited by the French emperor to live in France, and the painting was brought to France to complete. Around his death three years later, the painting was sold to Emperor Francis I. It ended up with Napoleon, who hung the painting in his bedroom. The *Mona Lisa* later became a national French treasure. That is why everyone wants to watch it today has to visit Louvre in Paris.

Leonardo died on May 2, 1519, near Paris at the age of 67.

A Rational Painter

There are two kinds of artists: sentimental and rational, which are manifested in the selection of subjects and painting styles. "Sentimental" here means non-analytical. For example, Su Shi (1037-1104) is a sentimental artist: He would get half-drunk before doing his calligraphy and paintings. Renaissance painters were employed to paint and had no freedom to choose subjects, so whether they are sentimental or rational has to be judged by their painting style. By this, Leonardo da Vinci is a rational painter.

Although oil painting is different from ink painting and can be constantly revised while being painted, generally oil painters have already thought about it for a long time before starting to paint. Leonardo is no exception, but he prepares more and deeper than others, more like a physicist conceiving and designing a complex experiment, and even some of the instruments have to be constructed in the lab. He once said: "Painting *is* science." He also said: "First study the science, and then practice the art which is born of that science." In other words, he paints from the perspective of a folk scientist, using many physical techniques he invented himself. He paints calmly and meticulously as a detached artist, much like a working physicist.

None of his paintings are signed. Like some good physicists, he did not produce much, only 15 are thought to have been painted in whole or mostly by him, and the rest may have been lost.

Things he designed on paper as an engineer included bicycles, helicopters, parachutes, machine guns, diving suits, tanks, robots, gravity perpetual motion machines, gliders, gearboxes, collapsible bridges, flamethrowers, scissors, and submarines. While some may have been made at the time, most remained on paper and were not realized until hundreds of years later. Imaginative, right?

There are two cool words to describe da Vinci's works: painting cold, techniques cool.

Discussion

1. Leonardo da Vinci's father married three times and had 12 children through marriage, none of whom were famous though. Therefore, in addition to his own efforts, da Vinci's achievements came from the genes of the peasant mother. The accomplished Mozi (c.476-c.390 BC) also came from a peasant family [6.6]. Therefore, don't underestimate the genes of peasants and don't underestimate the IQ of peasants' children.

2. The term "folk scientist" is not pejorative. Starting in Florence and continued in Milan and elsewhere, da Vince's notebook writings and drawings—some even carried out in secret, record his numerous scientific *observations*, *understandings*, and *innovations* about light, water flow, human and frog anatomy, botany, geology, mathematics, etc., most of which had nothing to do with his duties as an engineer. The notebooks, which contained 100,000 drawings and 6,000 texts, were intended for publication, but were too late to do so before da Vince died. So, his scientific work was carried out "civilly," not part of his job as an

engineer, although part of the time he did have an official (engineer) status. In other words, he meets the definition of folk scientist and is a *scientist*.

3. There are two important folk scientists in history: Stanford Ovshinsky (1922-2012, high school graduate [12.3]) in the United States, and Leonardo da Vinci in Italy (no formal education). Da Vinci had no scientific training and thus failed to raise his scientific observations to a theoretical level. His invention was too advanced, limited by the technological conditions of the time, and thus would not be made until hundreds of years later.

4. Da Vinci, like Newton (1642-1727), never married, no close woman friends, and had no children. Da Vinci was arrested for homosexual conducts, similar to charges as Alan Turing (1912-1954), except that the former was released for lack of evidence and the latter was convicted and punished. It may all be a coincidence, but it is an indisputable fact that the three have made significant contributions to humankind in the fields of art, science, and computer.

5. Leonardo da Vinci had no scientific training but he had said the following and practiced it. When he was about 61 years old, he said: "First, before I continue, I will do some *experiments*, because my intention is to first cite experience and then *reason* why this experience must work in this way. This is the real rule that should be followed to explain the effects of nature." This is completely the approach of modern science. Although a folk scientist, da Vinci was a pioneer who knew how to do science. The author of *The Science of Leonardo* (2007, left figure) argues that Leonardo da Vinci should be revered as the *father* of modern science.

6. Judging from his persevering personality, thoughtful working methods, and delicate works, da Vinci would have been a first-class scientist if he had the opportunity to enter a university for complete scientific training, could even be better at physics than Galileo Galilei (1564-1642) who was born 45 years after his death. If this were the case, modern science would have appeared a hundred years earlier. However, a hundred years

is nothing in the six million years history of humankind; science can wait. Moreover, art is more important than science [7.3]. And if one has to choose where to add one more Master, in art or science, better choose art over science.

Conclusion

The revelation of Leonardo da Vinci's life: A person can determine neither the time and space of his birth nor the environment in which he grows, but

- Be curious, keep your eyes open, look around, observe carefully, and remember them.
- Be Calm, tenacious and persevering.
- Think independently and ignore the mainstream.

Then one could be successful and, if lucky, could even become a Master of something.

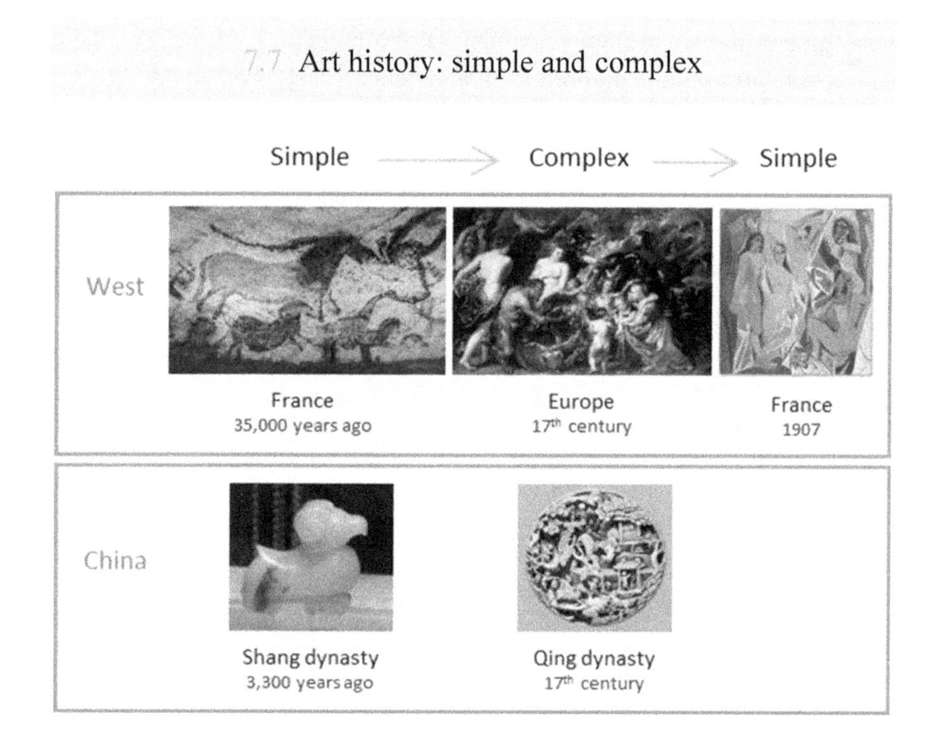

In art history, complexity of the art form and artistic thinking changed over time, which are different in the West and China. These changes are summarized and explained here.

Western art history: The art *form* went through simple → complex → simple but the artistic *thinking* is simple → complex. The cave painting in the West is simple and elegant, and the spirit of modern art is also simple and elegant, roughly speaking. The latter is related to the development of technology and the quickening of life's rhythm of the times [7.2]. Simple is not easier than complex, but simple saves materials, time, and energy.

Hong Kong and Taiwan were influenced by the West earlier than mainland China, and the art form there was simplified earlier. The trend of Japanese art is the same as in the West, and its traditional Zen ideas coincide with the simplicity of art, which is expressed in various aspects of Japanese architecture, utensils, and fashion clothing design.

Mainland China is different. The jade carvings unearthed from the tomb of *Fuhao* (妇好) of the Shang dynasty were minimalist and soul touching, but later art form became more complicated.

Chinese art history: Artistic *thinking* went through simple → complex → simple and the art *form* is simple → complex. The reason is that Chinese culture has a history of favoring complexity for over 2,000 years and was influenced by Western culture for only about 100 years. At present, Chinese art forms are still mostly complex than simple. Will they become simpler in the future?

Summary

Place	Art form	Artistic thinking
West	Simple → Complex → Simple	Simple → Complex
China	Simple → Complex	Simple → Complex → Simple

7.8 Aesthetics: old and new

Aesthetics is a discipline that studies the nature of beauty and taste, and its meaning, a branch of *philosophy* related to literature, art, psychology, linguistics, anthropology, history, etc. The aesthetics part of art is related to Aesthetics, but Aesthetics is not the same as the Philosophy of Art [7.2]. Aesthetics asks a wide range of questions; e.g., Why do some people find sunsets beautiful and others do not? Why is the color of temples red in China but brown in Japan?

Old Aesthetics

To understand the problem of sunset beauty, the *modern* research method is to first do a questionnaire to see how the sunset beauty is related to the age, gender, occupation, grown-up location, etc. of the viewer, and then do a brain scan of the viewer when watching, but that is not the general practice of philosophers.

Generally speaking, the early aestheticians go to literary works and artistic creation to find out the content related to beauty, read the expositions on beauty of many philosophers since Plato (including Kant and Hegel), discuss it, add their own experience, and call it finished. Later, aesthetics scholars such as Zhu Guang-Qian (朱光潜), will add analysis from a psychological point of view. Zhu has a complete aesthetic education and is recognized as China's aesthetics master.

Zhu Guang-Qian (1897-1986), aesthetic educator; BA from the University of Hong Kong in 1922; went to Europe in 1925, MA from the University of Edinburgh, UK; PhD from the University of Strasbourg, France. In 1933, he returned to China and taught at Tsinghua University, Peking University, Sichuan University, and Wuhan University. Since 1946, he has been teaching aesthetics and Western literature at Peking University. He is the author of *Twelve Letters to Youth* (1929), *On Beauty* (1932), *History of Western Aesthetics* (1963), etc.

The seventh letter in Zhu's *Letters on Beauty* (1980), "On beauty and aesthetic perception from a physiological perspective," written at the age of 82, bluntly points out that the basis of beauty is physiology; i.e., beauty comes from humans' physiological responses. At this point, the aesthetics philosopher can no longer do it alone (without the collaboration of physiologists, say), but can only stop. And that was exactly what Zhu did: He quitted aesthetics.

New Aesthetics

Some new approaches to aesthetic research.

1. *Experimental aesthetics* is oriented towards the natural sciences, using modern methods mainly from the field of cognitive psychology or neuroscience, resulting in *Neuroaesthetics*: The use of neuroscience to explain and understand aesthetic experience at the neurological level.

2. *Computational aesthetics* uses computer-science methods to predict, convey, and evoke an emotional response to a work of art. Aesthetics is not considered to depend on taste, but is a matter of cognition, so beauty can be obtained through learning. In 1928, mathematician George Birkhoff created the aesthetic metric, $M = O/C$ (the ratio of order to complexity). Since 2005, computer scientists have been trying to develop automated methods to infer the aesthetic quality of images. These methods typically use a machine-learning approach, using a large

number of manually graded photos to "teach" the computer which visual characteristics are related to aesthetic quality.

3. *Evolutionary aesthetics* uses evolutionary psychology to explain the benefits of aesthetics, arguing that *Homo sapiens*' aesthetic preferences are meant to enhance survival and reproductive success. For example, people prefer beautiful and open landscapes because they are conducive to a safe habitat for their ancestors. Another example: Good body symmetry and proportions are beautiful and attractive because it indicates that a person grows up healthy and is conducive to producing healthy offspring, and thus perpetuating genetics.

Conclusion

The way out for aesthetics is humanities-science synthesis [4.1].

7.9 Li Qing-Zhao: searching seeking

There is Curie in the West and Qing-Zhao in the East. Both are talented women.

Li Qing-Zhao (李清照, 1084-c.1155) lived from the Northern Song dynasty to the Southern Song dynasty, with a prominent background—maternal great-grandfather and maternal grandfather were both prime ministers. Her father is a *jinshi* (进士) and a Su Shi (苏轼) student. Li is a lyric-poet (词人)—the foremost talented woman in China's 3,000 years of cultural history. I think the most creative lyric-poets in Song dynasty are Su Shi and Qing-Zhao. Qing-Zhao surpasses his father and competes with his father's teacher.

This is Li's lyric (词，Ci') I learned at Clementi Middle School in Hong Kong:

《声声慢 • 寻寻觅觅》

寻寻觅觅，冷冷清清，凄凄惨惨戚戚。乍暖还寒时候，最难将息。
三杯两盏淡酒，怎敌他晚来风急！雁过也，正伤心，却是旧时相识。

满地黄花堆积，憔悴损，如今有谁堪摘？守着窗儿独自，怎生得黑！
梧桐更兼细雨，到黄昏点点滴滴。这次第，怎一个愁字了得！

Here is the English translation by Wang Jiao-Sheng:

A Long Melancholy Tune (Autumn Sorrow)
Despair

Searching, seeking.
  Seeking, searching:
What comes of it but
  Coldness and desolation,
A world of dreariness and misery
And stabbing pain!
As soon as one feels a bit of warmth
A sense of chill returns:
A time so hard to have a quiet rest.
What avail two or three cups of tasteless wine
Against a violent evening wind?
Wild geese wing past at this of all hours,
And it suddenly dawns on me
That I've met them before.

Golden chrysanthemums in drifts ——
How I'd have loved to pick them,
But now, for whom? On the ground they lie strewn,
Faded, neglected.
There's nothing for it but to stay at the window,
Motionless, alone.
How the day drags before dusk descends!
Fine rain falling on the leaves of parasol-trees ——
Drip, drip, drop, drop, in the deepening twilight.
To convey all the melancholy feelings
Born of these scene
Can the one word "sorrow" suffice?

I am not a Curie fan but a Qing-Zhao fan. Art is more important than science! Art makes the world worth living while science just makes living easier.

7.10 Luo Li-Rong: sculpture goddess

Luo Li-Rong (罗丽蓉), a Belgian-Chinese artist, was born in 1980 and graduated from the sculpture department of the Central Academy of Fine Arts, Beijing, in 2005.

Her sculptures have an elegant dynamic beauty, with great attention to detail, using the techniques employed by European sculptors during the Renaissance and Baroque periods.

Luo Li-Rong, goddess, super!

PART II

SCIENCE AND SCIENTIST

8

Science Basics

8.1 Defining science

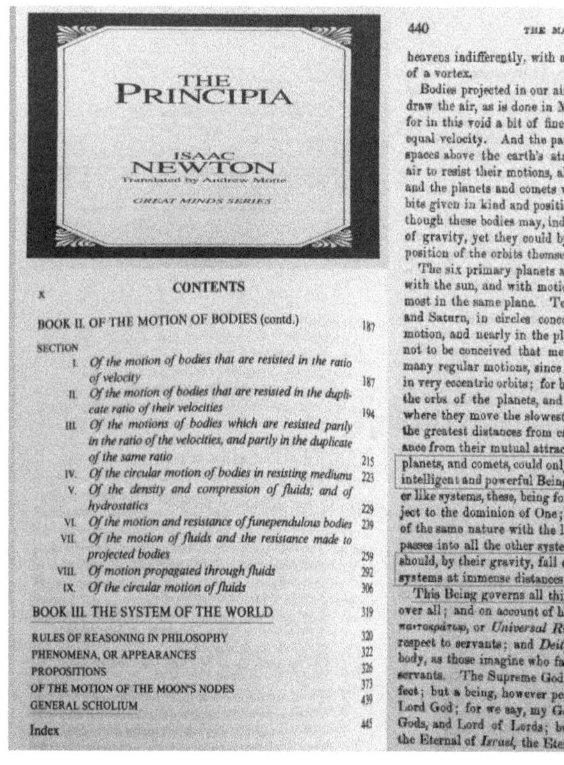

x CONTENTS

440 THE MATHEMATICAL PRINCIPLES [BOOK III.

heavens indifferently, with a freedom that is incompatible with the notion of a vortex.

Bodies projected in our air suffer no resistance but from the air. Withdraw the air, as is done in Mr. *Boyle's* vacuum, and the resistance ceases; for in this void a bit of fine down and a piece of solid gold descend with equal velocity. And the parity of reason must take place in the celestial spaces above the earth's atmosphere; in which spaces, where there is no air to resist their motions, all bodies will move with the greatest freedom; and the planets and comets will constantly pursue their revolutions in orbits given in kind and position, according to the laws above explained; but though these bodies may, indeed, persevere in their orbits by the mere laws of gravity, yet they could by no means have at first derived the regular position of the orbits themselves from those laws.

The six primary planets are revolved about the sun in circles concentric with the sun, and with motions directed towards the same parts, and almost in the same plane. Ten moons are revolved about the earth, Jupiter and Saturn, in circles concentric with them, with the same direction of motion, and nearly in the planes of the orbits of those planets; but it is not to be conceived that mere mechanical causes could give birth to so many regular motions, since the comets range over all parts of the heavens in very eccentric orbits; for by that kind of motion they pass easily through the orbs of the planets, and with great rapidity; and in their aphelions, where they move the slowest, and are detained the longest, they recede to the greatest distances from each other, and thence suffer the least disturbance from their mutual attractions. This most beautiful system of the sun, planets, and comets, could only proceed from the counsel and dominion of an intelligent and powerful Being. And if the fixed stars are the centres of other like systems, these, being formed by the like wise counsel, must be all subject to the dominion of One; especially since the light of the fixed stars is of the same nature with the light of the sun, and from every system light passes into all the other systems: and lest the systems of the fixed stars should, by their gravity, fall on each other mutually, he hath placed those systems at immense distances one from another.

This Being governs all things, not as the soul of the world, but as Lord over all; and on account of his dominion he is wont to be called *Lord God* παντοκράτωρ, or *Universal Ruler*; for *God* is a relative word, and has a respect to servants; and *Deity* is the dominion of God not over his own body, as those imagine who fancy God to be the soul of the world, but over servants. The Supreme God is a Being eternal, infinite, absolutely perfect; but a being, however perfect, without dominion, cannot be said to be Lord God; for we say, my God, your God, the God of *Israel*, the God of Gods, and Lord of Lords; but we do not say, my Eternal, your Eternal, the Eternal of *Israel*, the Eternal of Gods; we do not say, my Infinite, or

It is a common wisdom that when doing science, God/supernatural cannot be invoked. Why? Where did this cognition come from and how did this tradition begin? It turns out that science sans God/supernatural is a *conscious* act that happened in 1867, in England. And it is this act that cleared up the path for science to prosper. Unfortunately, this important fact is rarely known to professionals and laypeople alike, even today. Here, we trace the evolution of the word Science back to the 14th century, point out the limitation of the *second* definition of science of 1867, and present the 2007 and *third* definition of science that matches what we presently know about human and the rest of the universe. The importance of these two definitions, often overlooked, are detailed.

The Lack of a Definition

Everyone, according to their own experience or understanding, could and may have their own definition of the word Science. This will not pose a problem if they are *not* engaged in discussion or research related to the nature of science—such as on the history, philosophy, sociology, or communication of science. For those who do scientific research or those who only do case studies in science history, this does not pose a problem, too. But when discussing beyond case studies and discussing major issues such as the origin of science and the nature of science, the definition of science is an unavoidable issue. For example, to write a general history of science, what should be included? The situation in philosophy of science is similar, even more so.

Further examples could include two ongoing debates: Is traditional Chinese medicine science? Why modern science was not born in China (the Needham Question)? It is hard to imagine that without an agreed upon definition of science one can reach consensus on the answer to these two important questions.

Thus, the lack of consensus on the definition of science in the academic world is an urgent problem to be solved. It is a problem that needs to be faced head-on. Presently, in our opinion, progress in science history and science philosophy has been greatly hindered by the absence of this consensus.

In the absence of a unified definition of the word Science, scholars deal with this problem in two different ways:

1. A number of scholars take an *evasive* attitude. For example, David Lindberg's book *The Beginnings of Western Science* lists eight definitions of science, and then uses the words science and natural philosophy interchangeably in the book, which are wrong and has bad influence [2.1].

2. Some other scholars give *wrong* answers, such as equating science with modern science, which is obviously wrong: It is precisely

> because there was science before modernity that it is necessary to point out that this more recent part is *modern* science.

It is worth pointing out that George Sarton (1884-1956), the pioneer of the history of science discipline, never defined science and belongs to the evasive group.

The crux of the problem is that most people do not notice this: The English word Science has been defined twice in the UK, and only the second definition of Science is the "science" that everyone knows at present.

The First Definition of Science

The *first* time was in the 14th century, when Science referred to "established" knowledge (i.e., mainstream opinion), especially theoretical knowledge, covering all fields (including grammar and theology).

The Second Definition of Science

The *second* time, in 1867 (33 years after the word Scientist appeared), the word Science is defined as follows:

> Science is the study of *nonhuman* (material) systems in nature without introducing God/supernatural considerations.

The rationale to exclude God/supernatural is not hard to guess: Even if they exist, since no one can know exactly their properties, they cannot be relied on to build any dependable knowledge.

Many people misunderstand science as the definition of the first time, leading to various problems. Yet, this 1867 definition of science has long been documented:

1. The Oxford English Dictionary notes that "the dominant sense of the term science in modern use—branches of study that relate to the phenomena of the material universe and their laws, and which exclude reference to the *theological* and *metaphysical*—dates from April 1867." (See *Wrestling with Nature*, p 2.)

2. Raymond Williams's *Keywords* (1976/1983) says:

> We can find by 1867 the significantly confident, yet also significantly *conscious*, statement: "we shall...use the word 'science' in the sense which Englishmen so commonly give to it...as expressing physical and experimental science, to the exclusion of theological and metaphysical." The particular exclusion was the climax of a decisive argument, but the specialization excluded under that cover, many other areas of knowledge and learning.

The central idea of this second definition is to emphasize *decoupling* science from religion. In other words, science takes the initiative not to touch religious issues. Unfortunately, few science historians and science philosophers have noticed this, these two documents, and other formally published scholarly sources (see Discussion below).

Background of the Second Definition

Why is there a second definition? The background is that by the 14th century, the Philosophy founded by the ancient Greeks (which includes all disciplines today) has been divided into three parts: 'Philosophy' (single quotes), Theology, and Natural Philosophy. For religious reasons, the British believed that humans are different from other animals (that was before Darwin's theory of evolution), so they should be studied separately. The result is that study of human was grouped under 'Philosophy', and nonhuman systems (nonhuman animals, plants, etc.) belong to Natural Philosophy. However, the works of Natural Philosophy is divided into two broad categories: "invoke God" and "no God." The same goes for 'Philosophy' [2.1].

Britain's Industrial Revolution (1760-1820/1840) spanned the 18th and 19th centuries and boosted economics and education. Around 1800, the number of British universities expanded and increased rapidly. In particular, in the field of Natural Philosophy, a large number of scholars with teaching and research as their profession appeared. To describe this new group of professionals, William Whewell coined a new word in 1834: Scientist. Consequently, by 1867, the "no-God" part of Natural

Philosophy was greatly increased. To describe this new thing, the British called this part of the study Science. In fact, a new *word* was needed to describe a new *concept*: to understand nature without invoking God.

Unfortunately, they chose to use the existing word Science from the 14th century (instead of creating a new word) and gave it a *new* meaning—the essence of the second definition. And this caused a lot of confusion which continued to today.

After the Second Definition

With the new word and concept of science, people went back in history to see who and what research activities in the past met the definition of science. They agreed that the theory "all things are water," proposed by Thales (c.624–c.546 BC), the first philosopher of ancient Greece, about 2,600 years ago was the first theory about nature without invoking supernatural. By this, they honored Thales as the Father of Science, although he also believed that everything has a soul, a supernatural—this part is excluded from science [6.1]. Note that a scientific theory, like that of Thales, being right at that time, could be wrong later, which is pretty common in the history of science [8.4].

Similarly, people looked at the 1687 book *The Mathematical Principles of Natural Philosophy*, written by the natural philosopher Isaac Newton (1642-1727) more than 300 years ago. They attributed to science the parts of his book that had nothing to do with God (the three laws of motion and the law of gravitation), discarded the parts that mentioned God, and recognized him as a scientist (one of their own), although Newton was a devout Christian and had done a lot of serious research on the Bible. What all these mean is that scientists could be religious at the same time as long as their scientific works do not bring in God or supernatural.

In fact, according to the 1867 definition, Thales and Newton are only half scientists (partially introduced soul or God in their works) while Galileo is 100% scientist (who did not introduce God in his work) although he was also a devout believer.

The word Scientist was coined by William Whewell (1794-1866) in 1840. Strictly speaking, between 1840 and 1867, scientists, according to Whewell, are people who devoted themselves exclusively to the study of natural philosophy, whose work could invoke God. During this period, Newton was a 100% scientist. Yet, after 1867, scientists refer to people who is engaged in scientific research without invoking God. Thus after 1867, Newton became half a scientist. Quite intersting, right?

Deficiency of the Second Definition

Science's 1867 definition of limiting scientific research to nonhuman systems is a serious shortcoming, no longer in line with later scientific understanding [1.2]. According to this definition, medicine—the medical study on humans, is *not* part of science, let alone the so-called Social Science.

The Third Definition of Science

To solve this problem, there is a *third* definition of the word Science in 2007—the Scimat definition:

> Science consists of the process and results of human's effort to understand *all* things in nature without introducing any supernatural considerations.

Why the Third Definition Is Important

The last two definitions of science (1867 and 2007) are important, for the following reasons:

1. According to the 1867 or 2007 definition of science, the court could directly reject the request of intelligent-design people who insisted to teach creationism in secondary school *science* classes. That is, there is no need for the trial to happen in the first place.

2. Only with a *universal* definition of science, philosophers and historians of science can have a common language to conduct smoothly their discussion in meetings. And authors will know what to include in their general history of science, otherwise it

must be called "general history of nonhuman-system science" (if the 1867 definition is adopted).

3. According to the 1867 definition, the humanities were excluded from science, resulting in people's low expectation from these disciplines, which, of course, seriously affected the development of the humanities. Only by adopting the 2007 definition will the humanities become more scientific and can move forward.
4. The humanities include (political) decision-making, which is a key discipline that affects world peace. The low scientific level of the humanities is not good for all living beings.
5. Only with a universal definition of science can the Needham Question (Why did modern science not happen in China?) and the Chinese traditional-medicine question (Is it science or not?) be resolved [9.4, 9.5]. Common sense, right?

Discussion

Although everyone's understanding of the word Science may be different, there is a common understanding that science cannot refer to God/supernatural to explain any phenomena. This definition emerged gradually in the first half of the 19th century and was summed up by the British theologian William Ward (1812-1882) in the April 1867 issue of *Dublin Review* (see below).

There are at least *five* academic sources that mention the important fact of the 1867 definition that decouples science from religion. But, unfortunately, none of them have attracted enough attention from the science scholars:

1. Ward, W. G. [1867]. *Science, Prayer, Free Will, & Miracles: An Essay* (reprinted from the *Dublin Review* of April 1867). South Yarra, Victoria, Australia: Leopold Classic Library. (See footnote on p 1.)
2. Hayek, F. A. [1952]. *The Counter-Revolution of Science: Studies on the Abuse of Reason.* Chicago: University of Chicago Press. (See footnote 2 in Sec. 1.)

3. Cunningham, A. & Williams, P. [1993]. De-centring the "big picture": The origins of Modern Science and the modern origins of science. The British Journal for the History of Science **26**, 407-432.

4. Harrison, P., Numbers, R. L. & Shank, M. H. (eds.) [2011]. *Wrestling with Nature: From Omens to Science*. Chicago: University of Chicago Press. (See p 2.)

5. Lam, L. [2014]. About science 1: Basics—Knowledge, nature, science and scimat. *All About Science: Philosophy, History, Sociology & Communication*, Burguete, M. & Lam, L. (eds.). Singapore: World Scientific. (See p 15.)

Conclusion

The now accepted definition of science that no supernatural is allowed in science came from the 1867 definition. It was an intentional act and not an accident.

Dual use of a word, any word, causes trouble. Example: "Work" means "labor" in everyday parlance, but physicists define it as "work done by a force" (approximately, force times distance)—two very different things. The result is that ordinary people find it difficult to understand physicists' language and physics itself. Similarly, the second definition of the word Science, using an existing word, leads to confusion in communication for the same reason.

Imagine in 1867, if the British had not defined the word science a second time but instead created a real neologism (such as Scinece), many of the confusion about science could have been avoided.

Remarks

1. The Britons did not read Chinese; otherwise, they might find out that the ancient Chinese philosopher Guanzi (c.723 – 645 BC) had proposed "all things originate from water," about one hundred years earlier than Thales' "all things are water" [9.1]. Unlike Thales that no writings by him survive and no contemporary

sources exist, Guanzi's sayings are recorded in the book *Guanzi*. Thus, science, according to the 1867 definition, actually began in ancient China, not ancient Greece [9.2].

2. The *scientificity* (scientific level) of different disciplines is different: Nonhuman natural science is relatively easy and the scientificity is high; the human system is more difficult and the scientificity is low. Many people mistakenly take the former as the scientific standard, and therefore feels that the latter is not scientific enough or is not scientific. In fact, no matter how high or low, it is all science, except that they are at different stages of scientific development.

3. In China, some people divide science into "narrow science" and "broad science," the former referring to the science of nonhuman systems (generally called Natural Science) and the latter referring to the science of human system (mainly referring to Social Science). This classification is wrong. There is only one kind of science, although there are two types of research objects: nonhuman systems and the human system. Scientific research objects can also be divided into four categories according to their simplicity/complexity and certainty/probability [2.5].

4. Previously, Ren Hong-Jun (任鸿隽, 1886-1961) also divided science into two types: narrow and broad. In his opening essay for the inaugural issue of the Chinese magazine *Science* in 1915, he said:

   > The great name of science means knowledge and systematic. In the *broad* sense, all knowledge which separates into disciplines, each of which concentrates on one specific topic, is called science. In the *narrow* sense, if the knowledge is about a certain phenomenon, its reasoning is based on experiments, its observation is coherent, and it can be related to large number of examples, then it is science. Therefore, history, art, literature, philosophy, and theology are not science; astrophysics,

> physiology, and psychology are science. The so-called science today is science in the narrow sense.

This statement is equivalent to calling the *first* definition of science broad science and modern science as narrow science. This is wrong. Ren was just a chemist (with master's degree in chemistry from Columbia University), not a science historian or science philosopher. It is thus not surprising that he was wrong about science more than a hundred years ago. But it is a little strange that his wrong terminology or classification is still used by people today.

5. Note that in the second definition of Science in 1867, apart from theological/supernatural considerations, metaphysics was also excluded from science. Metaphysics is what is now called "philosophy" (double quotes [6.1]). This means that "philosophy" can be used as inspirations but not as validation of scientific theories. The validation for any scientific theory is the *Reality Check* (which includes experiments) [8.3]. As for the fact that science is inspired by "philosophy" or clouds in the sky, it varies from person to person, and it is actually not important. The reason is that "philosophy" theories are only philosophers' a priori *conjectures* after consulting the scientific knowledge available at that time [6.7]. Nature does not necessarily buy it. There are plenty of examples like this in the history of science.

8.2 Science not exact

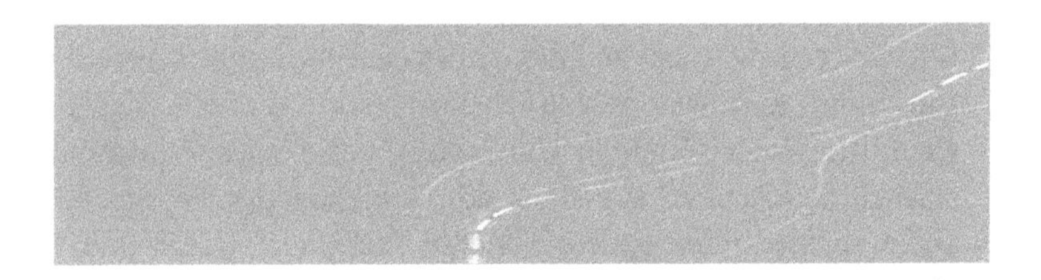

"*Exact* science" exists as a myth for the following reasons:

1. Every constructed scientific theory is an approximation. First, theories are models of real things. For example, Newton's three laws of motion talk about the motion of a *point* particle (*not* body), which is an ideal thing with mass but zero in size (see Remarks below for applying the laws to objects of finite size). Second, so far at least, a theory always is an approximation of a larger, later theory. For example, Newtonian mechanics is an approximation of Einstein's special theory of relativity (when the speed is small compared to light speed); the Standard Model in particle physics is a low-energy approximation of some future theory to be discovered.

2. Even if a theory is exact, the solutions obtained are usually approximate because exact solutions are rare. For example, of all the atoms, only the quantum mechanical model of the hydrogen atom has an exact solution.

3. Even if the theory and solution are exact, they need to be verified experimentally because nature has the final say in any scientific theory. But experiments involve instruments and measurements that always have limited resolution. For example, to check the equation A = B, we measure each quantity in A and B and try to prove that the left-hand side and the right-hand side of the equation are equal to each other. Assuming that our measurements yield $A = 2.5 \pm 0.1$ and $B = 2.5 \pm 0.1$, we can only say that this equation holds within the experimental-error range (due to the $\pm$ 0.1). In other words, the inevitable finite resolution of measurements prevents rigorous proof of any scientific theory.

In other words, strictly speaking, unlike mathematical proofs, science will never prove anything. There is no absolute, final thing in science.

Discussion

1. The conclusion that science is *not* absolute has its "philosophical" (double quotes [2.1]) meaning, and should have a profound impact on philosophers.

2. There is room for improvement in any theory, and new improvements *may* appear. Whether it will appear or not, only when it appears, we know for sure.

3. Approximations work in scientific research because, like getting lost in the forest, a rough map is what a person needs to get out of the forest.

4. It is precisely because of the use of approximation that science has advanced rapidly in the last few hundred years.

5. Without approximation, society cannot function. For example, two people make an appointment to meet at 2 pm. In fact, as long as you see the other person a few seconds around 2 o'clock (an approximation), you can claim that you two meet on time.

6. Approximation is inevitable, indicating that life is uncertain even if the uncertainty is very small. Dealing with others requires and can only be done with wisdom and humility in order for everyone to continue to live.

Conclusion

Science is inseparable from approximation. Science, though usually rigorous, is not exact. Science coexists with approximation and that is why science can move forward.

Remarks

When applying Newton's second law to an object, we first divide the object (abstractly) into many small volumes, and then *approximate* each volume as a point particle. After that, the effect of each volume is superimposed and take the limit of the volume size tending to zero.

8.3 Reality check

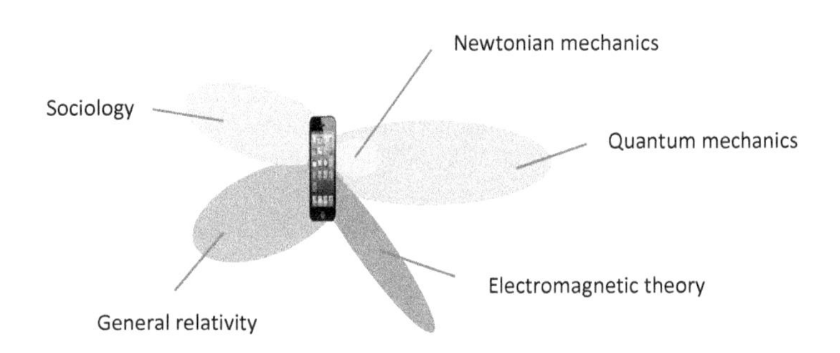

Since science is not exact [8.2] and scientists are human beings who could make mistakes [10.8] then why should everyone take science seriously?

Everyone should take science *seriously* for two reasons:

1. Science is *effective*. Our air conditioners and mobile phones, among many other good things, owe it to science.
2. Science is the *best* way to understand nature. As Newton said: "And although the arguing from Experiments and Observations by Induction be no demonstration of general Conclusions; yet it is the best way of arguing which the Nature of Things admits of."

Reality Check

Why does science work? This is because it must pass a *Reality Check* (RC), just like every legal driver must prove himself through a road test. RC means "confirmation" by experiment or practice (within the margin of error), or at least in agreement with the established data.

Due to the imprecise and incomplete data collected, sociological theories are difficult to be 100% confirmed. Confirmation is a necessary and crucial step for any theory to be recognized as part of the *Knowscape* [2.2]. It is the RC that makes scientific knowledge unique among all forms of "knowledge" and separates science from other approaches to understanding the world, such as meditation or religion.

Cell Phone Test

Of all the RCs, the *Cell Phone Test* (CP test) stands out because the working principle of the mobile phone depends on the validity of a large number of theories: Maxwell's electromagnetic equations, Newtonian mechanics, quantum mechanics, semiconductor theory, general relativity, sociology (i.e., enough people want to interact with others).

Any *new* theory, if it conflicts with any of the principles behind how mobile phones work, must explain why there is a conflict and why this new theory is better. A good and acceptable answer is that the new theory includes the old theory as a special case. We highly recommend the CP test to advocates of any new theory.

Discussion

1. Unlike (nonhuman) natural science, which has experiments as RCs, *humanities/social science* theories involve people. Due to ethical considerations, except for questionnaires, psychological tests, brain scans and other harmless tests, there are not many means of RCs in the latter, which, when available, often require decades to hundreds of years of waiting to know the results. Therefore, unlike physical theory, most social science theories have *not* passed the RC, and can only be adopted as a stopgap measure. They should be used with double care, not exhaustively, and should be modified in due course according to external changes (as practiced in economic forecasting). However, those who *think* they are masters of reading history and having good foresight will not change their theories or policies overnight. And that is a problem. Note that RC is at the core of the philosophy called *Pragmatism*.

2. RC is the *guardian* of science and is a *signature* of science. It is extremely important. But all historians of science (including George Sarton and Thomas Kuhn) do not mention or emphasize that science has and needs RC. The reasons are unclear but it could not be an oversight. What is known is that not mentioning the RC opens a back door for *postmodernism* and the *sociology of science* to talk nonsense about science, mislead sentient beings, reduces the trust and goodwill of ordinary people in science, and has a bad impact, which is a pity. A great pity!

8.4 Science not always right

Superconductivity was discovered in 1911 but the first correct theory did not appear until 1950, called the Ginsburg-Landau (G-L) theory, which is at the phenomenological level. It was only after the emergence of the microscopic BCS theory in 1957 that the mechanism of superconductivity was understood, and the G-L theory could be derived from the bottom up. (For the three levels of disciplinary research, see [4.3])

In fact, in the 46 years from 1911 to 1957, many theoretical articles (perhaps hundreds) were published, but only one remained correct. Those who got the theory wrong included eight Nobel Prize winners: Thompson, Einstein, Bohr, Bloch, Landau, Heisenberg, Born, and Feynman. All wrong, although there was no problem (i.e., not contradicting existing experiments) when the article was published.

This story tells us that while science is going on—sometimes more than 2,000 years (from Aristotle's mechanics to Newtonian mechanics) or 46 years (from experiment to theory in superconductivity)—many of the papers that published turn out to be wrong (hindsight) and cannot be used. Only those which have passed the *Reality Check* can stay [8.3].

It is not necessarily that the person who did the wrong theory is not smart enough (some people are, of course), but that the phenomenon that needs to be understood are too new and complex, and the key *guiding* experiments have not yet appeared. (In the superconductivity case it is the isotope experiment).

In addition, most of the equations in physics are nonlinear, with multiple solutions, and only experiments can tell us which one nature chooses. Therefore, good physicists know: Physics starts with experiments and ends with experiments. Superstring theory encounters this situation: too many solutions, no experiment.

It is very likely that about 90% of the experiments and about 10% of the theories will survive in journal articles. Papers that published, except those fraudulent ones, are all part of science. What is right is kept alive, used and written into textbooks. Those which are wrong are kept in scientific archives and become research materials for the historians and philosophers of science. All are useful!

Conclusion

When science is going on there are many theories and even experiments that turn out to be wrong later. So, science may not always be right, but the science that can stay must be right.

8.5 Science and reason insufficient

Human affairs cannot depend on science and rationality alone. Otherwise, sometimes they cannot be done; other times they are not effective; and in serious cases, they will get people killed. Why?

The Lack of Science

Science can develop because of the use of approximation. So, science is not exact [8.2].

In fact, science is only very successful or fairly successful for *simple* systems (such as the invention of air conditioner and mobile phone), and not so successful for *complex* systems (including the human system). For example, one cannot accurately predict the weather a week from now or the stock going up or down in minutes, let alone when one will win a lottery jackpot.

People's impression that science is very successful comes from not understanding or ignoring the fundamental differences between simple systems and complex systems.

In addition, for simple systems, only *classical* systems are deterministic (such as stones falling to the ground) while *quantum* systems are intrinsically probabilistic (although the equations are deterministic). This is because when linking the theoretical results of quantum mechanics to experiments or the real world, only the probabilities of an outcome can be talked about—the contribution of the Nobel laureate Max Born (1882-1970).

Although the complex system of humans is classical, it is also *probabilistic*, for reasons different from those mentioned above. Probabilistic because there are so many external factors (such as your mother's emotions) that cannot be included in the theory.

Because of the inevitability of probability, we live in uncertainty every day (such as not being able to predict tomorrow's Covid-19 test outcome), no matter how much science we know (or can know).

How to deal with yourself? It is recommended that you learn some basic probability knowledge (e.g., a high probability of success does not mean that it will succeed; there are enough people buying lottery tickets that someone is bound to win the jackpot in the end) and prepare for the worst scenario. As a scientist, add one more: learn to be *humble*.

The Lack of Reason

Reason (or rationality) alone is not enough because:

1. Rational thinking is never complete (e.g., when more factors are considered, irrational thinking becomes rational, and vice versa).
2. Human systems are probabilistic: Any predictions about humans can only be made with probabilities.
3. Therefore, decision-making based on rational thinking always involves *gambling* (i.e., incomplete information).

Since it is gambling, of course, other factors must be considered. For example, compassion for the vulnerable, empathy for the helpless, and the limit of personal tolerance when losing a gamble.

Conclusion

Both science and reason have their unavoidable limitations, and no one can or should depend on them alone in decision makings. Life is not easy. Each person has their own difficulties. Be kind to others!

9

Scientific Confusion

9.1 Guanzi: all things originate from water

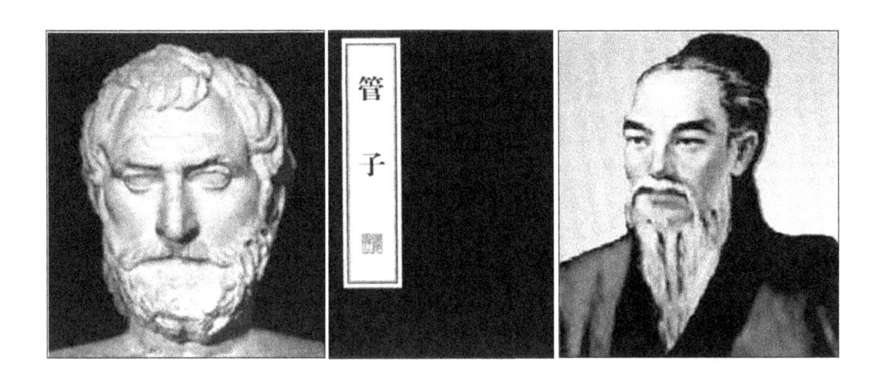

Thales (c.624–c.546 BC, left), the first philosopher of ancient Greece, proposed that “all things are water,” which does not use myths to explain natural phenomena and is recognized as the beginning of science [8.1].

However, before that in China, the philosopher Guanzi (c.723-645 BC, right) of the Spring and Autumn period (770-476 BC) had already proposed that “all things originate from water.” *Guanzi*: “The earth: the origin of all things, the root of all life, the birth of beauty and evil, virtuousness, and foolishness. Water: the flesh and blood of the earth, like what flows in the veins. Thus: water, materials also.” Also: “Human, water too. Vitality of male and female combine, and water manifold.”

Guanzi, a descendant of King Mu of Zhou dynasty; his father, a *daifu* (大夫) of the Qi state. His famous saying: “We know honor and disgrace when we have enough food and clothing.” He was the prime minister of the Qi state for 40 years, and soundly managed the country. Guanzi was also a scholar. He did his scholarly works while being a civil servant—pioneered the so-called “Guanzi model” for Confucius and others to follow, i.e., cultivate learning by official duties and apply learning to practice.

Guanzi’s *scientific* proposal that “all things originate from water” is about 100 years ahead of Thales’ “all things are water.”

9.2 China's four major leads

Left to right: Guanzi, Confucius, Su Dong-Po, Cai Yuan-Pei.

In addition to the four major inventions (papermaking, printing, gunpowder, and compass), China has four major leads:

1. Guanzi leads Thales: all things originate from water [9.1].
2. Confucius leads Plato: civilian education [6.4].
3. Su Dong-Po leads Cézanne: modernism in art [7.4].
4. Cai Yuan-Pei leads the United States: general education [4.2].

Among them, the first one, not using supernatural to explain natural phenomena, is the beginning of science.

The four leads involve three areas: science, education, and art. Unfortunately, the lack of successors made it impossible to pass down in China, and so was surpassed by the West.

The Guanzi case is that he was a full-time civil servant while doing scholarship part time, lacking disciples who could continue his works [6.3].

9.3 Einstein's letter

The New York Times.

TUESDAY, NOVEMBER 25, 1919.

A NEW PHYSICS, BASED ON EINSTEIN

Sir Oliver Lodge Says It Will Prevail, and Mathematicians Will Have a Terrible Time.

SPACE OF FOUR DIMENSIONS

In Which Gravity Ceases to be a Force and Becomes a Quality.

ATTEMPT TO MEASURE IT

C O P Y

A. Einstein,
112, Mercer Street,
Princeton,
New Jersey, U.S.A.

C O P Y

April 23, 1953

Mr. J. S. Switzer
3412 Del Monte Str.
San Mateo, Cal.

Dear Sir:

Development of Western Science is based on two great achievements: the invention of the formal logical system (in euclidian geometry) by the Greek philosophers, and the discovery of the possibility to find out causal relationship by systematic experiment (Renaissance).

In my opinion one has not to be astonished that the Chinese sages have not made those steps. The astonishing thing is that those discoveries were made at all.

Sincerely yours,
/s/ A. Einstein
Albert Einstein

In 1953, in a letter answering J. S. Switzer's question of why modern science did not arise in China, Einstein gave two reasons: The lack of formal logical system (in Euclidean geometry) and the practice of scientific experiments. This letter has been widely quoted but in fact is misread by scholars and laypeople alike. Here, we trace the origin of this letter, the background of the Stanford student J. S. Switzer, and explain why the letter was universally misread.

A Shining Star

In 1905, at the age of 26, Albert Einstein (1879-1955) published his doctoral thesis, plus four reports, across three different fields (Brownian motion, the photoelectric effect, special relativity), and became a shining star in physics. Fourteen years later, he became a world celebrity after making many newspaper headlines at the age of 40 (lower figure, left). It was 1919, eclipse observations confirmed the general theory of relativity he had proposed four years earlier (upper figure).

Since then, Einstein has had countless fans. In 1922, he won the 1921 Nobel Prize for the photoelectric effect, which made his fans doubly happy.

Four Letters

In 1953, two years before Einstein's death at the age of 76, J. S. Switzer, a fan who lived in the Bay Area of California on the west coast of the United States, wrote Einstein a letter (letter **A**), asking why China had not developed modern science. (The content of the letter did not show up in online search.) Einstein, who lived in Princeton on the east coast at the time, was a good celebrity who personally wrote replies.

Einstein's native language is German. His English is OK, but not as good as his German.

Very probably the reply letter was written in German (letter **B**), then translated into English and typed by the secretary, which was read and signed by Einstein, the author. This signed English letter (letter **C**) was sent out to Switzer. Note:

1. The English letter publicly available (letter **D**; lower figure, right) is *not* Einstein's handwritten original letter (letter B) which is supposedly written in German. It is *not* even the signed English letter (letter C) that Switzer received, but only a typed copy of letter C (unsigned; there was no photocopier at the time).

2. Einstein's letter B (as well as C and D) is a *private* letter written by a celebrity to a fan, *not* a scholar's officially published academic paper. Scholars are only fully responsible for officially published papers and for life.

First Paragraph of Letter D

Letter D has only two paragraphs. First paragraph:

> Development of Western science has been based on two great achievements: the invention of the formal logical system (in euclidean geometry) by the Greek philosophers, and the discovery of the possibility to find out causal relationship by systematic experiment (Renaissance).

The "Western science" mentioned here refers to modern science, and "logical system" and "systematic experiment" refer to the two means used by Galileo to make breakthroughs in the study of *simple* physical systems. Einstein was right, but not comprehensively. The two means he refers to are necessary but *insufficient*—lacking *poetic* thinking (analogy, imagination, intuition). All innovations in simple systems, including that of Galileo and Einstein, used poetic thinking.

Einstein was a very poetic man (could play the violin). Look at how he came up with the special theory of relativity, which is not deduced by logical thinking alone, but first there is a kind of comparison, imagination, conjecture, new concepts *before* constructing equations. The same is true when it comes to general theory of relativity. Of course, he knew that poetic thinking *is* important. He did not mention it in letter D, just because replying to fans' letters is impromptu, just like we reply to friends' private chats on WhatsApp—not thoughtfully writing rigorous academic papers. And people who read it cannot and should not hold him to it 100%.

In fact, Einstein said in his book *The Evolution of Physics*, co-authored with Léopold Infeld:

> The formulation of a problem is often more essential than its solution, which may be merely a matter of mathematical or

> experimental skill. To raise new questions, new possibilities, to regard old problems from a new angle requires *creative* imagination and marks real advances in science.

Here, "creative imagination" refers to poetic thinking.

Unfortunately, in response to the Needham Question, the first paragraph of letter D was used by many science historians to explain why modern science did not appear in China, and it was not answered correctly. (For the correct answer, see [9.5].) This is not conducive to scholarship and, more importantly, is unfair to Einstein, the author of the letter.

Last Paragraph of Letter D

The second and last paragraph of letter D says:

> In my opinion one has not to be astonished that the Chinese sages have not made *those* step. The astonishing thing is that *those* discoveries were made at all.

It is said that Joseph Needham (1900-1995) wrote the "those" in the second sentence of the last paragraph as "these." It may be a transcription error, or it may be that he has read letter B and translated himself the German word into English. After all, the English of letter D is the work of the secretary who is not a physicist; she could translate it inaccurately (writing Euclidean as euclidean implies this possibility). At the same time, Einstein himself is unlikely to take the time to proofread such a letter in detail, I guess.

In the second sentence, "those discoveries" has two different interpretations by Chinese scholars:

1. It will be astonishing should the Chinese discover logical thinking and systematic experimentation.
2. It is astonishing that in spite of not discovering logical thinking and systematic experimentation, the (ancient) Chinese did make their own great scientific discoveries.

The correct answer does not depend on whether "those" or "these" appears in the second sentence. And by carefully reading letter D it seems that both interpretations 1 and 2 are incorrect. The correct interpretation is:

3. The astonishing thing is that those discoveries actually happened, through the effort of the ancient Greeks (formal logical thinking) and Galileo (systematic experimentation).

Yet, the answer may be more transparent if one can see letters A and B. This is not impossible because letters A and B should still be preserved in the Einstein archives, and interested and serious historians can go and look for them.

However, all these effort in interpreting Einstein's private letter, word by word, may not be worth it. For the Chinese, doing your own things well is most important. Whether Einstein praised it or not is not important.

Conclusion

1. *Science* was first invented by Guanzi about 2,600 years ago, ahead of Thales [9.1]. And traditional Chinese medicine is a science [9.4]. So, science *did* exist in ancient China, as Joseph Needham asserted [9.5]. It is just that *modern* science (starting with simple systems) did not happen in China, as Einstein tried to explain in his letter.

2. However, historically and perhaps by accident, the Galileo breakthrough of modern science did happen in *simple* physical systems, in which logical thinking and systematic experiments are necessary but not sufficient. Poetic thinking is also required and could be more important. This was well recognized by Einstein but not pointed out in his letter, perhaps because the letter was written in a hurry.

3. After all, private letters are not published research papers. They can only be used as reference, not as evidence or support for serious academic arguments.

4. The ancient Chinese, like the ancient Greeks, did have poetic thinking, as demonstrated in *The Classic of Odes* (诗经) dating back 3,000 years. The reason that modern science was invented in Italy but not in China is

because the Chinese picked the wrong topics to study—they picked *complex* systems—not because of lack of poetic thinking [9.5].

Appendix: Looking for Mr. Switzer

According to Sue Lempert, former mayor of San Mateo, California, she bought the book *Science Since Babylon* by Derek J. de Solla Price, a professor at Yale University. She found in the References that there was such a letter from Einstein to San Mateo resident J. E. Switzer (Price misspells J. S. as J. E. in the book). Lempert was so excited that she set out to find Mr. Switzer. (Source: Sue Lempert, "Switzer mystery solved," *Daily Journal*, Feb. 1, 2016, updated July 12, 2017.)

Lemert found four J. E. Switzer who reside in Vista, California and, of course, none of them could be the Mr. Switzer she wanted to find. Fortunately, savior Chris McGuire showed up in time. McGuire is an amateur historian and genealogist who finally brought the truth to the world.

About Mr. Switzer

According to McGuire via Lempert:

> US Army Colonel John Singleton Switzer Jr., born 1895 in Fort Leavenworth, Kansas and died 1970 in Orange County, California. He lived for a short time at 3412 Del Monte St. … . It was while he was getting his master's degree at Stanford University that he wrote Albert Einstein. The original letter Einstein sent [Letter C] eventually ended up in the possession of his Stanford Professor—Arthur F. Wright who later moved to Yale University. The Colonel never really appears to have set down roots too long being a career military man.
>
> Wife no. 1: Edith Russell Switzer b. 1894 in Ann Arbor, Michigan and d. 1962 in Marin County, California. They were married in Washtenaw, Michigan in 1917. They had a daughter Ruth, born 1920 in Fort Benning, Georgia. Wife no. 2: Evelyn Wardall Knight, born 1907 in California and died 1982 in Orange County,

California. They were married in Marin County, California in 1963.

Switzer served in both World War I and World War II and is buried in Golden Gate Cemetery in San Bruno.

McGuire's Sleuth Work

According to Lempert, this is how McGuire found the clues:

> I [McGuire] used various search engines and genealogy sites to do the research and cross referenced multiple search terms. The break came when I found that Arthur F. Wright had the original letter [letter C]—with this new name I found the Einstein Archive that listed J. S. Switzer, then I hit pay dirt on a Chinese website—here is link: http://bbs.tianya.cn/post-free-2532845-109.shtml; this is where I found an actual copy of the letter [letter **E**].
>
> The rest of the info I found was through a basic genealogy search. I since found a few early school photos and a few other records ... John Jr. graduated in 1916 from the University of Michigan where he also competed on the varsity tennis team. It appears he also served as an associate professor at Cal Berkeley. He also wrote an ongoing article for the *Infantry Review* titled "The Champagne-Marne Defensive." He appears to have lived a full life.
>
> His father also John Singleton Switzer was a graduate of West Point and also was a colonel and is buried in Arlington National Cemetery.

Letter E

McGuire obtained letter E from the site: http://bbs.tianya.cn/post-free-2532845-109.shtml. Here is letter E, as reported by Lempert who obtained it from McGuire:

> Einstein
> 112, Mercer St.
> Princeton, NJ. U.S.A.
>
> April 23, 1953

Mr. J. S. Switzer
3412 Del Monte
San Mateo, California

Dear Sir: The development of Western Science has been based on two great achievements, the invention of the formal logical system (in Euclidean geometry) by the Greek philosophers, and the discovery of the possibility of finding out causal relationships by systematic experiment (at the Renaissance). In my opinion one need not be astonished that the Chinese sages did not make these steps. The astonishing thing is that these discoveries were made at all.

SY

/s/ A. Einstein

Albert Einstein

Note that letter E has a few words different from letter D (lower figure, right): using "these" and not "those," and "Euclidean" instead of "euclidean." It seems that letter E was copied from somewhere (which could be Needham's version), but definitely not from letter D.

Story Roundup

After retiring from the military, J. S. Switzer studied for a master's degree in history at Stanford University. He took Arthur F. Wright's class, which included a discussion on Chinese science. This led Switzer to write a letter to Einstein (letter A) and received the famous reply (letter C). Then Wright moved to Yale University, with Letter C in his pocket. There Dr. Price gave a series of talks in which he mentioned the Einstein reply Switzer received.

McGuire also mentioned that J. S. Switzer was in Fort Lewis in 1940 and was a neighbor of former President Dwight Eisenhower and his wife Mamie.

9.4 Chinese medicine is a science

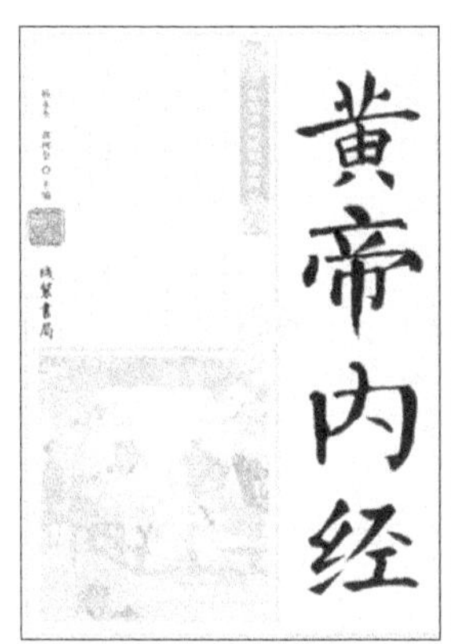

"Chinese medicine" here means "traditional Chinese medicine" (TCM). "Is TCM a science?" is a question that has been asked and debated for over 100 years, especially in China after the influx of Western science into the country. The debate is still ongoing and no universal answer has been reached yet. It is particularly important within China because this is not a purely academic issue, but also a question of cultural confidence and resource allocation. To settle this question, the first step is to clarify and agree upon on a common definition of science; the second step is to correctly understand the nature of science. Here, we set out to do exactly this and conclude that TCM is a science. But saying something is scientific only means that it was right (i.e., consistent with what was known) at the time proposed, irrespective of whether it is right or wrong later. Whether TCM theories are right or not *presently* is a question remained to be settled. More scientific studies are needed.

Science Is Mostly Wrong

A brief history of superconductivity: In 1911, Kamerlingh Onnes discovered (low-temperature) superconductivity; in 1950, Ginzburg-Landau (GL) phenomenological theory was published; in 1957, Bardeen-Cooper-Schrieffer (BCS) microscopic theory was published. In 1986,

Bednorz and Müller discovered high-temperature superconductivity. All won the Nobel Prize.

During the 46 years from 1911 (when superconductivity was first discovered) to 1957 (when a successful theory appeared) there should be a few hundred or more theoretical articles published. With the exception of the GL theory, all superconductivity theories before the BCS theory are wrong, but they are good science at the time of publication. That is, theories that were *right* at the time but were *wrong* later are all part of science, and their existence is the *norm* in the development of science [8.4].

In other words, science may not always be right, but the science that can remain and still being used must be right (i.e., consistent with the reality of nature).

In journals, experimental papers that survive test of time are much more (estimated 90%) than theoretical papers (about 10%). Theory is more difficult than experiment. Since theory is more influential than experiment, generally only the names of the theorists are widely known. Who can remember a few experimentalists? (See [11.7] for an example of exception.) This is also the norm.

TCM Research Is Scientific Research

Traditional Chinese medicine (like Western medicine) began in primitive society, and the theory of TCM was basically formed during the Spring and Autumn period and the Warring States period. The ancient myth and legend that Shen Nong (神农) tastes hundred herbs, has no credibility. But the *Inner Canon of the Yellow Emperor* (黄帝内经) is different. The 18 volumes of this book, the main part of which was formed in the Warring States period (475-221 BC) to Eastern Han dynasty (AD 25-220), summarizes the medical experience and academic theories from the Spring and Autumn period to the Warring States period, which absorbs previous knowledge in astronomy, calendar, biology, geography, anthropology, psychology and other disciplines before the Qin and Han dynasties. It uses the theories of yin and yang, the five elements, and the unity of heaven and man to make a more comprehensive exposition of the anatomy, physiology, pathology and diagnosis, treatment, and prevention of diseases of the human body.

Since then, the development of TCM has not been interrupted. Li Shi-Zhen (李时珍, 1518-1593), a medical scientist and imperial physician of the Ming dynasty, was born in a medical family—his father and grandfather were both famous doctors. At the age of 23, he studied medicine with his father (equivalent to a medical-school training), practice medicine at a pharmacy, and completed the *Compendium of Materia Medica* (本草纲目) in 29 years (equivalent to the current medical-school professor who is publishing papers while being a doctor).

Before the book was written, Li went to the mountains and countryside to observe and collect drug samples, gathered and *tested* herbs everywhere, visited famous doctors and scholars, and visited fishermen, farmers and other civilians to collect folk prescriptions. The book is divided into 16 parts and 52 volumes, with about 1.9 million words. It contains 1,518 kinds of drugs collected by various herbal experts, and 374 kinds of drugs added. It collects 11,096 prescriptions from ancient pharmacists and folks, with more than 1,100 pictures of drugs attached to the book.

It can be seen that TCM has experiments (testing medicine personally), theories (meridian theory, etc.), and academic publications (many great books). TCM has honestly gone through the four steps of the scientific research process: gathering information, asking why, guessing the answer, seeking empirical evidence. Only one conclusion can be drawn: TCM research is *scientific* research. This conclusion is still valid even if all its theories turn out to be wrong (see above). But in fact, there is no conclusive evidence showing TCM theories are wrong (see Discussion). Extended conclusion: TCM is a *science*. This conclusion can also be drawn from the following discussion.

TCM Is a Science

Ancient Chinese and Western medicine are about the same quality. The theoretical construction of the Warring States' *Inner Canon of the Yellow Emperor* and the ancient Greek's *Complete Works of Hippocrates* (400

BC) are very similar in several aspects: abolishing witchcraft, preserving medicine, overall concept, regulating balance, philosophical thinking, and clinical practice. In the 2nd century, Western medicine moved towards anatomy while Chinese medicine did not, and the two began to diverge. Since the beginning of the Renaissance in the 15th and 16th centuries, Western medicine has been combined with modern science and Chinese medicine has not, and the distance between the two has been widened.

However, since Plato, Western medicine, a study on humans, has been considered an applied study, not a natural philosophy, or a science as defined in 1867 [8.1]. But under scimat's 2007 definition of science, both Chinese and Western medicine are sciences, because they study the human body—part of nature and do not introduce any supernatural considerations [3.3].

Discussion

1. The discussion of whether TCM is a science should concentrate on the content and history of TCM, and the nature and definition of science. It should not involve other considerations (such as cultural pride). This is the approach followed by the above discussion.

2. TCM's terminologies, like meridians, acupuncture points, and qi (气), are descriptions at the *phenomenological* level [4.3], which should not be understood literally. As such, they are not necessarily things that can be directly observed in the human body.

3. With the advance of modern science, while TCM theory remains at the phenomenological level, Western medicine theory has entered the *bottom-up* level [4.3]. For any system (here the human body), how to connect the phenomenological level with the bottom-up level is not a simple problem. (For a successful example, see the case of superconductivity.) If the theory of TCM cannot be explained by Western medicine at present, it may be temporary, not implying that it must be wrong.

4. Not everything scientific has to be supported, just not opposed. Moral support is relatively easy, and material support should be selective due to the limited time, energy and financial resources of individuals and the limited resources of society. How individuals choose to support them is the freedom of the individual and does not need to be discussed. But how much society should invest in a project is something that should be discussed carefully and calmly, even for a long time.

5. Some people think that TCM theory is a holistic theory, which cannot be proved (because the existence of any so-called holistic system cannot be proved). Even so, holistic systems can be studied from the building blocks *and* their interactions, through computer simulations in some cases. Many complex systems (such as those in Sociology) have been studied this way with success.

6. TCM has basically stagnated for 400 years and needs to be modernized.

7. Even if the direction of integrating Chinese and Western medicines is correct, it is still not enough. What is needed is a leader who can integrate holism and reductionism, i.e., people who do complex systems.

Conclusion

1. Traditional Chinese medicine (TCM) research is scientific research.
2. TCM is a science.
3. There was TCM in ancient China, so there was science in ancient China.
4. TCM has accumulated thousands of years of valuable experience. Like Tang poetry and Song poetry, it is a valuable Chinese treasure. They form the basis for cultural confidence.

9.5 The Needham Question

The Needham Question of why modern science did not arise in China has garnered hot debates and numerous answers since it was raised in 1964. To answer this question correctly one has to learn about the definition of science and the differences between simple and complex systems. Unfortunately, it seems that none of these two issues had been entertained by the debaters and so all the answers are incorrect. Here, the correct answer is presented: The ancient Chinese had picked the complex system of human to study while breakthroughs in Western modern science happened in the study of simple systems.

The Needham Question

Joseph Needham (1900-1995) was a British biochemist, historian, and sinologist. In 1937, the married Joseph Needham fell in love with a Chinese woman, and then fell in love with Chinese culture, so he became a sinologist and compiled *Science and Civilisation in China*, a total of 27 books in 7 volumes (25 books are published).

In 1964, Needham asked two questions in his book *The Grand Titration: Science and Society in East and West*:

1. Why did modern science not develop in Chinese (or Indian) civilization, but only in Europe?
2. Why was Chinese civilization so much more effective than Western civilization in applying human knowledge of nature to human practical needs, from the 1st century BC to the 15th century?

The first question, called the Needham Question, is more fun than the second one, and there are many answers provided (including by Needham himself). But they are not convincing, and some are simply wrong. For example, one answer: Ancient China did not have science in the first place, but only technologies (such as the four great inventions) that were more advanced than in the West. This is obviously wrong, because there *was* science in the Ming dynasty and before: traditional Chinese medicine, which is the science of complex systems (human health) [9.4].

In fact, there may be a simple answer to the Needham question: The ancient Chinese picked the *wrong* topic to study more than 2,000 years ago, followed by bad lucks.

Breakthrough in Simple Systems

The scientific content includes both simple systems and complex systems [2.5]. The former represented by physics in the secondary school curriculum, and biology as the representative of the latter.

In the West, modern science began 400 years ago with the breakthrough in the study of *simple* systems (Galileo's free fall, pendulum, astronomical observations). This kind of breakthrough does require (mathematical) logical thinking, systematic experiments, *and* poetic thinking.

Ancient Greece and ancient China began from similar starting points, but the ancient Greeks (such as Aristotle) attached equal importance to simple systems and complex systems and studied them together, while ancient Chinese (such as Confucius) chose *complex* systems (human problems) to investigate and ignored simple systems.

The reason behind the different choices is that on the one hand, ancient Greeks had a lot of leisure that came with wealth, so they were free to ask all sorts of questions, from heaven to earth, from falling rocks to societal problems. On the other hand, ancient Chinese intellectuals had to look for government jobs to survive, which were mainly about governance of humans—a very complex system.

Complex-system problems involving humans are hard to crack, as witnessed by the relatively low scientificity of the humanities and social science today. This explains the slow progress in Chinese philosophies, and the absence of breakthrough in the humanities/social science—ancient or modern, East or West [3.1].

Breakthrough in Complex Systems

However, breakthrough in *nonhuman* complex systems is much easier to achieve and indeed happened. Examples are Darwin's evolutionary theory in 1859 and Mendel's hybrid inheritance of pea in 1865/1866, which came two hundred years more after Galileo. Note that these breakthroughs in biology are very low tech that do not involve mathematical logic, but only systematic experiments and poetic thinking, plus a lot of patience.

That modern science began with breakthrough in simple systems but not complex systems was entirely accidental. It has nothing to do with high-tech instruments (Galileo's telescope is not that high tech) but depend heavily on the emergence of capable individuals.

Imagine that after Confucius, as long as an ancient Chinese noticed the change in pea breeding 400 or more years ago, did not modern science appear in China first? It did not happen—bad luck for China.

Heaven and Man are One

More than 2,000 years ago, Chinese was rightly aware that "heaven and man are one" (天人合一). This means that:

1. The part of nature beyond human—the nonhuman systems—and human can interact with each other.
2. Both (self-) organized similarly.

These two cognitions had since been verified scientifically. For example, there are three universal organizational principles that cover both nonhuman and human systems: chaos, fractals, and active walk [22].

Since heaven and man are one, it is not necessary to study humans directly to understand human society. The same purpose can be achieved by studying ant society, say; after all, the latter is much easier.

The method of studying ants is simple: First keep a group of ants in captivity (without spending money), observe their social activities, and then according to their division of labor, use different colors of pigment on the heads of ants to identify, constantly observe and record. As long as a child or adult did it early, modern science appeared in China first.

This situation did not appear in history, perhaps because the curiosity of Chinese children was worn out early. They were forced at young age to read the *Four Books* and the *Five Classics* for the national examination—the way to get a government job for intellectuals in old times.

Circumvent Debate

Ancient Greece did not have many scientific instruments, and the research methods of science and other areas were mainly debates—the so-called *Socratic Method* [6.2]. In the West, this method has been passed down, and the improvement of all disciplines cannot be without mutual discussion or debate, just like to improve basketball skills, finding someone to intercept each other will be more effective than solo hard practice.

The tradition of avoiding debate in China may have begun with Confucius and followed by Confucianism, which is certainly not conducive to the development of science.

In China, after the so-called "science spring" year of 1978, when Yan Ji-Zi (严济慈) and Qian San-Qiang (钱三强), vice presidents of the Chinese Academy of Sciences (CAS), called for everyone to restore the fine tradition of public debate, Guan Wei-Yan (管惟炎) of the Institute of Physics and Hao Bai-Lin (郝柏林) of the Institute of Theoretical Physics took the lead and published their mutual dispute on a certain issue in physics in the *Nature Magazine* (自然杂志). About 20 years ago, the Science Weekend session of CAS' newspaper *Sciencetimes* (科学时报) featured polemics about scientism. At present, there are also tit-for-tat debates in the Chinese economic circles from time to time.

In any discipline, the words of any authority (including Einstein) do not count. One should seek truth from facts and think independently. To progress in any discipline, debate is indispensable.

Discussion

1. Guanzi (c.723-645 BC) invented *science* when he proposed "all things originate from water," ahead of Thales [9.1]. Unlike Thales, Guanzi's theory was not passed down because he was a high-level civil servant without disciples. Additionally, there definitely was science in ancient China since traditional Chinese medicine is a science [9.4]. Thus, science did exist in China, as Needham asserted. It is just that *modern* science did not happen there.

2. Historically, modern science began in the West with Galileo's breakthrough in simple *physical* systems, using three tools: (mathematical) logical thinking, systematic experiments, and poetic thinking [9.3]. But later, Darwin and Mendel's breakthroughs in complex *biological* systems did not require mathematics, only systematic experiments and poetic thinking, both of which China has.

3. Modern science could begin with breakthrough in either simple systems or (nonhuman) complex systems. That it happened with simple systems was purely accidental.

4. For China not inventing *modern* science, among the three reasons given above, the first one is the most important. Not taking advantage of the unity of heaven and man only means opportunities missed. Circumventing debate is only detrimental to development. But if you pick the wrong topic to study, no one can help you. In China's case, they concentrated on the complicated human system and missed the chance to invent modern science.

5. Even though Needham was not a trained historian of science, he did ask some profound questions about China's development in science, or lack of it. Needham was a good biochemist before he changed field and became a science historian. His assertion that science did exist in ancient China is correct while other science historians got it wrong because they do not understand the essence and nature of science. In one sentence, science is the study of all things in nature without mentioning God/supernatural [8.1].

6. Those who use Einstein's letter of 1953 to answer the Needham Question are wrong. See [9.3] and arguments presented above.

Conclusion

There was science in China since the ancient times even though modern science did not happen there. That is because the ancient Chinese picked the most complex human system to study. The human system is the most difficult topic to crack, as evidenced by the slow progress in the humanities in the West for more than 2,000 years since Plato.

Picking a research topic, like picking a companion, there is no inevitability. If you pick a wrong companion, it could immiserate you a lifetime. But if you pick a wrong topic, it will delay you for a thousand years.

Remarks

1. In the **UK**: Marriage, which is a legal relationship, is defined by law. Human interaction is a personal relationship, a private matter, defined by the persons involved.

Joseph Needham's wife, Dorothy Moyle (1896-1987), was a Cambridge colleague and professor of biochemistry. The Chinese woman Needham fell in love with was Lu Gui-Zhen (鲁桂珍, 1904-1991), an international student and a doctoral student of Moyle. The three got along well, resided on the same street, and often travelled together. Lu is Needham's long-term assistant and collaborator. In 1989, two years after Moyle's death, Needham married Lu. (See *The Man Who Loved China*, 2008.)

In **China**: In AD 101, three years after the death of his wife Wang Fu (王弗), Su Shi (苏轼, 1037-1104) of the Northern Song dynasty married Wang Fu's cousin, Wang Min-Zhi (王闰之),.

2. One of the arguments against the Needham Question is that one cannot ask why something did not happen. This is wrong. For example, two brothers went to climb a tree: One climbed up; one did not. Of course, you can ask the one who did not: Why didn't you climb up? Did you skip breakfast? You are not tall enough? Don't want to climb?

Or maybe the two brothers climbed two different trees. Of course, you can still ask: Why didn't they climb the same tree? Is it accidental? How do two people choose a tree to climb? Are the trees climbed about the same size?

Human ancestors came out of Africa, scattered in different directions, settled in different places, but their genes are almost the same. Hundreds of thousands of years later, some development was the same (such as 10,000 years ago there was agricultural settlement), some different (such as Chinese and Western philosophies [6.2, 6.3]). But regardless of the same or different, you can ask why, just like the two brothers climbing trees.

9.6 Faith aggregation

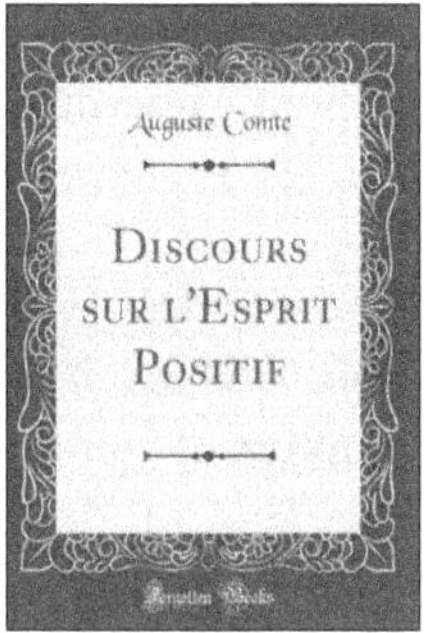

"Faith aggregation" is defined as a mass community based on common, *untested* beliefs. The classification of different kinds of faith aggregations is presented. Science theories usually have quick verifications and are not easy to produce faith aggregations. But the humanities/social science theories do not and are prone to produce followers building on faith alone, called schools of thought, which are actually faith aggregations.

Faith aggregation

Humans have two characteristics: thinking and sociality. If you can think, you will be curious, and finally you will ask about the meaning of life, for example. Sociality leads to group gathering, fear of falling behind, and there will be a herd effect.

Under the combined action of these two factors, thinking and sociality, there will be *faith aggregation*: a mass community based on common (untested) beliefs. There are two types of faith in faith aggregations:

A. Faith contains supernatural.

B. Faith does not contain supernatural (called "secular").

Type A example is Buddhism and type B example is Daoism. Type B is divided into two categories:

B1. Organized.

B2. Loose and disorganized.

By definition, organized faith aggregations are called *religion* (with or without supernatural). Types A and B1 are religions. Type B is often triggered by humanities/social science "theories" (hypotheses, in fact).

In contrast, natural-science theories, which could be confirmed or refuted more easily, are not easy to produce faith aggregations, except for complex-system theories. For example, climate-change theories: The grouping of those who believe in them, is close to be called faith aggregations; the same goes for those disbelieve in them [10.8].

Two examples of type **B1** (secular and organized):

1. In the East, Daoism originated from the *Tao Te Ching* (道德经), a literary book. Of course, in order to unite believers for a long time, in addition to a book, it is necessary to develop some additional things such as alchemy and rituals (even far-fetched immortal cultivation), as well as buildings for gatherings and Daoist temples.

2. In the West, Auguste Comte (1798-1857), the founder of Sociology, founded the Religion of Humanity three years after the publication of his book *Discours sur l'Esprit Positif* (Discourse on the Positive Spirit, 1844), and the following year the Society of Positivism. The Religion of Humanity has a hierarchy and rituals, and churches are built in France and Brazil.

Type **B2** (secular and loose) has no formal organization, is not called religion, and sometimes appears as "school of thought" in the humanities, like the followers of postmodernism.

Differences Between Humanities and Science

Science people often do not check the correctness of the science theory they want to use; they just pick it up and use it. For example, before using the formula F = ma, it is not necessary to look at what assumptions Newton made and how to derive it, and there is no need to do so because many experimental tests have proved this formula correct. So, people who use Newtonian mechanics are not based on faith, but on the *Reality Check* (which includes experiments) [8.3].

Similarly, humanities people also often do *not* check the humanities theory used, just pick it up and use it, or only make a cursory examination. Or some may do a detailed examination, but due to insufficient training, they cannot see the flaws in the theory. In either case, unlike the science situation, humanities theories lack quick verification. Consequently, humanities people need to have stronger *critical thinking* than science people to see whether the theory is right or wrong.

Unfortunately, on the contrary, in connection with the university curriculum, the critical-thinking training of the humanists is often less than that of the sciences, and many followers of humanities/social science theories are based on faith. Their grouping is often just a faith aggregation.

Conclusion

Generally speaking, science has quick (experimental) verifications, which are not easy to produce faith aggregation. The humanities lack quick verification; the followers and schools of thought are often a kind of faith aggregation, and some can even develop into religions.

9.7 Sciphilogy: Mach, Popper, Kuhn

Left to right: Ernst Mach, Karl Popper, Thomas Kuhn

To understand why many philosophers of science could be so wrong but still attracted numerous followers and exerted out-of-proportion influences, in academia and beyond, we suggest that one should look at their personal life in some detail and examine the societal conditions that allowed it to happen. This approach is named *Sciphilogy*—the human science of philosophers of science. Here, the case of Ernst Mach, Karl Popper, and Thomas Kuhn are presented as an example of how Sciphilogy works.

Sciphilogy

Philosophy of science, a branch of philosophy, belongs to the humanities. There is no love for no reason, no hate for no reason, and similarly, no science philosophy theory for no reason.

Sciphilogy is a new discipline I created in 2014. Starting from the growth and life of the science philosophers, Sciphilogy studies the ins and outs of science philosophy, and discusses what is right and what is wrong, involving science history, science philosophy, psychology, sociology, era background, history, and culture. Sciphilogy is the "human science of science philosophy" and the "study of science philosophers." Sciphilogy, first discussed in Chapter 2 of my book *All About Science* (2014), is a discipline of Scimat [3.3].

Here we talk about three important science philosophers: Mach, Popper, and Kuhn. The three were insiders or outsiders of science. Science here refers specifically to physics, because these three people are involved in physics-related matters. Mach and Popper are Austrians; Kuhn, American. Mach is 64 years older than Popper, and Popper is 20 years older than Kuhn.

Mach

Ernst Mach (1838-1916), at the age of 22, received his doctorate in physics from the University of Vienna and attended medical physiology for a semester. He did physics in universities, 3 years in the Czech Republic followed by 28 years in Austria, and then 6 years at the University of Vienna as a science philosopher. He retired at the age of 63 and died at 78. In addition to being an accomplished physicist, he also contributed to physiology and psychology. Mach is an *insider* in physics.

He was always looking for a unified view of the science of inanimate systems (physics) and living systems (psychology). He proposed the philosophy of phenomenalism/positivism. He said he did not believe atoms existed. One of his central ideas was that in science, everything that cannot be observed should be excluded from physical theories, because he believed that physical theories should be limited to descriptions of things one can sense and measure.

Mach had enormous influence, and his many admirers included the young Albert Einstein (1879-1955). Mach is credited with giving birth to the (failed) *Vienna Circle*. What caused Mach to fall off the philosopher's platform was a "small" but fatal mistake he made. He insisted that atoms/molecules would never be observed and therefore not allowed in theory. Then, he lost. The lesson here is this: Betting on the future of anything is not a sure thing and should be done with caution.

Here is what happened. Current scanning tunneling microscopes can magnify atoms by a factor of 100 million, allowing us to "see" individual atoms directly. But as early as 1905, Einstein explained Brownian motion with molecular theory and made predictions. Soon Jean Perrin (1870-1942, Nobel Prize 1926) experimentally confirmed the predictions. Theory plus experiments indirectly but convincingly prove the existence of atoms. Thus, Mach was wrong on the issue of atoms.

Popper

Karl Popper (1902-1994), born in Vienna in 1902, at the age of 17, was attracted to Marxism and joined the Austrian Social Democratic Workers Party, but abandoned the ideology in the same year. He was a supporter of social liberalism throughout his life. He received his doctorate in psychology at the age of 26 from the University of Vienna. In 1937, when Nazism was on the rise, he immigrated to New Zealand at 35. A year after the end of World War II, he moved to England at 44 and worked at the London School of Economics (part of the University of London) until his retirement at 67. He died in England in 1994 at the age of 92.

Popper established the Department of Philosophy, Logic and Scientific Method at the London School of Economics in 1946, contributing to the establishment of science philosophy as an autonomous discipline within philosophy. Popper is a 100% physics *outsider*.

Popper is known for his demarcation of science: The criterion for determining a theory as a scientific theory is its *falsifiability*, not its verifiability. The idea was conceived in the "winter of 1919-1920," when he was only 17 years old, presumably after becoming fascinated and abandoning Marxism, before training in psychology. The falsification criterion was officially proposed in the book *Logik der Forschung* (Logic of Research, 1934; English translation *The Logic of Scientific Discovery*) when he was 32. This idea looks good on paper, but for three reasons, it is not difficult to see it could not be correct and is impractical:

1. There is a **logical** problem with falsification. Strictly speaking, no scientific hypothesis can be falsified empirically, as Popper hoped, as Pierre Duhem (1861-1916) pointed out when Popper was four years old: Behind any *explicit* scientific hypothesis (such as the upper limit of the speed of light in special relativity) there is always a bunch of *implicit* assumptions (such as conservation of energy). Therefore, when the (reliable) experimental results do not agree with the predictions of the

explicit hypothesis, it is impossible to rely on logical inferences alone to determine whether the explicit hypothesis or the implicit hypothesis is wrong. Unfortunately, while Popper did know about Duhem's work he chose to turn a blind eye or not fully grasp it.

2. There is a **technical** problem with falsification. Even if the assumption that implicit assumptions are always true, it is technically impossible to rigorously negate the explicit assumptions of scientific hypotheses through experimentation. The reason is: All experiments use instruments, and instruments have an error, so all experimental "proofs" are only proofs within the error range, not absolute proofs. In fact, all scientific judgments are the judgments of scientific people based on experiments and experience, and are non-absolute judgments with probabilities. Furthermore, science history tells us that not all implicit assumptions are always true. One example is the 1957 experiment that confirmed parity is not conserved in weak interactions [11.6].

3. There is a **time** problem with falsification. Let us say that there is a theory that predicts that a comet will fly by or the Earth will destroy in 10,000 years, which is falsifiable, but it will take 10,000 years. Is this theory a scientific theory?

But in the face of many simple and obvious objections, why did Popper persist for so long? And why was he able to do so for so long? These are academic questions worth exploring (see below).

Kuhn

Thomas Kuhn (1922-1996), BS in 1943, MS in 1946, PhD in physics (solid state theory) 1949, all Harvard University; several theoretical papers on physics were published around 1949. But before and after he obtained his PhD, for eight years (1948-1956), he was a lecturer in general education and later an assistant professor of general education and science history at Harvard. In 1956, after a failed promotion at Harvard, Kuhn was hired full-time by the History Department at the University of California, Berkeley, with part-time in the Philosophy Department. In 1961, he was terminated by the Philosophy Department, but promoted to full professor by the History Department. Unlike

Popper, Kuhn is an *outsider* with a PhD in physics (but did not do physics after the degree, unlike Mach).

Kuhn's rise to fame came from his book *The Structure of Scientific Revolutions*, published in 1962 (2nd edition, 1970; 3rd edition, 1996; 4th edition, 2012; called *Structure* below). Two years later, he joined Princeton University as Professor of Science and History. In 1979, he moved to MIT (not Harvard) as a professor of *philosophy*, where he remained until his retirement in 1991 at the age of 69. Kuhn was diagnosed with lung cancer in 1994 (reportedly smoking 5 packs of cigarettes a day) and died in 1996 at the age of 74.

He published three science history books during his lifetime. But it was not these books (which was rated mediocre) that left him famous, but his book *Structure*. Since its publication, he has spent the rest of his life revising it—not making it more and more clearly but instead, making both his critics and friends more confused. Kuhn ended his life as an apparently unpleasant professor. Understanding why a man with rigorous physics training failed to articulate his mind for 34 long years (1962-1996) is an interesting topic worth analyzing. But first, let us look at the simple mistakes Kuhn made in *Structure*.

First, Kuhn soon discovered that the term "paradigm shift" he coined in 1962 was problematic. Yet, instead of abandoning the *concept* like a mature physicist, he abandoned the *term* and replaced it in the second edition of the book in 1970 with two new words he coined: "disciplinary matrix" and "exemplar." In other words, he started playing with words.

Another concept, "incommensurability," he proposed in 1962, was similarly in trouble. This error is easy to see. For example, Newtonian mechanics and Einstein's special theory of relativity do not have so-called incommensurability: The mass in the former is the "rest mass" in the latter—the mass when the velocity of the object tends to zero. (See next section for more discussion.)

Philip Anderson (1923-2020), a Nobel laureate and Kuhn's physics classmate at Harvard, commented on *Structure*: Fortunately, many scientific revolutions did not happen the way he said, except quantum mechanics.

How Mistakes Are Made

Both science insiders and outsiders could make serious mistakes, but the reasons for which they are committed are different and of different nature.

Mach

Mach forgot the lessons from science history of the invention of new instruments. For example, the invention of optical microscope can turn something previously invisible, such as a single-celled amoeba, into something visible. Similarly, atoms are super small, and if they cannot be seen at the time, it does not mean they will not be seen later when new instruments like the scanning tunneling microscopes are invented. Furthermore, by Mach's epistemological thinking, quarks within the protons and nucleons should not be proposed at all, but as history showed, this is absurd.

As Einstein remarked:

> [T]he prejudices of these scientists against atomic theory can be undoubtedly attributed to their positivistic philosophical views. This is an intersting example of how *philosophical prejudices hinder a correct interpretation of facts* even by scientists with bold thinking and subtle intuition.

Popper

Popper had no physics training, and he did not know enough and in detail about past cases in science history. Just as philosophers like to shoehorn the *world* into bottles they like [6.7], science philosophers like to shoehorn *science* into bottles they like. The two thousand more years of science development since ancient Greece is like a person going through adolescence. Science philosophers' understanding and prediction of how science is or should be is like predicting a person's future based on her adolescence years—this is only immortals can do. When mortal humans do it, it is easy to make mistakes and often off the mark.

Popper's main basis for falsification is the case of physics. Non-physical theories are sometimes difficult to give predictions, even if the theory is good. For example, the biological theory of Darwinian evolution did not give any precise predictions when it was first proposed, and while it did predict that the Earth must be very old (so biological systems had enough time to evolve), it could not say how old it was or what animals would be the fittest to survive. Popper thus erroneously concluded that evolutionary theory was not a scientific theory, but later in life revised this view to come to the opposite conclusion.

In addition to being unwise in his logical or critical thinking, Popper's basic problem is that he completely ignored or was unaware of the fundamental difference between the theory of deterministic simple systems (e.g., Newtonian mechanics) and the theory of probabilistic complex systems (e.g., sociology and psychology). It is wrong to make the same demands on both (such as mathematization or short-term verification) and ask the same questions. For example, you cannot ask which year a person will die but only what is the probability of dying in that year.

Kuhn

Kuhn made mistakes for a different reason than the previous two. According to what Kuhn told his MIT colleague and Nobel laureate Steven Weinberg, while studying Aristotle physics at Harvard in 1947, he suddenly realized that Aristotle's mechanics was not "bad physics" at the time, although it later found to be "wrong physics." According to Weinberg, the (obvious) paradigm shift from Aristotelian mechanics to Newtonian mechanics may have led Kuhn to come up with his later ideas. If so, then this is the key to understanding the Kuhn incommensurability fallacy.

The concepts, terms, and results of Aristotelian mechanics are nowhere to be found in Newtonian mechanics because the former failed the *Reality Check* (RC, i.e., consistent with the real world, including experiments [8.3]). In contrast, Newtonian mechanics passed the test, so some of its concepts, terminology, and results were retained or accepted by Einstein's theory of relativity. In other words, any theory that *fails* the RC will eventually disappear and will be incommensurate with the valid theory that emerges later. On the contrary, anything that *passes* the RC is preserved and can be identified later in some way, and there is no incommensurability with later theories.

In fact, the RC is the most critical thing that distinguishes science from other forms of human exploration, such as meditation. Curiously, the RC is not mentioned in the discussion of science by Kuhn and many of the philosophers, indicating that they have not grasped the essence of science and have only a partial understanding of the nature of science.

Why Wrong Theories Can Be Maintained

Humanities theories are different from science theories. The latter usually have experimental tests quickly, and if they are really wrong, they can only be abandoned. The former generally has no experimental tests, and if they are wrong, it can still be constantly argued with opponents, so that the theories will endure. But there are two premises:

1. The proponent of the theory has a teaching position in a prestigious school—an effective platform.
2. The theory has enough followers.

There are also two tricks in defending a theory: clearly and vaguely. Note that the trick of avoiding debate, applicable in China, say, cannot be used in the international academia.

Mach's attitude towards atoms is this: "Atomism is a good working hypothesis for the study of chemistry; it must be used with great care on studying and working in science; but it is extremely dangerous as a noetic theory."[1] When other physicists and chemists began to believe that atoms really existed, he did not change his opinion until his death.

Remark: When faced with strong skepticism, Popper and Kuhn chose not to retreat. Both have prestigious faculty positions and many followers, but the former used the hard-shoulder tactic—defending it clearly, while the latter used muddy-the-water tactic—defending it vaguely. These two different approaches reflect, respectively, their personal upbringing, societal experiences, and personal characters.

Where the Followers Come From

Humanities/social science theories lack quick empirical verification, and their followers and schools are often a mass community based on beliefs: faith aggregation. Popper and Kuhn's followers come from the B2 type of faith aggregation (i.e., secular and loose [9.6]), but from *different* sources.

Mach

Mach's followers were the intellectuals in Vienna, who in the 1920s organized themselves into a philosophical association named Verein Ernst Mach (Ernst Mach Society). It was transformed into the Vienna Circle promoting logical positivism. Some of these people moved to the United States in the 1930s when the Nazi party in Germany annexed Austria. Compared to Popper and Kuhn, Mach's followers are mostly academia people; their number is sizable but small.

Popper

Popper's life (1902-1994) covered all the Soviet Union years (1917-1991) and the entire Cold War period (1947-1991). He was introduced to Marxism and Darwinian evolution at the age of 10, and at 17 worked for a psychoanalyst, a Freud's disciple. His life, from the time he joined and left Marxist organizations at 17, was devoted to anti-totalitarianism.

The falsification idea was formed in Popper's early years (published in German in 1934; in English, 1959). It was Popper's best-known scholarly work, but *not* his most important. His academic fame comes from his two books on political philosophy: *The Poverty of Historicism* (published in a series of periodicals in 1944/45; completed in 1957) and *The Open Society and Its Enemies* (1945). He fought the left liberalism and the right authoritarianism with equal enthusiasm. He believed that freedom was more important than equality, and advocated a progressive democratic society, with followers in the Western camp of the Cold War, including the influential Friedrich Hayek (1899-1992, Nobel Prize in Economics in 1974).

In fact, falsification is not popular in the philosophical community, and its supporters mainly come from some scientists, a few science philosophers, and believers outside the academia. Most of these scientists are contemporary physicists (including Einstein, Bohr, Schrödinger), and none of them are science philosophers. The falsification theory is consistent with their daily work experience, but that was the innocent years in physics *before* parity nonconservation was discovered in 1957 [11.6].

Why do some people find the unrealistic falsification idea attractive? Because it provides a simple-to-use tool that meets the needs of these people, like a portable "good guy/bad guy discriminator" (albeit an ineffective instrument). In fact, the demarcation of anything makes little sense because everything has gray areas, including the good guys/bad guys and the laws (that is why there is a jury system).

Note that non-scientific things are also very important to the world, such as Su Dong-Po's (苏东坡) poem/prose and calligraphy/painting [7.4], Leonardo da Vinci's *Mona Lisa* [7.6], Shakespeare's plays, and Li Qing-Zhao's (李清照) lyric [7.9]). Moreover, scientific things are not necessarily good; there can be bad science, wrong science, and some fuzzy, stagnant science like traditional Chinese medicine [9.4].

Later, Popper finally learned that his falsification idea was untenable, and changed his words to say that it is only an ideal state that scientific theories can be falsified.

Kuhn

Kuhn wrote *Structure* because he was dissatisfied with his predecessor and colleague George Sarton's (1884-1956) one-dimensional glorification of science history.[2] In addition to the two polemical views of "paradigm shift" and "incommensurability," the focus of the book is to emphasize the societal nature of the scientific process—the human factor. It is this latter that has made *Structure* a bestseller, which was translated into more than 20 different languages and sold more than a million copies. Why is it a hot seller? There are two reasons.

First of all, there are four conditions for an *academic* book to become a bestseller:

1. The technical threshold is not high or there is none (no equations), so it is easy to read.
2. Not too thick, easy to finish reading.
3. Happens to be able to meet the needs of a large number of readers.
4. The writing is vague, and after reading, it can be interpreted differently by different people to ensure debates.

Structure may be the only academic book that can meet all four of these conditions at the same time.

Second, the social conditions and happenings of the 1960s provided the book with a large number of general readers outside of science. There were two "anti-science" movements in the United States in the 20th century:

1. The first occurred in the 1930s, following the Great Depression of 1929. There are two aspects to complaining: (1) Science (in fact, technology) has been blamed for the dehumanization of society (machines replacing workers). (2) The great triumphs of physics in the early 20th century (the emergence of relativity and quantum mechanics) made some scientists hot-headed and casually expressed laypeople-level opinions on social affairs. These scientists were sought after by the media, causing the masses to resent science, in fact scientism.

2. The second occurred in the 1960s—the anti-Vietnam War and student-movement years. Adding to these were the two atomic bombs that ended World War II in 1945 that killed countless people, turning anti-nuclear into anti-scientist and anti-science. More importantly, *postmodernism*, which asserts equality for all and no distinction of good or bad, became popular in the 1950s,

grew in the 1960s, and spread from France to American campuses, including the University of California at Berkeley (where Kuhn was employed). In short, the 1960s were a fiery era of anti-war, anti-authority, and anti-establishment. By chance or not, the *Structure* precisely emphasized the impediment of academic authority to scientific progress.

Thus, before the *Structure* came out in 1962, a large number of potential readers were waiting for such a book. With a PhD in physics from Harvard University and a full professorship at the University of California, Berkeley, Kuhn is a scientific insider in the public eye—someone who knows what he is talking about science. Unfortunately, Kuhn is actually an outsider (even though this point is not that important as shown in Mach's case). Kuhn confirmed to them what they had always suspected: Scientists are human beings and have selfish times, that their work is not entirely objective and credible, and that the scientific *process*, like any other human enterprise, is influenced by societal forces.

All of this is actually *obvious*. But it was previously not valued, or, more likely, no one considered it worthy of writing a book. In fact, there are three basic facts about science that are usually overlooked:

1. Science has *two* parts—the scientific process and the scientific results [3.3].
2. Personal and societal factors affect only the scientific process and not the scientific results.
3. The scientific results, not the scientific process, are the part of science that can really affect society in the long run.

Unlike previous books on science, *Structure* focuses on the scientific process and *rightly* points out that the scientific process is societal (a piece of cake; see Remarks), but it takes a big step forward by *mistakenly* extending sociality to scientific results, arguing that "when paradigms change, the world itself changes with it." Does it mean that as soon as Newtonian mechanics appeared, the previous world *itself* would change at the material level? *Structure* is deliberately vague on this important issue, extracting the objective credibility of scientific results that pass the Reality Check (such as the law of gravitation), and providing "scientific" support for postmodernism and relativism. And here lies the greatest disservice *Structure* provides to the academia and society.

If *Structure*'s arguments are wrong, Kuhn's reputation in the academia will be affected. But the damage outside the academia is more important. The damage does not come from whether the "paradigm shift" or "incommensurability" are correct or not, but the strong implication in the book that scientific *results* have no objective validity (i.e., they are artificial constructs). This is a fatal effect, but unfortunately, it is also the basis on which Kuhn became famous.

In the late 1960s, the *Structure* became increasingly popular, especially among radical students who felt the book liberated them from the tyrannical tradition. Elsewhere, Kuhn-style "anti-science" relativism was used against scientism, which uses poison to destroy pests and leaves a toxic environment behind when successful.

Kuhn's Hurt

Historian Robin Collingwood (1889-1943) argued that latecomers must "reconstruct" the thought processes of historical figures by placing themselves in the position of the historical figures based on historical data [5.1]. According to this, to understand Kuhn's behavior patterns, it is necessary to put yourself in his shoes and understand how he thinks.

According to Kuhn biographers Wes Sharrock and Rupert Read: "Thomas Kuhn died in 1996, convinced that his lifework had been misunderstood, and failing to complete a categorical restatement of his position before his death."[3]

A professor at a prestigious university who became famous inside and outside the academia and sold an academic book for more than a million copies ended up in a depressed mood. Why? A Harvard PhD in physics has failed to make his academic claims clear for 34 years. Why? Is there something unspeakable? What is Kuhn's unspeakable secret? To answer these questions, let us start with Kuhn's "philosopher complex," from his student days.

During his first year at Harvard, Kuhn took a year of philosophy classes. In the first semester, he studied Plato and Aristotle; in the next semester, he studied Descartes, Spinoza, Hume, and Kant. He also audited several of Sarton's science history lessons, but found it boring. As a PhD student in physics, Kuhn taught a science history course for senior undergraduates in 1948 at the invitation of the president, a chemist. He continued teaching this course until 1956 when he left Harvard. Before completing his PhD, he wanted to change field and enroll in the philosophy department, but gave up because it was too difficult, so he had to barely finish his PhD in physics. Since then, becoming a philosopher has been an important goal in Kuhn's life.

Kuhn became an assistant professor at Harvard in 1952. In 1956, Harvard felt that his *The Copernican Revolution* (published in 1957) was not sufficiently scholarly and rigorous, and *refused* to offer him a tenure appointment. That year, when he was hired in the history department of the University of California, Berkeley, he asked for a part-time job in the philosophy department and got it. Five years later, in 1961, he was *dismissed* from the philosophy department, leaving Kuhn extremely angry and seriously hurt. When he transferred to Princeton University in 1964, he was Professor of Science History and Philosophy and in 1968 Chair Professor of Science History. It was only in 1979 that he transferred to MIT and became a professor of pure philosophy, although he only did philosophy of science.

Although Kuhn openly dissociated himself from "Kuhnism" (so as not to be laughed at by other science philosophers), he refused to clarify what he meant by "world change" when he said that "when paradigms change the world itself will change with I." Does it mean a change in the material world or a change in people's concept of viewing the world? This question is actually not difficult to answer. It is equivalent to when a person sees a beautiful woman on the street and comes home to find that his wife has become ugly. Does it mean that the molecular arrangement on his wife's face has changed and become ugly? You bet!

Why did he refuse to answer this simple question directly and instead, continue playing with words and concepts? The reason is that if he had replied that the material world had not changed, but only that people's ideas had changed, there would be little left of *Structure*, and his treasured philosopher status would have been greatly devalued.

Kuhn's biographers put it this way: "Arguably, Kuhn (although not a 'proper' philosopher) has been not just the most influential philosopher of science in the second half of the twentieth century, but the most influential philosopher, full stop, and tended to be at his worst side when *trying* to be one of those."[4]

Kuhn's UC Berkeley colleague Paul Feyerabend (1924-1994) put it this way: "I venture to guess that the ambiguity is *intended* and that Kuhn wants to fully exploit its propagandistic potentialities."[5] What this means is that Kuhn enjoyed the global fame of the vague book *Structure* so much that he decided to prolong the aura by keeping it vague.

Academia advances through debate, so academic books must be written clearly, otherwise the more debate, the more confused. People who cannot write clearly can try writing poems. Kuhn's dilemma is the phrase in Shakespeare's tragedy *Hamlet*: "To be or not to be." In Kuhn's case, it is "to be clear or not to be clear." The essence of the problem is that Kuhn's capacity as a philosopher does not match his fame, and his selfish ego does not match the truth-seeking tradition laid down by the ancient Greek philosophers.

Popper was able to abandon his falsification idea because of his *integrity*, and because in addition to the falsification, he had several other important books. Kuhn, on the other hand, has only one "enduring" book. If he gives it up he will be remembered merely as the science philosopher who had enjoyed his "15 minutes of fame," which, of course, is too far from the name of philosopher he wanted. And so, he can only pretend to be the only one who is sober while everyone else is drunk, prolonging it until death called.

Being a person and a scholar, the first and utmost important thing is to be honest. Socrates is the scholar who did not go against his will in court;

Galileo did. These human reactions could be understood. What is more complicated is the phenomenon that academics make deceiving statements without any coercion. This phenomenon occurs in both the humanities and science professions, seemingly more often in the former than in the latter. This is a topic that deserves to be explored in depth.

Conclusion

1. The academic claims of the humanities are not as easy to verify as those of the natural science, which leaves a lot of room for argument. But it is not impossible to distinguish between good and bad (or right and wrong) theories. It depends heavily on the researcher's skill. Not all claims are the same, although they can often coexist temporarily.

2. Due to the lack of quick dependable Reality Check, humanities theories are often based on beliefs, and their aggregation is a kind of "faith aggregation." The Kuhn phenomenon falls into this category.

3. Humanists are more likely to be fooled by their own theories. In the face of clear evidence or arguments against their own theory, science philosophy's insiders and outsiders often react differently. Take the three people mentioned here as an example: The insider Mach faced them silently; the outsider Popper persisted but eventually admitted his mistake; the outsider (with a Harvard PhD in physics) Kuhn chose to hold on and persist on mistakes to the very end. It is dictated by personal characters and upbringings.

4. Science is like an ancient Greek tragicomedy that has been performed on the stage for 2,600 years. Science philosophers are like drama critics who came in late, showing up in the last hundred years. If the critics do well, they can help to promote the development of drama. If it is not done well, the critics will mislead the audience and become a small circle for self-amusement.

Notes

1. Brauner, B. [1924]. Einstein and Mach. Nature **113**: 927.
2. Kuhn criticized the old-fashioned writing of science history in his 1962 book *Structure*, but did not name Sarton. He made it even clearer in 1984

when he talked about the professionalization of science history. See Kuhn, T. S. [1984]. Professionalization recollected in tranquility. Isis **75**(1): 29-32. See also Pinto de Oliveira, J. C. & Oliveria, A. J. [2013]. Kuhn, Sarton, and the history of science. http://philsci-archive.pitt.edu/10078/.

3. Sharrock, W. & Read, R. [2002]. *Kuhn: Philosopher of Scientific Revolution*. Malden, MA: Blackwell. p 1.

4. Ibid, p 209.

5. Hoyningen-Huene, P. [2000]. Paul Feyerabend and Thomas Kuhn. *The Worst Enemy of Science? Essays in Memory of Paul Feyerabend*, Preston, J., Munévar, G. & Lamb, D. (eds.). Oxford: Oxford University Press. pp 109-110.

Remarks

1. Sciphilogy can be broadened to form "human science of scientific culture," but the study of science philosophers is more fun.

2. It seems Kuhn and others' understanding of paradigm shift is equivalent to "first-order phase transition" in physics: A characteristic quantity (called *order parameter*) undergoes a *discontinuous* change at the phase transition point (e.g., water vaporizes to form steam). In fact, there is also a "second-order phase transition": The order parameter changes *continuously* at the phase transition point, but the slope of the curve is discontinuous (e.g., ferromagnet loses magnetization when warm up). It is possible that before and after major breakthroughs in physics (e.g., the Copernican revolution, the emergence of quantum mechanics), the number of relevant papers—the order parameter—may appear as a second-order phase transition, rather than the first order. A quantitative analysis is needed to see whether this is true. It can be done easily by searching some key words online and see how the number of papers changes in time. The results would have great impact since, after all, Kuhn's book has sold more than a million copies.

3. On July 13, 1965, at the International Conference on Science and Philosophy held at the University of London, the budding 43-year-old Kuhn and the already famous 63-year-old Popper attended the conference at the same time. At that time, Kuhn's *Structure* was published for three years, and Popper's *The Logic of Scientific Discovery*

was published in English for six years. The meeting of two science philosophers of disparity was arranged by Imre Lakotas (1922-1974). However, the occasion, known as "Kuhn vs. Popper," was not actually debates between two protagonists. A collection of essays *Criticism and the Growth of Knowledge* was published after the meeting. See Fuller, S. [2004]. *Kuhn vs. Popper: The Struggle for the Soul of Science*. New York: Columbia University Press.

4. *Scientific American*'s review of the book *Structure* is "much ado about very little." John Horgan, author of *The End of Science* (1996), has two old interviews on the magazine's blog Cross-Check: "What Thomas Kuhn really thought about scientific 'truth'" (May 23, 2012) and "The paradox of Karl Popper" (August 22, 2018). Popper was 90 and Kuhn was 69 at the time of the interview, both in their later years.

10
On Science

10.1 Science and religion

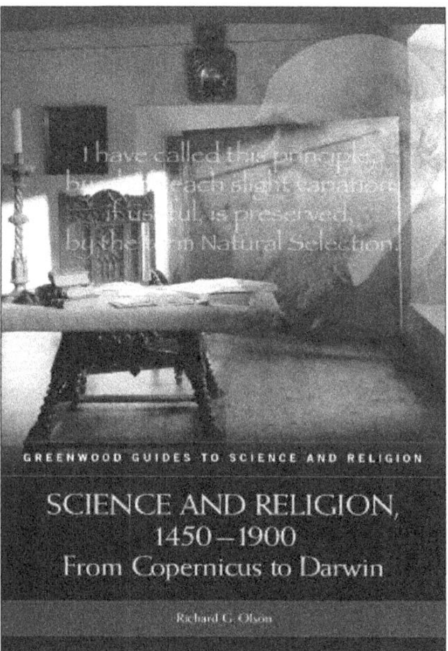

Religion existed 9,000 years ago while science only emerged 2,600 years ago [8.1]. Religion offers three things: ready-made answers (such as the meaning of life and where to go after death), placebo, and the circle of friends, the first two of which are the shortcomings of science. The premise of religion requires the abandonment of independent thinking, and science encourages independent thinking. The approaches of the two are diametrically opposite to each other.

In a place with *theocratic* government (state and religion are united into one), science, as a latecomer, is like a tenant whose welfare depends on the kindness of the house owner—religion. Such was the case in Italy 400 years ago (Galileo's time) and 350 years ago in England (Newton's time). The relationship between the owner and the tenant is mutually beneficial, but the owner will prevent the tenant from acting unfavorably against him, so from time to time there will be some disputes between the host and the guest. And conflicts between the two arise.

Where church and state are *separated*, both science and religion are tenants, and there is competition between the two: a relationship of competition for resources and influence. However, in ancient China, Daoism greatly promoted the development of science such as Chinese traditional medicine and alchemy.

Why Conflict?

Several hundred years ago Galileo was punished by the Vatican when his theory conflicts with the official version. But that happened in Italy when the state and religion was one. Today in places that state and religion are formally separated why there still is conflict between the two? There are two reasons for this:

1. Religion did not *retreat* fast enough when science advanced. As long as religious descriptions of nature contradict known scientific findings, the two are in conflict. (For example, the Bible says the universe was created in six days, while scientists say it took 13.7 billion years to evolve from the big bang.) Otherwise, there would be no contradiction between science and religion.

In other words, when science moves forward and religion is reluctant to take a step back, conflict will occur. A famous example is Galileo Galilei (1564-1642), a Roman Catholic. He was punished by the church for refusing to abandon his conclusion that the center of the solar system is the Sun and not the Earth.

Fair to say, historically, when science advanced quickly in the modern-science era, started by Galileo 400 years ago, religion has removed natural phenomena from its concerns, but not quickly or completely enough. For example, the Vatican still maintains its own observatory.

Scientists deal with this factual contradiction between science and religion personally by adopting the scientific view in their works and skipping the religious view when they go to church. Not a perfect solution, of course.

2. There are *misunderstanding* of the very nature of science, by the religious people and even some scientists. For example, unaware of the 1867 definition of Science that science consciously decouples itself from God/supernatural, some creationists insisted that the intelligent-design theory be taught in science classes (in the United States) and lost the case in court [8.1]. On the other side, some scientists are unaware that science, which confines itself to the physical world (the universe), can never prove the non-existence of God (or any supernatural) which are non-physical, and say things in the opposite.

To remove the conflict, Stephen J. Gould (1941-2002) in 1999 proposed his principle of NOMA—nonoverlapping magisteria (domain of authority). He argues science could help us to understand the physical world but not more abstract things like morality and the meaning of life, which should be left to religion. This is wishful or mistaken thinking by Gould since no one can show abstract things (thought up by the brain) are beyond science, and religion will not give up physical things, like the origin of the big bang. Furthermore, religion, the self-proclaimed authority of morality and meaning of life, usually bases its teachings on an ancient book and nothing else, which sometimes is against human nature.

Dialogue

Although there are occasional conflicts, religion and science feel the need to communicate with each other for different reasons, and there indeed is dialogue between the two from time to time.

On the religion side, religion is an enterprise at the organizational level, often a global enterprise. Many religions offer membership (called believers). Like any business, even a non-profit organization must meet the requirements of the people it serves—members or customers. Many of these members owe modern science for their comfortable living (e.g., air conditioning, electrical lighting, and cell phones) and, for most of them, even their jobs. These people would like to see their religion more in harmony with science.

On the science side, since many religious people are taxpayers who can influence the government funding of science, science people would like to help them more in understanding science. In fact, in the United States, more than 90% of the population believes in a personal God.

Seeing mutual benefits, both the National Academy of Sciences and the American Association for the Advancement of Science have set up projects to promote a dialogue between science and religion. After all, with science on one end and religion on the other, there lies a spectrum of common enemies—pseudoscience, antiscience, antireligion and pseudo-religion—between them, the two greatest enterprises on earth.

What these dialogues could achieve is not to convince the other side but to remove misconceptions about each other, and hopefully, increase the respect of each other. Furthermore, it is a good opportunity for both sides to understand the proper limits of science as well as that of religion—good for world peace.

Three Religious Tactics

In order to retain believers who trust science, the religious side has taken actions to deflect or minimize the factual contradictions between religion and science. The solution is actually very simple. There are three tactics commonly used by religion:

1. The **gap** method. There are many unknown things in the frontier of science—the "gaps" in scientific knowledge. When an important new gap appears, religious people will come out and say that God did it, and the existence of this gap proves the existence of God. This method is called the "God of the gaps." As old gaps are closed in science, new gaps will surely appear. So, this method will always work.

2. The **Newton** method. This method was demonstrated by Newton. Newton cleverly used science to explain some things in the Bible and make it consistent with scientific understanding. For example, the biblical book of Genesis says that God created Earth on the first day and the Sun on the fourth day. Newton first noted that the Earth rotating against the Sun once is called a day. And if there is no Sun, one cannot

define a day. So, the first three days in the Bible are not the three days that we generally understand. It could be any number of years and so the Bible does not conflict with science. And it works, in a certain sense.

3. The **apology** method. When all other methods failed, religion resorted to apologies, retreat, and then re-explain. In 1992, for example, the Pope publicly apologized to the world. He said: We were not very polite to Galileo at that time, and we did not handle it well. Four years later, he said: We agree with the evolutionary theory. He acknowledges that man is a kind of animal, and like other animals, it evolves, but he adds that God takes special care of man, and that before each person is born (in the womb or immediately born), God puts this soul in his brain, but does not do it to other animals. The Pope explains the peculiarities of man, and you cannot say he is wrong, but you do not know whether he is wrong or not, because there is no evidence.

Meaning of Life

What is the meaning of life? This question implies several questions: Where did I come from? Where will I go after death? What is the purpose of living? These questions could be answered from the perspective of philosophy, science, theology or metaphysics The answers depend on the life history and cultural background of the individual. But there are no definite answers, not yet.

The scientific answer is represented by that of physicist and Nobel laureate Steven Weinberg (1933-2021) who writes in his book *The First Three Minutes* (1977): "The more the universe seems comprehensible, the more it also seems pointless." Subsequently, in response to numerous negative reactions, in the book's second edition (1993), he added that in a pointless universe, one can still find meaning by doing her own things.

The religious answer is simple and clear: It is God's will that one finds herself in this world, to bring glory to God and spread His words. Of course, to make it works, there are hell and heaven (punishment and reward) going with it.

In fact, there is a third answer if life on Earth is brought in by aliens. Then, humans' question of meaning of life will become the aliens' question. And we can stop worry about it; instead, the aliens should.

As a practical matter, the final answer, if exists, may not come during one's lifetime. And one should be patient and life has to go on. After all, during the 46 years between the discovery of superconductivity in 1911 and the correct (BCS) theory in 1957, no scientist killed himself. That means, one has to live healthily and happily. The good news is that after death, one will know for sure whether God and heaven/hell exist or not.

Before that, one can adopt a positive view of life by enjoying the world since one is already here (like finding oneself in the Disneyland, like it or not). Work hard, play hard. Help those near you (that will excite the happy hormones in your brain). And leave a better world behind for future generations who might be smart or lucky enough to find the ultimate answer. For intellectuals, invent something or write a good book, leaving a name in history—the equivalent to living forever, which is better than making babies.

Finally, one has to admit that religion is the quickest way to inject morality in a population when enough number of people believe in it as long as it is peaceful and harmless, like taking daily vitamin pills.

Thank Religion?

It has been said that without Christianity there would be no modern science (i.e., science since Galileo). It is meaningless to talk about history in this way. Because for any two things in history, A and B, as long as A occurs before B, one can always connect B to A through enough intermediate links. Therefore, it is always right to say that there is no B without A, but to say it is to say it in vain.

Take an example. It could be say that without B's dad (A) there would be no B in this world. Of course, this sentence is true, and the middle link is B's mom. But without B's dad, B's mom is very likely to find another man, borrow some sperm from him, and give birth to another child who could be stronger and smarter than B.

Moreover, the so-called Christianity of Galileo's time was actually a theocratic and brutal Catholicism. It did contribute, involuntarily and half-heartedly, to the emergence and development of modern science (hiring Galileo as a tenured professor was positive). Involuntarily because the Vatican could not stop people's natural curiosity in understanding what they observe. Half-heartedly because it did not allow science for science's sake: to question anything (like the existence of God) and overthrow any theory that does not agree with experiments/observations. Science was allowed to proceed under the pretext that its purpose is understand how God works.

Indeed, when Galileo was not tied up and burn to dead, everyone has breathed a sigh of relief. And we should be thankful?

10.2 No pseudoscience

Any topic (including quantum communication) about nature or artificial materials (e.g., semiconductors, plastics) can be legitimate scientific research. What is currently unknown or sounds impossible may be possible in the future. Science does not exclude this in principle, and there are many examples of this in the history of science.

Therefore, from the perspective of basic research, *any* topic can be done. Yet, there are other factors to consider whether it should be done or publicized with great fanfare.

In particular, whether quantum communication is an illusory industry depends on whether the participants are honest and how much investors are willing to gamble.

The above statement is true for any topic, including the establishment of a research Institute for the Non-conservation of Energy, depending on who is the director; if Einstein is the director, don't laugh, hurry up and see if you can buy some of stocks. Topic selection/investment is to bet on what its success rate is and how long it will take to succeed.

Recently, *Scientific American* (July 13, 2021) reprinted an article: Star Trek's warp drive leads to new physics. It is about a junior high school student who watched *Star Trek* and wondered if such faster-than-light travel is feasible. This person grew up to become a theoretical physicist and was trapped at home during the Covid-19 epidemic. With more free time, he remembered the issue of faster-than-light travel that excited him as a child. Going back to it and through careful consideration, he finally wrote an academic paper based on general relativity, which was published in a regular physics journal.

Even this kind of topic is legitimate scientific research, and it can be seen that there is no forbidden area in the selection of scientific topics. Thus, any honest research is not pseudoscience.

Discussion

1. In fact, experts have studied carefully, debated fiercely, and published several books, concluding that there is no way to draw a line between pseudoscience and true science. Technically speaking, *demarcation* between science and pseudoscience does not exist.

2. As individuals, if the discourse of a topic comes from an unknown (whether the author is honest or no), in order to save time, we can choose not to read it. As a journal editor, one can reject a manuscript on the grounds that it is not good enough. As the manager of government funding, one can bet on its success rate and decide with taxpayers' money in mind. In all these three cases, there is no need to *label* any work as pseudoscience.

Conclusion

There is no pseudoscience.

Remarks

For more, see my 2014 book *All About Science*, Chapter 1.

10.3 No antiscience

Science itself, so beautiful, so useful, can anyone object to it? Has anyone ever objected to science? Indeed, there are self-proclaimed or former "antiscience" people. But are they really antiscience?

Historically, there were two revolt-against-science movements in the United States in the 20th century (see *The Physicists*). The *first* occurred in the 1930s, following the Great Depression of 1929. There are two complains: 1. Science (in fact technology) is blamed for dehumanizing society (such as machines replacing workers). 2. Some scientists cross borders, offering simple-minded opinions on social issues, encouraged by excessive respect for these opinions by the media.

The *second* occurred in the 1960s: the anti-Vietnam War and the years of the student movement. The reasons for opposing science were the same as the first time, only this time it added the revulsion against the exaggerated influence of physicists who were revered for their success in building the atomic bomb. Related to this, postmodernism became all the rage in the 1950s and spread from France to American campuses in the 1960s. In short, the 1960s were an era of anti-war, anti-authority and anti-establishment turbulence. In this matter, both scientists and critics are responsible, and there are some misunderstandings:

1. The education of scientists is very *narrow*. For example, a superconductivity expert has virtually no say in black-hole problems, let alone climate change or social issues. Moreover, contrary to the claims of the physicist and Nobel laureate Robert Millikan (1868-1953) and others, the critical thinking acquired by these scientists through their experience with simple nonhuman systems (in mathematics and physics) helps, but does not automatically make them experts in dealing with human problems. Human problems are complex systems: Each problem is nonlinear and has multiple solutions; each outcome involves multiple factors and cannot be accurately predicted [2.3]. Furthermore, every human problem depends on history, and the only system that most physicists have learned in their profession that is history-dependent is ferromagnets, the complexity of which cannot be compared to the complexity of humans.

2. Many scientists do not even realize the *basic* difference between the (deterministic) simple systems they are familiar with and the (probabilistic) complex systems to which human problems belong, and inadvertently cross the line when making public comments on the latter. Some of their comments bordered on *scientism* and angered the public.

3. Physics achieved several major victories in the early 20th century (invention of quantum mechanics and two theories of relativity) and later contributed significantly in ending World War II (atomic bomb). The media's excessive attention to star scientists is not the fault of the scientists. This has to do with the nature of the media, driven by a strong public interest in any type of star figures. Additionally, the media and the public are ignorant of the difference between simple systems and complex systems because complex systems are not included in high school (or even college) courses. Fortunately, the situation has improved. For example, the media no longer asks Nobel Prize scientists what they

think about clothing fashion, nor does they ask Lady Gaga what she thinks about gravitational waves.

4. The adverse effects of science on society come from people who decide when and how to use scientific results. And human *decision-making* belongs to decision science—a discipline of the humanities. Therefore, no matter how much the natural science of nonhuman systems advances, science cannot harm humankind. In fact, science can harmlessly benefit mankind only when the development of the humanities receives sufficient attention. Accordingly, the so-called antiscience is actually *anti-humanities*, caused by the rapid development of science but underdevelopment of the humanities.

Discussion

1. Those who destroy science from within are those who violate scientific *ethics* and commit scientific fraud. They can be called *anti-scientists*. They violate the first credo of doing science: integrity.

2. Some Chinese historians and philosophers of science claim themselves to be (or have claimed to be) antiscience. More than 30 years ago, in order to oppose the leftists' *scientism*, they uncritically took Kuhn's discourse on science [9.7], jumping from anti-scientism to embracing *relativism* and adopting the rightists' antiscience. It is time to reflect on the personal academic claims of the past and remember the scientific spirit: Seek truth from facts and keep pace with the times.

Conclusion

Antiscience people are actually opposed to some aspects of scientific application—decision-making, resource allocation, and scientific omnipotence—which belong to the humanities. Therefore, the so-called anti-(natural) science is actually anti-humanities. In the final analysis, there is no such thing called antiscience.

Remarks

For more, see my 2014 book *All About Science*, Chapter 2.

10.4 Folk scientists dare

Opposition to Einstein's theory of relativity from an ideological or philosophical point of view has been seen before, both in China and abroad, which had died out. Yet, it has recently been seen in China again. Why? There are four basic reasons:

1. Progress of science in China is insufficient, which for a long time failed to come up with important breakthroughs (in physical theory).
2. Some people in the society can't wait.
3. Folk scientists are brave, without much to lose, and quick to act, seeing a vacuum to fill.
4. Most folk scientists, by definition, have no formal scientific education or training and misunderstand the nature of physical theories.

Item 1 comes from the basic contradictions at the societal level and the disadvantages of coming from behind in science development. And, unfortunately, the more the science-management system is corrected, the more wrong it is, such as counting papers, three-part wages, no tenure system, under-table hiring contracts that demand unrealistic number of quick papers. Good news is that there are partial solutions, hard to implement though.

In item 4, folk scientists' misunderstanding is that an experimentally proven physical theory will always be surpassed by newer theories in the future (e.g., Newtonian mechanics surpassed by Einstein's special theory of relativity), i.e., in new applications (e.g., inside black holes), the old theory may not work while the new theory will work. But, importantly, the new alternative theory must be able to explain *all* the experiments/phenomena that the old theory has explained, *and* be able to explain and predict new phenomena (and finally be confirmed). Insiders call new theories *surpassing* or *generalizing* the old theory, while laypeople (such as folk scientists) call them "overthrowing."

Different language hinders communication, but it is not the most important, as long as both parties know. If the two sides do not know, the more they talk, the more confused they are.

The important thing is that the folk scientists do not understand the requirements of science for new theories, and the more courageous they are, the more "breakthroughs" they produce.

10.5 Science knows borders

Scientific research is divided into two parts: basic science and applied science. The phrase “science knows no borders” is correct if it is meant that the validity of (natural) science is universal across country borders. If it is meant that the spread of scientific knowledge should ignore country borders, then it is wrong. In that case, it should be replaced by “basic science knows no borders; applied science knows borders.”

For example, theory of special relativity is basic science, which is available worldwide *after* publication, including $E = mc^2$. Neutron reactor research in the manufacture of atomic bombs is applied science, which was initially kept secret and could not be used by enemy countries, even though after more than a decade in peace time, it was unsealed and made available worldwide.

Similarly, applied science with commercial use have national borders and are protected by patents, and patents are not sold to enemy countries. But patents have a term limit (like drug manufacturing). When the limit is reached, it will change from with-border to no-border. This is the current Western way of resolving the border/no border problem related to applied science, but it is not necessarily recognized or respected by all countries.

10.6 The sci-tech relationship

Science and technology are related, a mother-child relationship or a brotherhood relationship, but science and technology are completely different things. Science pursues *understanding*, while technology pursues *usability* (even without understanding).

Brotherhood

Brotherhood relationship is technology invented first, then the scientific explanation. Science is only 2,600 years old [8.1] and of course there are many technological inventions before, but it is not until there is science that there is a theoretical explanation. For example, the wheel was invented thousands of years ago but it was only in the last few hundred years that we understand how the wheel works, involving the three laws of Newtonian mechanics and friction. Another example: Big data, deep learning of AI, etc., are technologies, which have not been scientifically understood and thus have not yet entered the brotherhood regime.

Mother-child

Mother-child relationship means technology appears after the relevant scientific theory is discovered. Many modern technologies have emerged because of the maturity of science, e.g., mobile phone after semiconductor theory. Having a mother does not necessarily have a child, but once there is a child one can talk about the mother-child relationship. Of course, one can also talk about the child's personality and life without mentioning the mother—history of technology. In contrast, in the case of brotherhood, we can talk about the mutual influence between brothers—history of science and technology.

Sci-Tech

There is a modern Chinese term called *Keji* (科技), meaning science-and-technology, translated here as *Sci-Tech* or Scitech. Some claim that this term reflects the Chinese ignorance of the basic difference between science and technology, not merely a short-hand convenience. If this is true, it explains why there is no parallel word in the Western dictionary.

Historically, even though science was independently invented by Guanzi in China and Thales in ancient Greece (with the former about 100 years ahead of the latter [9.1]) and in spite of the appearance of the first-rate scientist Mozi [6.6], the development of science in China greatly slowed down after Confucianism was adopted in the Han dynasty in 134 AD [1.4]. Only 100 years ago modern science invented in the West became a hot pursuit in China and more so in the last few decades. As a late comer to build a strong country, the implicit official position is to emphasize technology more than science. Or it could be due to misunderstanding of the very nature of science that science and technology are grouped together as one entity. In any case, the presence of this term sci-tech causes all kinds of confusion in the academia and society in China.

Naturally, all societies and cultures have the freedom and right to create new words, as well as the right to hit walls and self-harm, which is fully respected. For example, counting papers, frequently evaluating professors, and selecting a thesis supervisor before and upon application to graduate schools (without chance of changing after admission) all belong to this category. The consequences are that there is no way to innovate and graduate students commit suicide What is sacrificed is independent innovation, the dream of a strong country, and national rejuvenation.

10.7 Human-blind

Can simulated things completely replace the real thing? Can the metaverse replace the real universe? Answer: Instrumental things, yes; other things, no. This is a blind spot for many technologists: Thinking that 100% simulation can replace the real thing. Wrong.

To give a few examples:

1. The Jane Fonda character in the science fiction movie *Barbarella* (1968). She is ready to have sex with a handsome guy, takes out two pills, and says: You eat one; I eat one (the contactless sex she knows). The handsome guy says no, using the old style (the current practice). The camera skimmed by, and after a few seconds, Jane lets out a long sigh: Wow! All fans of the metaverse and technology should watch this movie once, and again.
2. A da Vinci painting can sell for tens or hundreds of millions, but a realistic copy or a faked one (can't tell the real from the fake) is only worth tens of thousands. Why?
3. An abstract painting can sell for tens of thousands dollars or more, but if one is told it was painted by a monkey, it is only worth a few hundred or thousand dollars. Why?

4. The fastest human 100-meter race is about 9 seconds and the horse runs 100 meters much faster, but everyone will only be excited by the speed of human, not the horse. Why?
5. Suppose you have a child. One day someone comes in, takes her away, and stacks a fake one for you. Will you love it the same? Would you want to get the real one back? Should you get her back. Why?

The answer to all these questions: Because humans especially value and appreciate the achievements and experiences of humans (same species of animal).

Conclusion

Overzealous technology fans are not science-blind, but human-blind.

10.8 Trust scientists? Trust science?

I believe in scientists. I don't believe in scientists. Both sentences are problematic.

I believe in science. I don't believe in science. Both sentences are also problematic.

The first two sentences are tantamount to saying that I believe in men or not. Of course, it is wrong because the answer should depend on who the man is and what the man is saying. The last two sentences are wrong for not asking what kind of science it is and when it was done.

The following discussion refers only to honest scientists and science that has not been falsified. Dishonest scientists and science certainly cannot be trusted.

Scientist

Scientists, like writers or artists, have different levels of proficiency. So, can masters such as Nobel laureates or academicians be trusted? Not necessarily.

If he talks about his award-winning work and story, it is 99 percent believable. If he is talking about progress he has not participated in after years of the award-winning work, that is from a half-outsider; don't believe it. If he is talking about something outside his field of expertise (such as condensed matter people talking about high-energy physics, high-energy physics people talking about social issues), that is from a 100% outsider; don't believe it. However, whether the real masters are right or wrong, their words are *enlightening* and worth listening to.

Science

More than 80% of the scientific results of the 2,600 years since Thales, the father of science, were wrong, but they were right at the time, which is inevitable in the long history of scientific research [8.4]. So, *most* of the science is *not* credible.

Apart from outdated science, any academic paper that has just been published at the forefront of research cannot be trusted.

1. Simple systems

Experimental results should wait for the third party to verify (preferably by two or more), and theoretical results should wait even longer, until the dust has settled. For example, the theory of low-temperature superconductivity has to wait for decades, and the theory of high-temperature superconductivity has been waiting for 33 years without ends in sight.

The *verified* scientific results of *simple* systems can be and should be trusted. For example, apples will fall to the ground instead of to the sky; a lifted stone and falling towards your feet will hit your own feet; the trajectory of the rocket is predictable (so that rockets can be fired, landing on the moon, and going to Mars); $E = mc^2$ (the atomic bomb has been made; no way not to believe it).

2. Complex (human) system

Complex systems (such as medicine) are more complex. Due to its complexity and low reproducibility, some medical results are not overturned until decades later. For example: Smoking is harmless; butter and egg yolks are harmful to health—these were overturned. Yet, because medicine involves health and life, one cannot wait long for the results to be certain; one can only believe it temporarily (here we are mainly talking about Western medicine).

Social science is more troublesome. Any scientific verification of social theories is difficult, and in most cases even impossible (unlike the case of studying simple systems such as electrons in physics). Personal things can be verified in a short period of time later, but social things may not be known until many decades later (such as China's one-child policy and the harm of counting papers), or they may not be known until hundreds of years later. Therefore, in applying social science theories one can only be humble and cautious, learn while believing temporarily, and doubt it all the time, just in case.

3. Complex (nonhuman) systems

The science of probabilistic nonhuman complex systems is relatively simpler than that of the human system, but is often misunderstood. Misconceptions come from a perception error of computer models and probabilistic outcomes. For example, the science of *climate change*.

First, the complexity of climate models increases with the following: the spatial dimension of representation, the resolution of these representative dimensions, the extent of climate system components and processes included in the model, and to what extent these processes are given a realistic rather than simplified form (see plato.stanford.edu/entries/climate-science). The model contains N coupled variables, and as long as one is missed or the initial conditions measured by the instrument are not accurate enough, the result will be different—very different since it is a chaotic system [22]. According to this, prominent scientists who have questioned the results of the climate-warming computer results include Freeman Dyson (1923-2020) and Nobel laureate Phil Anderson (1923-2020). The former has done a little work in this area; he questioned the climate computer models in their details. The latter has first-hand experience seeing the ineffectiveness of economic computer models.

Second, climate models are probabilistic systems. Assuming that the computer results are 100 % accurate (in fact, it is impossible), how will the results be *interpreted*? Correct interpretation requires two correct understandings:

1. Computer results by themselves *don't* count; they only count if God (i.e., mother nature) approves.
2. Very small probability events may occur (called "black swans"; winning the lottery jackpot is an example).

Even a professional group like the American Physical Society (APS) forgets the first point. That is, forget that scientific conclusions come from experimental confirmation, not from the consensus of many scientists' theoretical results (i.e., mainstream opinion), and predictions of climate change *cannot* be experimentally confirmed. As a result, APS declared in 2007 that "there is undisputed evidence that global warming is occurring" (abandoned in 2015), prompting Nobel laureate Ivar Giaever (b. 1929) to withdraw from the society in 2011 for disagreeing with the discourse.

The meaning of the second point is not understood by many people. When the computer says that the Earth will not warm if human pollution of air not stopped, is a small probability event, this could actually happen. That is: Even if humans do nothing, the Earth will not warm up too fast. (But it is better not to bet on this because we will regret losing the bet. Besides, making the air clean is good for health.)

This example shows that even if the science of complex systems is credible, interpreting it wrong has consequences.

What to Do?

Most scientists cannot be trusted, and most science cannot be trusted. What to do?

At the *individual* level:

1. When encountering scientific problems, one should consult independent, private professional societies (which are maintained by membership fees), or top hospitals on issues related to medicine, online, rather than trusting the media.

2. Don't take unnecessary risks (e.g., bungee jumping), no matter how safe they are told to be.
3. Don't take risks you cannot afford (e.g., bouldering without safety gears).
4. Find the best doctor and the best hospital if you are seriously ill (unless it is your time to go, it is the doctor and the hospital that decide you live or die).
5. Buy a lottery ticket every month (will solve the problem if you win; affordable if you lose).

At the *collective* level:

1. Decision makers should consult widely (including professionals and ordinary people) before making a decision, because human affairs are probabilistic [2.4] *and* the decision is made on behalf of the people—a collective "bet," the consequences of which will be borne by all the people.
2. Decision makers should refer to and be familiar with *world* history and be able to make more accurate predictions. Therefore, the implementation of any social practice should be taken in small steps, the more humble and cautious, the better.
3. Decision makers make decisions on behalf of many ordinary people. They should be picked by the people with extreme care.

Conclusion

Scientists are not necessarily credible, not because they are bad, but because of the nature of science. Don't waste time arguing something is unscientific because what is scientific is not necessarily credible. Instead, we should be concerned with distinguishing between good science and bad science, outdated science and science that is still valid.

The application of any scientific result (e.g., whether to drop an atomic bomb or not and where to drop it) is decided by the decision-maker. Decision-making considerations belong to the *humanities*, not natural science. Natural scientists will neither take any credit nor be held responsible.

10.9 Science: a summary

A summary:

Definition of Science: 1867 (UK)

1. Many people, including historians of science, mistake science for only the study of nonhuman systems (nonhuman animals + plants + inanimate systems), i.e., universities' so-called science disciplines minus mathematics, also called "natural science." This is wrong and has its historical factors. This definition of science comes from the second definition of the word Science in English (1867 [8.1]).

2. The essence of the 1867 definition of science was the conscious decoupling of science from religion: the removal of considerations of God/supernatural from the domain of science *and* the exclusion of the study of human (because of British religious factors: humans are specifically created by God).

3. With this definition, people go back in history and recognize Thales of ancient Greece as the *father of science* for proposing "all matters

are water." They also exclude the God part of Newton's *Principia* and, by ignoring his study of the Bible (according to which he predicted the destruction of the world in 2060, a prediction that can be tested) and posthumously recognize him as a *scientist*.

4. However, Guanzi of China's Spring and Autumn period, about 100 years ahead of Thales, had proposed that "all things originate from water." That is, Guanzi is the *first* one in history who invented science [9.1].

Origin of Science

5. Science originates from *curiosity*. Greek mythology has a complete explanation of the world but it cannot make predictions, and thus cannot satisfy the curiosity of Thales and others. *Decoupling* from Greek mythology is the first step for science, taken by Thales.

6. Science is not born of the pursuit of freedom. Thales was a free man, and his theory of "all things are water" had not be suppressed by the authority.

7. To talk about the *origin* of science, one should start with Thales, not with Plato.

8. It is also *wrong* to think that Science = Modern Science. According to this statement, there was no science even in ancient West, and in fact there is no mainstream literature to say so. If so, there is no need for the term Modern Science. That there is such a term indicates that before modern times there was science and there were scientists (such as Thales and Archimedes).

Definition of Science: 2007 (Scimat)

9. In these contexts, the 2007 Scimat definition of science is: Science consists of the process and results of human's effort to understand *all* things in nature without introducing any supernatural considerations [3.3].

10. What distinguishes science from other modes of knowledge gathering is the *Reality Check* (which includes experiments whenever possible) [8.3].

11. According to this definition, science has *two* parts: scientific process and scientific results. The scientific process is done by scientists, but it is the historians and philosophers of science who study the scientific process. There are *two* parts to the scientific results: what is still in use, and what was right then but wrong now (the majority of them).

12. According to the scimat definition, scientific content includes *all* material systems, i.e., includes all studies on humans (which are made up of atoms): the humanities, social science, and medicine (collectively referred to as Scimat—science of human).

13. This view is fine in continental Europe, but there are still some people in English-speaking countries and China who do not adapt and disagree.

14. However, this view was agreed with by Karl Marx [3.1].

15. According to the scimat definition, Western medicine is scientific, and Chinese medicine is also a science (with experiments and theories), so there *was* science in ancient China [9.4].

16. That all disciplines (except Theology) are part of science has circumstantial proof: The so-called scientific spirit (seeking truth from facts, being meticulous, correcting mistakes) applies to all disciplines.

Scientific Thinking

17. It is beneficial to divide the object of scientific research into two categories: simple systems and complex systems. The nature of the two categories is very different, and the breakthrough of modern science came from Galileo's study of the former.

18. Scientific thinking has *two* types: 1. Poetic thinking (analogy, imagination, intuition). 2. Logical thinking. Both are important.

Western science relies on (Euclidian) logical and poetic thinking [9.3].

Science and Philosophy

19. The word philosophy has *three* meanings: Philosophy (the one in ancient Greece), ‘Philosophy’ (the one from the 14th century), and “Philosophy” (the one in the current university), distinguished by quotation marks. The distinction, not yet recognized in the academia, is very important; otherwise it is impossible to carry out meaningful discussions [6.1].

20. Philosophy (love of wisdom) includes all learning, which is later divided into three parts: Philosophy = ‘Philosophy’ + Theology + Natural Philosophy. ‘Philosophy’ deals with all aspects of human (except the biological aspect, which is medicine) while Natural Philosophy studies nonhuman systems and Theology assumes the existence of God.

21. Both ‘Philosophy’ and Natural Philosophy have *two* parts: “no God” and “invoke God.” Later, the no-god part of Natural Philosophy was named Science in England in 1867; the no-God part of ‘Philosophy’ is divided into *two* parts: ‘Philosophy’ = Humanities + Social Science, while Humanities contain “Philosophy,” which focuses on the basic problems of human and the essential problems of other disciplines.

22. So, Philosophy > ‘Philosophy’ > “philosophy.”

23. By the scimat definition, all disciplines (except Theology) are part of science, then Science > Humanities > “Philosophy”; i.e., “Philosophy” is part of science.

24. But ‘Philosophy” is not part of “science” (according to the 1867 definition of science).

25. Note: All disciplines, including science, are separated from Philosophy (of ancient Greece).

26. Since all disciplines (including "Philosophy") are part of science, which forms an important part of *humanity*, is there anything other than science in humanity? Yes, such as religious exploration and meditation, are all part of the humanity. (Note that humanity and humanities are two different words and have different meanings.)

Science Is Neutral

27. Scientific results are neutral, for satisfying curiosity and for fun.

28. The application of any scientific results, including $E = mc^2$ that leads to atomic bombs, requires decision-makers. Decisions determine life and death but decision science is a part of the humanities, so the humanities (including art) are the *most* important disciplines.

29. Science philosophers or so-called "antiscience" people often get this wrong; they are actually "anti-humanities." Science is not responsible for their mistakes.

30. Whether or not and on what occasion one should trust scientists and science have to be determined carefully, case by case [10.8].

11
Scientist

Newton: a sum up

The life of Isaac Newton (1642-1727) can be summed up as follows:

1. Newton's theoretical works are excellent but his experiments are also very good and innovative.
2. He was quite interested in art when he was a child but it did not show it when he grew up.
3. The unworkability of alchemy is known only in the last 100 years after nuclear physics emerged. In Newton's time it was science, not "pseudoscience." Newton's alchemy research was a rational, scientific choice, and if successful, it would be equivalent to making room-temperature superconductivity at present. His topic was at the forefront of science then, bold enough and super important.

4. In his time, more than 300 years ago, the modern meaning and concept of the word Science—that scientific works cannot refer to God/supernatural—was absent, which appeared only about 150 years ago in 1867. In his day, people were doing natural philosophy, and it was no problem to appeal to God and the Bible in their research [8.1].

5. Newton reached the pinnacle of physics at age 24 and it is natural that he left physics and tried to do something more or equally important. Studying Bible to find out the ultimate fate of humankind and doing alchemy fit into this aim. He merely changed topics but not changed fields in research—all the topics he studied are within natural philosophy.

6. Unlike the Chinese tradition (started by Guanzi [6.3]), Newton did not do research and work as a civil servant simultaneously. He finished all the research he wanted to do before entering public office, and soon quit his Cambridge chair professorship to work full time in government.

7. Newton was a pure *rationalist*, not a rational romantic. It is a bit pity!

11.2 Einstein: a human being

After Albert Einstein (1879-1955) died in the 1950s, his manuscripts and private letters, stored mainly at the Hebrew University of Jerusalem and Princeton University, were not easily accessible.

It was not until 2006, when 1,400 Einstein private letters donated to Hebrew University by his stepdaughter Margot were made public (einsteinpapers.press.princeton.edu) that we had a more real understanding of Einstein's qualities as a husband, father, and man. We learned that Einstein cared about his children, had more than a dozen lovers—sometimes more than one at the same time—and did not hide it from his wife. Also, when he was 40 years old, he first proposed to his 20-year-old female secretary, got refused, and then married her mother.

The 10-episode TV series *Genius*, aired on the National Geographic Channel in 2017, has a detailed portrayal of this part of his life. The actor who played Einstein is Geoffrey Rush (left figure), who won the Academy Award for Best Actor in 1997 for the movie *Shine* and played captain Barbossa in the *Pirates of the Caribbean: The Curse of the Black Pearl* in 2003.

Einstein did a super good job in physics and had average violin playing skills. He has a passion for people and the world: He frequently replied to fans' letters and agreed to make an atomic bomb to save the world. All these are easy to understand because he was a *rational romantic*.

Apart being a physicist, Einstein is a human being just like ordinary people.

11.3 Gödel and Einstein: what's the chat?

Kurt Gödel (1906-1978), born in the Czechoslovakia in 1906, grew up with excellent academic performance and a lot of curiosity which earned him the nickname "Mr. Why." He contracted rheumatic fever at the age of six and was particularly careful about his health and food throughout his life. Gödel entered the University of Vienna in 1924 and soon abandoned physics for mathematics. Because of his talent, he was invited to join the Vienna Circle while still a university student, but he did not share the Circle's positivistic philosophical views. Instead, he was a Platonist: He believed that, in addition to objects, there was a conceptual world that humans intuitively had access to.

He had few close friends due to shyness. But "he did, however, like the company of women and was apparently quite attractive to them" (see plus.maths.org/content/goumldel-and-limits-logic).

Gödel's two papers established him as the foremost mathematical logician of the 20th century. One was his doctoral dissertation "The integrity of the axioms of first-order functional calculus," submitted to the University of Vienna in 1929 and published the following year. Another was his paper "Non-deterministic propositions on mathematics and related systems" published in German in 1931. These two works are summarily called "Gödel incompleteness theorems." How important is it?

Let us put it this way: The incompleteness theorems ruined the lifelong dreams of the great German mathematician David Hilbert (1862-1943) and the British philosopher Bertrand Russell (1872-1970), ended the Vienna Circle, and influenced the entire idea of artificial intelligence later. Its shock in the mathematical community is equivalent to the shock of Einstein's theory of relativity in the physics world.

Albert Einstein (1879-1955) immigrated to the United States in 1935 while Gödel immigrated in 1940. Einstein is 27 years older than Gödel; both worked at the Institute for Advanced Study located in Princeton (not Princeton University); their mother tongue is German; both came to escape the Nazis in Europe; both are at the top of their respective fields; and they are good friends with each other. How good?

On December 5, 1947, Gödel went through an interview for his application of U.S. citizenship. He took two close friends with him; Einstein was one of them. After the interviewer knew that Gödel was from Austria, he asked what kind of government Austria has. Gödel: It was a republic, but it turned into a dictatorship. The interviewer: This can't happen in the United States. Gödel: No, I can *prove* that this could happen. Fortunately, the interviewer was a friend of Einstein, and in order to get Gödel through, he quickly terminated the interview. On April 2, 1948, Gödel was sworn in as a US citizen.

Gödel and Einstein did not watch movies or football games, but they both loved beautiful women. When the two of them talked about physics and mathematics all day at work, they should have talked enough about these, right? So, when they walk home together after work (left figure), guess what they talked about? Ordinary people guess wrong. My guess: They were not talking about physics and math or philosophy; they were talking about women. After all, male scientists are also men. And (most) men love women.

In Heaven

Gödel, who died 23 years after Einstein's death, starved himself to death for fear of food poisoning. It is ironic that a man who believes in reason dies because he is scientifically illiterate. This means that the mathematician did not really understand science and was not a thorough rationalist. Gödel was a *semi-rationalist* while Newton a *rationalist* and Einstein a *rational romantic*. The fact that the former two had a nervous breakdown while the latter did not is probably related to this.

In Heaven, apart from physics, Einstein and Newton can talk about human's fate, their other common interest. On the other hand, apart from mathematics, Einstein and Gödel can talk about their other common interest: women.

Gift

Einstein, Gödel, Fermi, and von Neumann were the four great gifts that Europe gave to the United States before and after World War II.

In contrast, more than 20,000 Tsinghua University graduates working in Silicon Valley can only be regarded as a small gift from China.

114 Rabi, Oppenheimer, Yan: to Europe

The development of physics in China and the United States is only a hundred years old. The starting point was different, but the starting route was the same, i.e., to study in Europe. (The center of physics in the world at that time was in Europe where quantum mechanics and relativity were invented). I. I. Rabi and J. R. Oppenheimer from America and Yan Ji-Ci from China are the three physicists who took that route. After returning home, the three became the founder of physics in their own country. In particular, Yan became the father of Chinse physics. This is how it happened.

History is of course made by people, but at turning points in history, there are always some names that will be remembered.

Rabi

In the United States, Isidor Isaac Rabi (1898-1988; commonly known as I. I. Rabi), born in Austria, came to the United States at age one. He got his BS in Chemistry from Cornell in 1919 and went to Columbia University for a PhD, partly to pursue a woman. After earning a PhD in theoretical physics from Columbia in 1927 he spent two years in Europe, working at different times with Sommerfield, Bohr, Pauli, Stern and Heisenberg. He returned to the United States in 1929 to teach at Columbia. His work led to the creation of the molecular-beam magnetic-resonance detection method.

When I was studying for a PhD in physics at Columbia University, I always met a small old man in the corridor who was doing nothing, whistled as he walked. I thought it was a bit funny. I only knew his name as Rabi, and only learned his grandfather status in American physics years later (there was no Internet to check at that time). After Rabi's death, the department retained his office and connected it to the department library.

Oppenheimer

J. Robert Oppenheimer (1904-1967), with Harvard bachelor's degree, went to Cambridge University in the United Kingdom to study for a PhD in 1926, transferred to Germany to study with Max Born (1882-1970; 彭桓武 Peng Huan-Wu's PhD mentor, 黄昆 Huang Kun's postdoc mentor), and made many friends in Europe (Heisenberg, Jordan, Pauli, Dirac, Fermi, Teller). A year later at the age of 25 he took a PhD in 1929, and was hired by both the University of California, Berkeley and the California Institute of Technology. He became a full professor at the age of 32.

Rabi and Oppenheimer began their college teaching career in the same year. Rabi in the East Coast raised the level of experimental physics in the United States and won himself the Nobel Prize in 1944. Oppenheimer in the West Coast raised American theoretical physics to a new level and made a name for himself as the father of the atomic bomb. Two masters in a hundred years, contributed fundamentally to experimental and theoretical physics that helped to make America's later hegemony in physics possible.

Yan

Chinese physics only began in the Republic of China, bought back by Chinese students who studied aboard and returned to the motherland. Among the returnees about 100 years ago, the most outstanding was Yan Ji-Ci (严济慈, 1901-1996). He graduated from Nanjing Higher Normal University in 1923 and received a bachelor's degree from the National Southeast University (now Nanjing University). In the same year he went to Paris to study for a PhD, mentored by a member of the Collège de France. He also did research with Madame Curie. In 1931 Yan became the director of the Institute of Physics, National Academy of Beiping (former name of Beijing).

The Institute has published more than 80 papers (including 51 published abroad by Yan), accounting for 1/3 of the country's physics papers, laying a solid foundation for Chinese physics. People there were full-time researchers without teaching duties. From an academic point of view, Yan's Institute of Physics was the most important physics research place at that time. By the number of researchers or the number of papers published, the Institute overwhelmed completely the universities. It is fair to say that Yan is China's father of physics [11.5].

When I was in high school in Hong Kong, I read Yan's physics textbooks. When I worked at the Institute of Physics, Chinese Academy of Sciences (CAS) (1978-1983), Yan was the vice president of the Academy and the head of the Graduate School of the CAS in Beijing. (He later became the president of the China University of Science and Technology in 1980.) During those six years I met Yan several times and had once visited him at the 301 Hospital.

It is a pity that Chinese physics has only really recovered in recent decades after it was disrupted by the Japanese in 1937 and beyond. Unfortunately, in recent years, it has suffered from the self-imposed system of "three-part salary," over assessment, frequent time-consuming form-filling, and other mismanagement practices. For Chinese physics to reach the top there is still a long way to go.

11.5 Yan Ji-Ci: China's father of physics

43 / 55

- 1923 留学法国巴黎
- 1927 法国国家科学博士
- 1931 北平研究院物理研究所所长(6年内发表53篇文章，除2篇外全在国外发表)
- 1935 法国物理学会理事
- 1937 抗日战争爆发，离开北平

严济慈一生以“捍卫科学”为己任，容不得半点弄虚作假。(新华社)

- 1913年，中国物理学高等教育开始
  北京大学开办物理学課
- 1920年代末，中国物理研究开始
  1930年国内只有4篇物理论文发表，1931年19篇，多来自严济慈领导的北物所

中国物理学初创时的三条轨跡

1. 清华大学1926设物理系培养科研人才，著眼“提高”(叶企孙系主任，教授有吳有訓、萨本栋、周培源、赵忠尧、任之恭等，学生有王竹溪、钱学森、马大猷等)
2. 燕京大学物理系著眼“普及”(教会学校，培养中学物理教师，主任謝玉銘)
3. 北平研究院物理研究所开辟研究道路(1929成立。1931严济慈任所长。1929政府颁布大学组织法。不少大学因而开设物理系，人才短缺。该所只能收到大学毕业生，培养后送英、法、美，有陸学善、钱临照、钱三强等十余人。该所全依研究者兴趣，目的是使中国有研究。严：“打基础，辟道路”。不兼差、不兼課。是最重要的物理研究机构，完成论文80余篇，佔全国的1/3，标志着中国物理学研究的发端)

抗战使萌芽的中国物理学夭折，打断发展的锐气和连读性。1949前中国物理研究只有10年左右，不到20年的历史。从研究角度来说，严济慈是“中国物理学之父”。

Image from Lui Lam's seminar on "Physics innovation: Basic principles and experience inheritance" at the Department of Physics, Nanjing University, December 28, 2016

A brief biography Yan Ji-Ci (1901-1996) is given in [11.4].

1. No one from Yan's generation had a masterpiece in physics research. The experimental conditions were too poor—this easy to understand; the lack of good theoretical work can be attributed to the Sino-Japanese War—no way to study peacefully.

2. Yan is an experimentalist. His doctoral work is very good. When his doctoral supervisor was elected as a member of the Collège de France, he talked about Yan's work instead of his own when he delivered the mandatory public presentation (at that time, the doctoral supervisor did not coauthor student's articles). And the French and Chinese newspapers promptly reported; so, Yan became famous as soon as he got his PhD. When he returned to China by ship the reporters were already waiting for an interview at the dock.

3. A hundred years ago, the world's physics center was in Europe, not in the United States. It was then more reasonable to study in Europe (such as Peng Huan-Wu 彭桓武 and Huang Kun 黄昆) than to go to the United States. But the United States did have scholarship for Chinese students coming from the Boxer Indemnity, so of course there were people who did go and study in the United States.

4. Yan was the director of the Institute of Physics, National Academy of Beiping (equivalent to the Institute of Physics of the Chinese Academy of Sciences) for eight years. During that period, he published 53 papers, with 51 published abroad. Knowing the below-average condition of experimental equipment available at that time in China, he was doing okay.

5. The Sino-Japanese War broke out in 1937 and Yan left Beiping (called Beijing today). Yan himself said that he had not published a research paper since 1939. In 1945, when World War II was won, Yan was 44 years old. He was 48 years old when the People's Republic of China was founded in 1949. It was too late to resume research after interruption of so many years, not to mention that the experimental conditions were not available at that time. In the early fifties, all the people of his generation and of his caliber were arranged by the state to do organizational work (such as establishing research institutes) or assigned to do atomic bombs.

6. After graduating from Peking University, Qian San-Qiang (钱三强) worked at Yan's Institute of Physics. A year later, Yan took him to Paris to study nuclear physics under Madame Curie's daughter (also a Nobel Prize winner). Later, Qian became the "father of China's atomic bomb" and so Yan could be called the "*grandfather* of China's atomic bomb." Those Chinese people who appreciate the atomic bomb have to appreciate the contribution of not just Qian San-Qiang but also Yan Ji-Ci.

7. Yan is the benefactor of Qian, so Qian's lifelong courtesy, respect, and gratitude to Yan as a disciple should be understandable.

8. In 1913, Yanjing University (forerunner of Peking University) started teaching physics, marking the beginning of physics *education* in higher institutions. At the end of the 1920s, physics *research* in China began. Specifically, there were only four physics papers published in 1930; followed by 19 papers in 1931, mostly coming from Yan's Institute of Physics (IoP). In the 1920s-1930s, there were clear division of labor in physics education/research in China: (1) Tsinghua University's physics department concentrated in educating/producing research people. (2) Yanjing University's physics department educated middle-school physics teachers. (3) Yan's IoP concentrated in physics research.

9. As far as research is concerned, Yan's Institute of Physics was the largest in size, the one publishing most papers and employing the largest number of researchers (without teaching duties). China's universities at that time were incomparable in terms of number of researchers, research time, and equipment. Therefore, according to international practice and judging from the scale and impact of *research*, Yan is the "China's father of physics" (see Chapter 12 of my 2014 book *All About Science*).

10. If who the father of Chinese physics is, is determined by educational achievements then the Italian Jesuit missionary Matteo Ricci (1552-1610) of the Ming dynasty who introduced science to China, who lived in China for nearly 30 years and died in Beijing, would be the father of Chinese physics. This is obviously absurd.

11.6 Sacred no more: parity broken

Basic Concept in Physics Is Reported Upset in Tests; Conservation of Parity Law in Nuclear Theory Challenged by Scientists at Columbia and Princeton Institute THEORY IN PHYSICS IS REPORTED UPSET K Mesons Led to Doubts

By Harold M. Schmeck Jr.

Jan. 16, 1957

The New York Times.

See the article in its original context from January 16, 1957, Page 1 Buy Reprints

New York Times subscribers* enjoy full access to TimesMachine—view over 150 years of New York Times journalism, as it originally appeared.

SUBSCRIBE

There are two madams in physics: Madame Curie (1867-1934) and Madame Wu (1912-1997). Madame Wu is Chien-Shiung Wu (吴健雄), the first female president of the American Physical Society and the first recipient of Israel's prestigious Wolf Prize (1978). Wu was associate professor at Columbia University when she did the experiment that decidedly showed parity is non-conserved in weak interactions. The story of this pivotal experiment and its impact in physics are discussed here.

A Press Conference

At 2 pm on January 15, 1957, Columbia University held a press conference in Pupin Hall, the physics department building, nine years before I arrived at Columbia. Veteran Professor I. I. Rabi took a sabbatical at MIT and was recalled as chair. At the meeting, it was announced that the experiments done by Chien-Shiung Wu and others confirmed that parity was not conserved. The next day, the *New York Times* front-page headline was "Basic Concept in Physics Is Reported Upset in Tests; Conservation of Parity Law in Nuclear Theory Challenged by Scientists at Columbia and Princeton Institute ..." There are two announced experiments:

1. Wu collaborated with four scientists (two Britons and two Americans) to do it at the National Bureau of Standards (NBS) in Washington, DC, a four-hours' drive away.
2. Professors Leon Lederman (1922-2018) and Richard Garwin with a graduate student did it at Columbia.

However, this press conference was unusual in two aspects:

1. The two experimental papers were submitted to *Physical Review* on the same day as the press conference and had not been reviewed.
2. Somehow, there were misleading hints that the experiment, a collaboration between Columbia and NBS, is "Wu's experiment." (Note that there is no mention of the NBS in *New York Times*' headline.)

Unlike Princeton University, which is located in a small town, Columbia is located in New York City, which has a population of more than eight millions. The American Physical Society was founded at Columbia in 1899; the journal *Physical Review*, founded in Cornell in 1893, was taken over by the Society in 1913, and by 1957 the editorial office had moved to the Brookhaven National Laboratory on Long Island, just outside of New York City. Therefore, it goes without saying that Columbia has a close relationship with the journal.

Loss of Innocence

Before 1957, physics was forced by experiments to abandon many basic assumptions, such as the change of energy in atoms from continuous to discrete (quantum mechanics), and space-time from absolute to relative (special relativity), but none was more fundamental and shocking than abandoning the left-right symmetry (although only in weak interactions). The reason is that every day when we look in the mirror, or at the face of our girlfriend, we see the left and right symmetry of the body (approximately).

The left-right symmetry in physics (called parity conservation) was considered natural and fundamental by everybody (including the genius Wolfgang Pauli). This faith in intuition was then shaken by the parity nonconservation experiment. However, the scientific spirit is, in the face of reliable empirical evidence, bow your head to accept the new (maybe cry first), just go back and revise the theory. And that was exactly what happened in 1957.

The loss of innocence in 1957 was a watershed in physics. From that point on, no one wants to guarantee that any of the so-called "fundamental" assumptions/principles of physics will be forever true, including the conservation of energy.

Conclusion

Regarding that press conference, with a full score of 10, Columbia gained 8 and lost 2; Wu gained 2 and lost 8. What Columbia lost was a prestigious school's demeanor, and Wu lost the Nobel Prize [11.7].

Remarks

In fact, since 1905, the physical principle of "conservation of energy" has been replaced by "energy-mass conservation" in nuclear reactions (because of Einstein's discovery that $E = mc^2$).

11.7 Wu's hurt: Nobel Prize missing

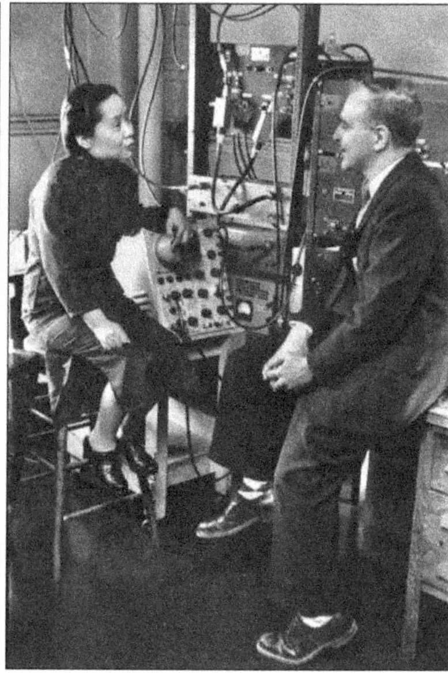

Left: Chien-Shiung Wu. *Right*: Chien-Shiung Wu and Ernest Ambler (1956).

In the corridors and elevators of Columbia's physics department, I often met a woman in a simple cheongsam, who was not walking fast, looking gentle and calm. At Pupin Hall, where the physics department is located, everyone knew that she was the only female professor in the department, called Chien-Shiung Wu. Students called her Madame Wu, and acquaintances (including Wolfgang Pauli) called her Gee Gee (sister). It was said that once a student called her Gee Gee, and Wu glared at her.

Wu's father worked in the Chinese revolution that overthrew the Qing dynasty. At one point, he took refuge in his hometown and established an elementary school. Wu studied there. And after finishing high school, Wu graduated from the National Central University at the age of 22. In 1936 she took a ship to San Francisco to visit the University of California, Berkeley (the tour guide was later her husband Yuan Jia-Liu,

袁家骝) before going to the University of Michigan for a PhD. But she was so impressed by Berkeley and decided to stay.

Wu took his PhD in 1940 and taught at a girls' school and Princeton University, and never left Columbia after working there in 1944. Wu's lab is on the top 13th floor of Pupin Hall, where I passed outside when I went to the rooftop to see the view. Before 1956, Wu's experiments at Columbia were all done by her and her subordinates, but the most important parity experiment in her life was done in cooperation with four male peers in the National Bureau of Standards (NBS, which later became the National Institute of Standards and Technology) in Washington, DC, a four-hours' drive away.

The Experiment

The experimental design is not complicated, but there are two technical barriers:

1. The cobalt-60 spins should be aligned at extremely low temperature (0.003 K). And NBS's Ernest Ambler is the one who pioneered the new technology needed to make this happen while he was at the University of Oxford a few years ago. That is why the experiment had to be done at NBS, not Columbia.
2. There needs to be a large enough crystal to protect this cobalt-60 (to prevent disruption of the spin alignments from heat). Wu's group at Columbia did not know to do that and checked all the crystal-growth literature in vain. Finally, Marion Biavati, a graduate student of Wu, accidentally succeeded in growing a one-centimeter crystal while cooking at home.

In Spring 1956, Wu started planning the experiment. Wu and Ambler agreed to do the joint experiment in September 1956 and the experimental measurements began in October, the following month. Wu still had teaching duties at Columbia University and could mainly on weekends go to Washington, DC, but she did send a graduate student full time there at NBS. Finally, the parity-nonconservation experimental data started appearing on Thursday, December 27, 1956, in Wu's absence.

And the group convinced themselves that the data were real on January 9, 1957.

The Paper

The published paper has only five authors (Wu, Ambler, Raymond Hayward, Dale Hoppes, Ralph Hudson; the latter four belong to NBS, alphabetically by last name), and there are *no* graduate students. This absence of graduate students who made crucial contributions may be in line with the house rules of Wu's laboratory, but is not in line with the rules of the physics profession. In fact, the final paragraph of the Wu et al. paper merely says "The inspiring discussions held with Professor T. D. Lee and Professor C. N. Yang by one of us (C. S. Wu) are gratefully acknowledged"—no mention of the graduate student who grew the large-size crystal. Throughout the paper, not even a footnote refers to the contribution of the graduate students.

In contrast, in Leon Lederman's paper there are three authors (Richard Garwin, Lederman, Marcel Weinrich), all alphabetically, and the third is a graduate student. Lederman was Wu's colleague who spent four days rushing out another parity-nonconservation experiment using Columbia Nevis Lab's cyclotron, after learning Wu's results on January 4, 1957.

Physical Review (PR) received two papers from Columbia on January 15, 1957, the date of Columbia's press conference [11.6]. Two days later, the journal received another paper from Valentine Telegdi (1922-2006), University of Chicago, whose experiment also shows parity nonconservation. Columbia's two papers were published on February 15, 1957 (PR, vol. 105, issue 1413), but Telegdi's paper was first rejected (apparently due to insufficient data) and was published after more data were added, on March 1, 1957 (PR, vol. 105, issue 1414), two weeks later than Columbia's.

Note that in this Jerome I. Friedman and V. L. Telegdi paper, the graduate student's name appears first and the names are in alphabetical order, too, like the Lederman paper.

The Nobel Prize

The experiment done at NBS was proposed by Wu. She actually proposed this to her colleague T. D. Lee who, together with C. N. Yang, included it as the first possible experiment to test parity violation in weak interactions in their prize-winning theoretical paper of 1956 (see the 1996 book *Madame Wu Chien-Shiung* by Chiang Tsai-Chien, 江才健). Unfortunately, Wu's suggestion of this Co^{60} experiment in β decay is *not* acknowledged in the Lee-Yang paper (not even by a footnote), violating the tradition of respecting colleague's original ideas and courtesy practiced in the physics profession. In fact, Wu, along with three other physicists (M. Goldhaber, J. R. Oppenheimer, J. Steinberger), are merely thanked for "interesting discussions and comments" in the final paragraph of acknowledgments. This is an issue of academic *ethics*, which lies at the core of the question of why Wu did not get a Nobel Prize (see below).

Back to the experiment

After Wu approached Garwin for help she was told NBS is the proper place to do it. And when NBS' Ambler accepted Wu's invitation to collaborate they knew that more experts are needed and so three more NBS experts were added to the team: Ralph Hudson (cryogenics), and Raymond Hayward and Dale Hoppes (radiation-detection). According to the American Physical Society's historic physics site report on NBS, the NBS team "devised and performed experiments analyzing beta decays" resulting in the overthrown of parity symmetry in weak interactions.

That is, this is not a Columbia experiment; it is a Columbia-NBS experiment; it was done at NBS and could not succeed without the four NBS scientists' major contributions. This point is sometimes overlooked by some people, due partly to the misleading Columbia press conference of January 15, 1957 (see below), but is emphasized by Nicholas Kurti and Christine Sutton in a *Nature* commentary in 1997, and by the Nobelist Val Fitch in 2002 (see: Hargittai, M. [2012]. Credit where credit's due? Physics World, Sept. 13, 2012).

Back to the press conference

The Columbia press conference of January 15, 1957 was misleading, to say the least. The two experiments announced: the major one, the Co^{60} experiment, was done at NBS, a collaboration between equal peers from two institutions; the secondary one by Lederman et al. was done at Columbia [11.6]. To be fair, the press conference, if needed to be called, should be a joint conference called together by Columbia and NBS. As it happened, it resulted in the misconception, even among some physics people, that it was a "Wu's experiment," not a "Columbia-NBS experiment," that discovered parity nonconservation. When the truth became more known, the Nobel Prize committee is not amused, I guess.

On this issue, the knowledgeable physicist and Nobel laureate Philip Anderson (1923-2020) has this to say. In the *APS News*, February 2002, Letters session, there is a piece under the (editor added) title "Wu's leadership role questioned," submitted by Anderson:

> In your otherwise well-written historical piece in the December 2001 issue of *APS News* about parity non-conservation, it is unfortunate that you perpetuate an injustice which seems to have become permanently embedded in the "history" books. You refer to an NBS "team led by C. S. Wu" as having done the crucial experiment.
>
> Entirely aside from the fact that an NBS team cannot have been led by a non-NBS scientist, there was no question of any formal leadership role for Dr. Wu, to my knowledge, and the NBS scientists, specifically Eric [Ernest] Ambler, have repeatedly chafed at the dominant role given to her in the histories. This injustice may have happened because histories tend to be written by theorists, and she was the one who communicated with the theorists, but she appears to have made no effort to correct it. That communication may have been strengthened by the fact that Lee and Yang are also Chinese as well as by academic snobbery which was at that time alive and well at Columbia.

Back to the Prize

1. Wu herself believed that she deserves the Nobel Prize. In 1989, at the age of 77, Wu said in a private message to his former colleague Jack Steinberger (1921-2020) who just won the 1988 Nobel Prize:

> I treasure and value your praise, coming from a modern-day physicist with such a critical mind, more than any prize and honor in science. I dedicated my whole life to, and found happiness in, research in weak interactions. Although I did not do research just for the prize, it still hurts me a lot that my work was overlooked for certain reasons.

Wu won all the major awards (including the inaugural Wolf Prize from Israel) except the Nobel Prize. The "prize" she said meant the Nobel Prize.

2. Wu's experiment was her own idea, not taken from the Lee-Yang article. Instead, the experiment was suggested by her to Lee, who included it in the Lee-Yang paper (see above). The Lee-Yang article does not predict whether parity is conserved or not, and the Nobel laureate Murray Gell-Mann (1929-2019) had teased them for lack of courage. For many years afterwards, Lee constantly stressed that they did not make any predictions and it was Wu who discovered that parity was not conserved, so she deserved the Nobel Prize. Steinberger also said that Wu deserved the Nobel Prize, so Wu in her message, first thanked him for his support, and then followed with the sentences quoted above.

3. It is stated explicitly in the will of Alfred Nobel (1833-1896) that the Prize is to be awarded "to those who, during the *preceding* year, have conferred the greatest benefit to humankind" [3.2]. Accordingly, since all the three experimental papers on parity nonconservation were published in 1957, none of the authors could be considered for the 1957 Nobel Prize, unlike Lee and Yang who's paper appeared in 1956.

4. Furthermore, the Nobel Prize committee needed time to sort out the credits due the authors of the three experimental papers (Wu et al., Garwin et al., and Friedman and Telegdi). The Columbia-NBS team

started their experiments in September and the Telegdi team in October, both in 1956, independent of each other and using two different approaches, while the Lederman team started on January 4, 1957, after learning about the NBS results from Lee. Therefore, a few years later, if the Nobel committee still wants to give the experimentalists the Prize, they could pick Wu, Ambler and Telegdi, say. But, after considering the messiness of the press conference and the publication process of the whole affair, the committee may no longer like the taste of this bowl of soup, or soup opera, I guess. After all, not all deserving scientific discoveries received the Prize which is given annually and to a maximum of three persons only for each prize.

5. According to Chiang Tsai-Chien, Wu's biographer, "the most probable reason why Professor Wu was passed over was her tense relationship with her collaborators at the National Bureau of Standards." See: Tsai-Chien Chiang [2015]. Wu Chien-Shiung: A brief biography. *AIP Conference Proceedings 1697*, 040004 (2015).

6. It is absurd to even consider the Nobel Prize discriminates women given the fact that Madame Curie got it two times, even though gender bias did or does exist in some university campuses. There is story allured to the case of Wu at Columbia when her promotion was under review before her success in the parity experiment.

The Lesson

Be fair to your co-workers and competitors, especially the graduate students. After all, like the Olympic Games, the science "game" invented and passed down by the ancient Greeks, is about the search for knowledge about nature. It is fun to win the game but it is the heritage and tradition of the game that should be treasured and preserved. And the way to do that is to educate the students properly, show them how to give credit to your colleagues and collaborators, and help them to succeed. This practice seems to be glaringly lacking on the parts of Lee and Yang as well as Wu.

The game must go on. And graduate students are the key that the game will go on. They are our future!

11.8 Physics competition: against the clock

The Woodstock of Physics

On Wednesday evening, March 18, 1987, at the Hilton Hotel in Manhattan, New York City, I squeezed into a large room of the American Physical Society's March Meeting. Starting at 7:35 pm, five speakers gave their talks, each 12 minutes, followed by 20 minutes of discussion after all the presentations. The speakers: Alex Müller (Switzerland), Shouji Tanaka (Japan), Paul Chu (朱经武, USA), Zhao Zhong-Xian (赵忠贤, China), and Robert Cava (USA). There were 1,200 people in the room, and 1,000 people watched TV broadcast outside the room. This interim report on high-temperature superconductivity will be known as the Woodstock of Physics. (The 1969 Woodstock Music and Art Fair was attended by 450,000 people.)

The competition for the over-90-K high-temperature superconductivity was fierce. In order to be the first one to publish, Bell Labs' paper by Cava et al. was submitted directly to the editorial office *Physical Review Letters* in Long Island on March 5 1987, by a two-hour drive from New Jersey.

Race Against the Clock

Going back 30 years, in January 1957, *Physical Review* received three manuscripts: two from Columbia University's Chien-Shiung Wu and Leon Lederman on the 15th, and one from the University of Chicago's Valentine Telegdi on the 17th. At 2 pm on January 15, 1957, Columbia held a press conference in the physics building, announcing or implying that the "Wu's experiment" overturned parity conservation and shocked the world [11.6]. A month later, on February 15, Columbia's two papers were published while the University of Chicago's was blocked by the first review and published two weeks later on March 1. Therefore, history records that Wu was the first person to overthrow the conservation of parity [11.7].

Lederman is a colleague of Wu. His experiments were started on January 4, 1957, when Wu's initial positive results were made known to him by T. D. Lee, also a Columbia colleague. On the other hand, Telegdi's experiments began in early October 1956 after he read the Lee-Yang paper preprint, without knowing anything about Wu's experiments. According to Laurie Brown:

> The work was also slowed because Val had to leave the country in late fall when his father died in Milan, Italy, and he went there to help his mother cope. Still, upon his return, he found that Friedman had 1300 measured events and clear evidence for parity nonconservation. The result was definite but preliminary (they were aiming for 2000 events) but the competition was fierce and they decided to publish.
>
> Friedman and Telegdi communicated their result to *Physical Review* [*Letters*] for fast publication. However, while the editors decided to publish the Columbia-NBS experiment and the Columbia cyclotron result the same week, they delayed the Chicago announcement until the next issue, where it appeared with a note (after protest) stating that it had been delayed "for technical reasons." This event caused Val great unhappiness that was only partly compensated by his receiving the Wolf Prize in 1991, in large part for this important work. ("Valentine Louis Telegdi,

> 1922-2006: A biographical memoir by Laurie M. Brown," National Academy of Sciences, 2008).

The "technical reasons" mentioned here are that Telegdi's first draft was rejected by the editorial office on the grounds of "insufficient data," and only after revision was it approved.

Conclusion

Taking high-temperature superconductivity and parity nonconservation as examples, the competition in physical experiments (and theories) is sometimes fierce and has to race against time. Would you rather win the race by not going to your dad's funeral? Are you willing to attend your dad's funeral and lose the Nobel Prize?

Afterward

Valentine Telegdi (1922-2006), born: Jan 11, 1922, Budapest, Hungary; Enrico Fermi Distinguished Service Professor of Physics (1971-1976) at the University of Chicago. On the 1957 incident that his parity paper appeared in *Physics Review* two weeks after the Wu et al. paper, this is his reminiscence on March 4, 2002, in his conversation with Sara Lippincott:

> LIPPINCOTT: After you and Friedman did this, did others do it, too?
>
> TELEGDI: No, other people did it in parallel. Those people who did it were under the direct influence of Lee and Yang, which I was not. And in fact, because of these circumstances, some of my competitors got into print before me—which was really very bad.
>
> LIPPINCOTT: Is that—
>
> TELEGDI: Miss [Chien-Shiung] Wu, at Columbia—the Dragon Lady.
>
> LIPPINCOTT: Did you know her?
>
> TELEGDI: And how!
>
> LIPPINCOTT: She was a dragon lady?

TELEGDI: Well, only for us. For the Chinese, the dragon is a symbol of good luck.

Source: Valentine L. Telegdi (1922-2006), interviewed by Sara Lippincott, March 4 and 9, 2002, Archives, California Institute of Technology, Pasadena, California.

In 1976, Telegdi moved to ETH Zürich (his Alma Mater and Einstein's) in Switzerland, being professor there from 1976 to 1989. He chaired CERN's Science Policy Committee from 1981 to 1983.

He was awarded the Wolf Prize in 1991, finally in par with Chien-Shiung Wu—his competitor and inaugural recipient of the Wolf Prize in 1978.

On April 8, 2006, Telegdi died in Pasadena, California.

11.9 Wu's English biography

There are many famous ethnic Chinese physicists, including six Nobel Prize winners: Tsung-Dao Lee (李政道), Chen-Ning Yang (杨振宁), Samuel Ting (丁肇中), Steven Chu (朱棣文), Daniel Tsui (崔琦), and Charles Kao (高锟). But there is only *one* English biography, *Madame Wu Chien-Shiung: The First Lady of Physics Research* (World Scientific, 2014), translated from Chiang Tsai-Chien's Chinese book *Wu Chien-Shiung* (1996), translated by Tang-Fong Wong (黄腾芳).

Here is the first paragraph of the Translator's Preface:

> In January 2007, Patricia Cladis was interested in the life story of Madame Chien-Shiung Wu. Surprisingly, she found no detailed biographical books written in English on Wu in the market, which remained the case until now despite Wu's tremendous accomplishments. Lui Lam introduced me to Pat, who just got hold of a copy of Tsai-Chien Chiang's book published in 1996 through her contact in Taiwan, a book she couldn't read. By May of 2007, both Pat and Lui raised the idea of translating Chiang's book into English. After several professional translators we

approached could not find time, I (reluctantly) committed to give it a try.

In fact, there do exist Chinese biographies about T. D. Lee, C. N. Yang and others. I hope that there will be people who will translate them into English or directly write new ones in English.

Notes

Patricia Cladis (1937-2017) was born in Shanghai; did not understand Chinese; worked at Bell Labs all her life and discovered the "reentrant phase" in liquid crystals.

Tang-Fong Wong graduated from Pak U Secondary School (柏雨中学) in Hong Kong, earned PhD in physics from Brown University, and retired from operations research at Bell Labs.

11.10 Feynman, the authentic

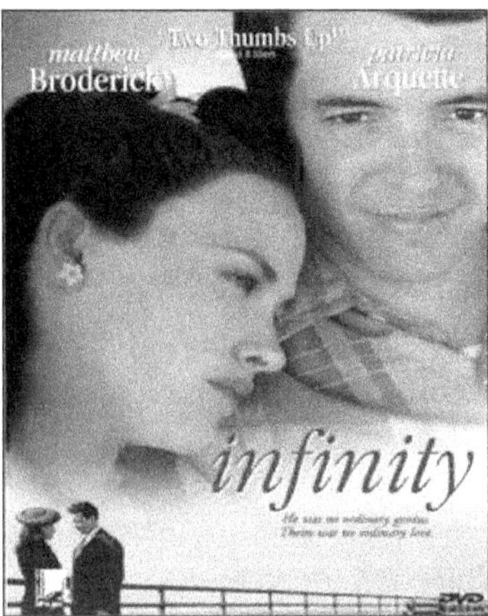

Richard Feynman (1918-1988) was the PhD mentor of my PhD mentor. He was a native of New York City who applied to Columbia University after high school. At that time, there was a quota limit for admitting Jews at Ivy League universities. Columbia belongs to Ivy League and rejected Feynman due to insufficient slots. Feynman ended up entering Massachusetts Institute of Technology (MIT). He initially majored in mathematics but switched to electrical engineering after finding mathematics too abstract. And after finding electrical engineering too non-abstract he switched to physics, which he claimed was somewhere in between. After completing his bachelor's degree at MIT, Feynman received his PhD from Princeton University in 1942.

His student Philip Platzman (1935-2012) worked on polarons with him (using Feynman's path integral method) and became famous before working at Bell Labs all his life. I was a PhD student at Columbia University but my thesis was done with Platzman at Bell Labs [11.11].

Feynman married three times. His first wife is Arline Greenbaum (1919-1945, left figure).

With Arline

Arline was Feynman's classmate and lover in high school. After Feynman received a bachelor's degree in physics from MIT, he went to Princeton University for a PhD with a scholarship. And one of the conditions of the scholarship was that he could not marry until he graduated.

Despite this, he continued to date Arline and resolved to marry her after graduation, knowing that she had cervical lymph node tuberculosis which was a terminal illness at the time. The illness is contagious; apparently she was unable to have sex and expected to live no more than two years.

In 1942, Feynman graduated with a doctorate from Princeton University. On June 29, they took a ferry from Manhattan Island to Staten Island (two boroughs of New York City) and married at City Hall. The ceremony was attended by no family and friends, and the legal witnesses were a pair of strangers. Feynman could only kiss Arline's cheek. After the ceremony, he took her to Deborah Hospital in Browns Mills, New Jersey, and visited her every weekend.

In 1943, Feynman was drafted to Los Alamos to work on the atomic bomb. The project's leader J. Robert Oppenheimer made a long-distance call to Feynman from Chicago to tell him that he had found a nursing home for Arline in Albuquerque, the largest city near the Los Alamos hills in New Mexico. Feynman and Arline were among the first to travel to New Mexico, arriving by train on March 28, 1943. The railroad provided Arline with a wheelchair, and Feynman booked her a private room on the train for an additional fee.

The salary for making an atomic bomb is not high. Feynman's monthly salary of $380 was only half of what he needed to live on and pay Arline's medical expenses. So he was forced to dip into her $3,300 savings to make up for it. Every weekend, he borrowed his colleague Klaus Fuchs' car and drove two hours to Albuquerque to see Arline. (It turned out that Fuchs, a German, was a spy who provided atomic bomb secrets to the Soviet Union. He was convicted in 1950 and spent nine years in prison before continuing to do theoretical physics in East Germany.)

On June 16, 1945, Arline became critically ill and Feynman drove down the mountain to Albuquerque, where he sat with her for several hours until her death. Arline died at the age of 25 and did not see her husband win the Nobel Prize (1965). On October 17, 1946, Feynman wrote a love letter to his deceased wife and sealed it. The letter was unsealed after Feynman's death in 1988 (https://lettersofnote.com/2012/02/15/i-love-my-wife-my-wife-is-dead/).

Feynman never put on a face to do physics (or anything). He laughed a lot. The only exception was when he was at the bedside of his dying wife—he cried. The story of Feynman and Arline was made into a movie in 1996: *Infinity*.

With Physics

Feynman was Jewish. In front of his Jewish friends he had said things that were politically incorrect. As told by Platzman, Feynman once said something like this: A person doing physics is like a child buying something in a small shop owned by a Jew: He should pay attention to whether the shop owner is honest in the deal and count every penny in the change. What Feynman meant is this: Do physics with extreme attention to detail and don't be fooled by God. Feynman and other physicists like him, including Platzman [11.11], inherited the meticulous tradition of the ancient Greek philosophers who loved wisdom. (As a corollary, physicists should never even think about faking their works. Those who did should be kicked out of the profession immediately.)

Atomic-scale physics and technology began with Feynman's lecture, "There's plenty of room at the bottom: An invitation to enter a new field of physics," given at the 1959 Annual Meeting of the American Physical Society, held at the California Institute of Technology. Here, the word bottom is a pun, which, together with "black hole," are the two most erotic terms in the history of physics.

Feynman frequented the bottomless bar (the maid is naked below waist but not above it) near Caltech for a while, to kill time and do physics. He once drew a sketch of the maid (copy included in *The Art of Richard P. Feynman: Images by a Curious Character*, compiled by his daughter Michelle). The original was given to his student Phil Platzman, my PhD mentor, and he gave me a copy upon my request when I visited him at Bell Labs.

Loving beauty is one of Feynman's hobbies, although his hobby of beating drums is more known. Feynman is not afraid of controversies. The title of his "plenty of room at the bottom" talk means there are many possibilities if one looks down there (at the atomic scale). But the word bottom also means somewhere in the lower part of a body. Such a title is not entirely politically correct these days. But it is one of the things that makes Feynman different from other scientists. He is *authentic*—doesn't hide, doesn't deny.

Conclusion

Feynman loves beautiful women, but has never committed sexual harassments. Loving beauty and sex are two different things.

Doing history of science or science popularization cannot take things for granted; one must pay attention to details and be careful to verify them. Instead of hiding "shortcomings" for dead scientists, the human side of them should be exposed, too. Unless great scientists are presented as real persons, no one can learn from them.

For example, Einstein's love history and list of lovers have become known in the past few years. But his contribution to physics has not been affected. Instead, it makes us feel that he is extremely human, just like the people around us [11.2].

Scientists are human. Popular science should not only popularize scientific knowledge but to convey this point to the public.

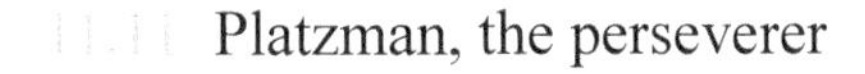

Platzman, the perseverer

Columbia University's Department of Physics focused on particle physics and nuclear physics in the 1960s. T. D. Lee and C. S. Wu were among the top professors in these areas. At that time, there was only one professor of condensed matter physics in the department who did experiments and one who did theories (Joaquin Luttinger). The two obviously could not take in too many graduate students. In 1969, I was a PhD student and wanted to do condensed matter theory; so, I decided to cross the Hudson River and went to Bell Labs in Murray Hill, New Jersey, to find a PhD thesis supervisor. On that particular day, I had the chance to pick one out of two supervisor "candidates"; i.e., I went to interview them. One smiled a lot; the other didn't. I picked the one who smiled.

The smiled one was Philip Platzman (1935-2012), born in New York City on May Day 1935, received a BS in physics from the Massachusetts Institute of Technology in 1956 and a PhD in physics from the California Institute of Technology (Caltech) in 1960. His PhD supervisor is Murray Gell-Mann (1929-2019), and his co-advisor is Richard Feynman (1918-1988), both Nobel Prize laureates. Platzman's doctoral dissertation

"Meson theoretical origins of the non-static two nucleon potential" was unremarkable. After graduating, he left particle physics and spent his life studying condensed matter at Bell Labs. In 1997, Platzman shared the Arthur H. Compton Prize of the American Physical Society with Peter Eisenberger.

Thesis

Platzman is a theorist, but leads an experimental group. Eisenberger, a year after graduating from Harvard, was hired to conduct experiments to study the distribution of electron momentum in solids using X-ray Compton scattering. At that time, Eisenberger went to Japan to buy back the most advanced X-ray machine and measured highly accurate Compton profile of sodium and lithium metals. Platzman came up with an explanation scheme to do numerical analysis on a computer, and he did not want to do it himself, so he asked me to do it. I didn't want to do it either. And within a month I rigorously proved that the solution would not work, and wrote him a report of several pages. He agreed with the results, wanted to publish it, then calmed down and thought that it was too low to produce an article with only negative results, so he gave up. After that, neither he nor I knew how to continue.

At the end of 1970, the Baodiao movement started, which was initiated by Chinese students in the United States but soon spread to overseas Chinese all over the world [11.15]. In 1971, I suspended my thesis work, closed down the *Chinese Language Movement* magazine (in support of people of Hong Kong) that I initiated with friends, and went to live in the Chinatown in Manhattan, to be close to the masses. We established the Chinatown Food Co-op to "serve the people" (see Remarks).

After that, embarrassed to receiving Columbia University's monthly stipend as a graduate student but doing nothing in return I moved back to the vicinity of the university, solved the Compton-profile problem, and wrote up my doctoral thesis in 1972. Coincidentally, my proposed solution used the works of acquaintances of Feynman, Platzman, Luttinger, and Hohenberg [11.12]. According to the results of my thesis, an experimental and theoretical paper (1972) with Eisenberger and

Platzman, and two theoretical papers (1974) with Platzman were published—a total of three. The first theoretical paper contains results called Lam-Platzman Theorem and Lam-Platzman Correction—names given by other physicists who cited the paper.

I have been with Platzman for three years, and he has never called me in to see him. It was I who took the initiative to meet him two or three times a year, at Bell Labs. To get from Manhattan to Murray Hill, one needs four means of transport (Manhattan subway, cross-river tunnel subway, train, taxi), and it takes one hour to be there. When we met the first thing he always asks was: What's new? I thought he asked me what was going on in the community and always replied: Nothing. It was many years later I realized that he was asking what new discoveries I had made in my research. But my answer wasn't wrong: I made no progress in my research (sometimes because I was not doing it).

I guess his words and mentoring style are copied from Feynman, who is famous for hands-off in advising students. The students brought up this way would be really independent and can go out to fight in the real world once they graduate—if they can really graduate.

Master and Apprentice

After my solution came out, Platzman arranged for me to attend the American Physical Society's March Meeting, an annual gathering with thousands of condensed-matter people (currently 11,000) from all over the world. This meeting is pretty unique: The invited talks are 30 minutes each, including Q&A time; the invitee's registration fee is not waived; other reports are 10 minutes each, plus 2 minutes of Q&A; the abstract is accepted without review as long as it is submitted or signed by an APS member.

Before the meeting, Platzman asked me to go to the lab at a weekend to rehearse, and that was fine. To save money or not waste money, he asked me to share a room with him during the meeting. Although in separate beds, that was the problem: I never shared a hotel room with a foreigner. So I went to buy a set of pajamas, which is kept to this day. The meeting was held in Hilton Hotel, New York, and it was the first academic presentation I gave in my life.

Platzman loves Chinese food, knows what is delicious, and is more correct and flexible with chopsticks than I am.

I started working at the Institute of Physics (IoP), Chinese Academy of Sciences, Beijing, in 1978 [11.15]. In January 1983, Platzman wrote to me at the IoP, saying that his family of four would visit Tokyo for a month in June, and that they wanted to visit Beijing for a week, and hoped that our side would follow the Japanese practice and cover his family's air tickets and local expenses. I wrote back that his visit is welcomed, but asked him to pay for his own tickets. He didn't come.

A few years later, I started working in San Jose in 1987. In September 1997, Platzman emailed me that he intended to visit China, but did not mention any air ticket this time. In December, the IoP sent him a letter of invitation agreeing that he and his wife would visit Beijing for two weeks in November 1998 and that the IoP would pay local fees. Platzman postponed his visit to China to the next year, but told me in April of 1999 that it would not take place for the time being, but for the future.

In this way, Platzman never visited China, otherwise we could have asked him everything about Bell Labs.

Platzman joined Bell Labs in 1960 and retired in 2001. Before retiring, the lab held a symposium for him on October 27, 2000, to celebrate his 40th anniversary of service, which I attended (see photo).

Ten Years

After retirement, Platzman continued going to the lab every day to work (without pay, I guess). But the next year he had an accident, paralyzed from the neck down, and only his head could move. According to him, it was the result of a fall down the stairs from the second floor at home. (I did not dare to ask how it happened.)

He hired a male nurse to live at home, took care of him 24 hours a day, and continued to live a "normal" life, i.e., going to the lab every day to work as usual, for ten years. He published 20 papers (with co-authors) in the last ten years and died at home on February 7, 2012, at the age of 77 years—seven years longer than Feynman.

Both Feynman and Platzman, master and apprentice, had physical problems in the last ten years of their lives (Feynman had cancer, with multiple surgeries). But they both insisted on working normally.

End

Platzman was not very wise in two ways:

1. Worship his own supervisor (Feynman) too much.
2. The family house has stairs and he lives upstairs.

The first point is not fatal; it just affects his physics. The second point is not fatal either; it just slows down his physics. But life is short; physics is not everything. The important thing is to live well and leave the world without regrets. Whether he had regrets, I don't know. He didn't tell me.

Phil Platzman, the perseverer, sticking to physics, sticking to the End.

Remarks

1. Joaquin Luttinger (1923-1997), PhD in Physics, MIT, 1947, made his name for "Luttinger liquid" and Luttinger Theorem. I did my doctoral thesis with Phil Platzman at Bell Labs but my nominal PhD supervisor in Columbia's physics department is Luttinger. Luttinger Theorem is used in my PhD thesis.

2. Arthur H. Compton (1892-1962), American physicist, the doctoral supervisor of the Chinese physicist Wu You-Xun (吴有训), won the 1927 Nobel Prize in Physics for the discovery of the Compton effect in 1923, which proved the particle properties of electromagnetic radiation.

3. Peter Eisenberger, BS in physics (honors) from Princeton University in 1963, PhD in physics from Harvard University in 1967, stayed at Harvard for one year as a postdoctoral fellow, doing research on biophysics and polaron theory. He joined Bell Labs in 1968 and left in

1981 as director of the physical sciences laboratory at the newly founded Exxon Research and Engineering Company. In 1989, he became Professor of Physics and Dean of the School of Materials at Princeton University. From 1996 to 1999, he was Vice Provost of the Earth Institute at Columbia University and Director of the Lamont-Doherty Earth Observatory. He is currently a professor of earth and environmental sciences at Columbia University.

4. In 1971, when I was a PhD student at Columbia, I joined my friends to establish the Chinatown Food Co-op in Manhattan's Chinatown, a mass organization with a political agenda, the purpose of which is to serve the masses, narrow the distance with the masses, raise their political awareness, encourage them to actively participate in local political activities (such as fighting for Chinese immigrants' rights), and identify with progressive political ideas. The Co-op has more than 10 members of men and women, all volunteers, including doctoral students from Columbia University, medical students from other schools, immigrants from Hong Kong, and local natives; the common language is English.

Every Friday at 5 am, I was responsible for driving an old truck, going to the vegetable wholesale market in Bronx with a few members. We bought boxes of vegetables and eggs, transported them to a church basement in Chinatown, put them away, went to school or work, and then met and tidied the goods up in the evening. At 11 am on Saturday, every member arrived at the basement, and because the goods were sold wholesale prices, the neighbors would come. I spoke in Cantonese for 10-15 minutes in the beginning, and the goods are sold out in an hour. The members of the Co-op ate together at noon in my Chinatown apartment, reviewing while eating, which could last four hours.

It was in the Co-op that I improved my spoken English and developed organizational skills, which was very helpful later in my physics career, in leading research groups and establishing the liquid crystal societies [13.5]. The Co-op was quite successful, and some neighbors wanted to marry their daughters to us—highly educated intellectuals with kind hearts. One of the female members later became mayor of Oakland, a California city.

Hohenberg, the eloquent

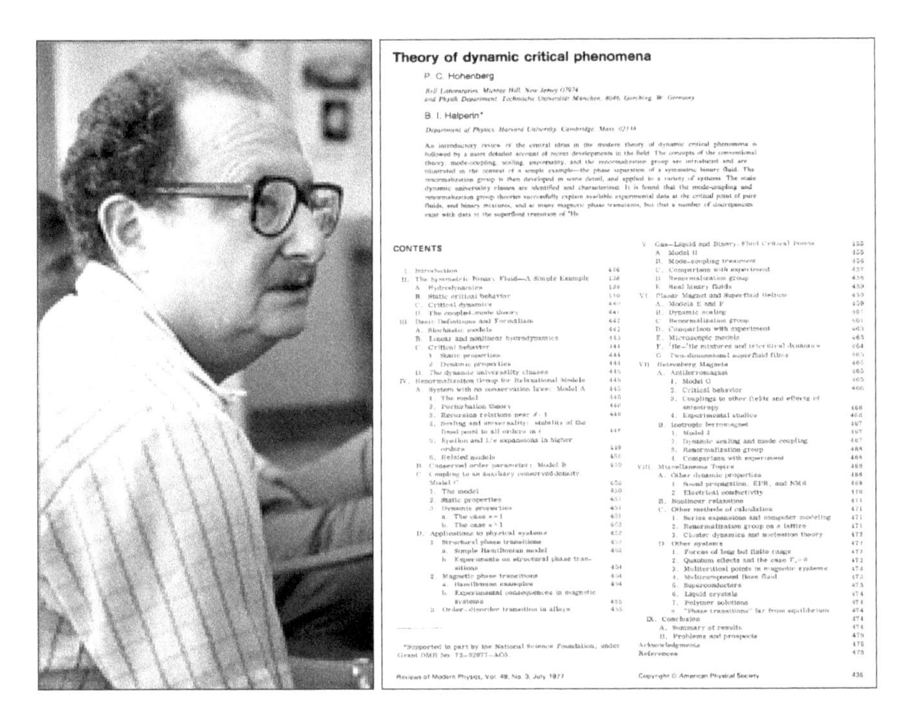

Theory of dynamic critical phenomena

P. C. Hohenberg

Bell Laboratories, Murray Hill, New Jersey 07974 and Physik Department, Technische Universität München, 8046, Garching, W. Germany

B. I. Halperin*

Department of Physics, Harvard University, Cambridge, Mass. 02138

An introductory review of the central ideas in the modern theory of dynamic critical phenomena is followed by a more detailed account of recent developments in the field. The concepts of the conventional theory, mode-coupling, scaling, universality, and the renormalization group are introduced and are illustrated in the context of a simple example—the phase separation of a symmetric binary fluid. The renormalization group is then developed in some detail, and applied to a variety of systems. The main dynamic universality classes are identified and characterized. It is found that the mode-coupling and renormalization group theories successfully explain available experimental data at the critical point of pure fluids, and binary mixtures, and at many magnetic phase transitions, but that a number of discrepancies exist with data at the superfluid transition of ⁴He.

CONTENTS

*Supported in part by the National Science Foundation, under Grant DMR No. 72-02977-A03.

Reviews of Modern Physics, Vol. 49, No. 3, July 1977 — Copyright © American Physical Society — 435

Pierre Hohenberg (1934-2017), born in 1934 in Neuilly-sur-Seine, France, is a French-American theoretical physicist who specializes in statistical mechanics. When he was young, he wanted to become a professional chef, but he did not become one, so he went to Harvard University to study physics, where he received his bachelor's degree in 1956 and his doctorate in 1962. In 1962-1963, he worked as a postdoc at the Institute of Physical Problems in Moscow; 1964-1995, Bell Labs (1985-1989, Director of the Department of Theoretical Physics); 1995-2003, Deputy Provost for Science and Technology, Yale University; 2004-2011, Senior Vice Provost for Research at New York University.

Hohenberg was a Fellow of the American Academy of Arts and Sciences (from 1985); member of the National Academy of Sciences (from 1989), the American Philosophical Society (from 2014) and the New York

Institute for the Humanities (from 2016). He received the Fritz London Prize in 1990, the Max Planck Medal of the German Physical Society in 1999 and the Lars Onsager Prize of the American Physical Society (APS) in 2003.

Hohenberg was also politically active: A postdoc in Moscow for a year is an unusual choice. He chaired a committee of the APS for the freedom of scientists in 1983, and in 1992-1993 served on an APS committee in support of scientists of the former Soviet Union.

Doing Physics

In 1964, Hohenberg and Walter Kohn (1923-2016) proved two theorems, providing a solid theoretical basis for density functional theory. Kohn won the Nobel Prize in Chemistry for this theory in 1998, and I did my doctoral thesis based on this work in 1972 [11.11].

Hohenberg wrote two famous review articles in the *Reviews of Modern Physics*: "Theory of dynamic critical phenomena" (with Bertrand Halperin, 1977) and "Pattern formation in nonequilibrium systems" (with Michael Cross, 1993). The publication of a review article is a signal for the author to retreat after successfully conquering a topic. [The exception is Leo Kadanoff (1937-2015), whose review article signals the entrance to a topic.]

In 2014, 80-year-old Hohenberg teamed up with Richard Friedberg (one of T. D. Lee's two "genius" students; the other is Norman Christ) to propose a new interpretation of quantum mechanics based on consistent histories.

Debating Physics

Hohenberg is a literate physicist who works in New Jersey but lives in Greenwich Village, an urban village in Manhattan, New York. The village is known as an artist's paradise, the capital of bohemia, the birthplace of the Beat, and the East Coast of the counterculture movement of the 1960s. The Bitten End bar in the village is where singer and literary Nobel laureate Bob Dylan debuted. Famous cultural figures

Thomas Paine (1737-1809), Mark Twain (1835-1910) lived in the village, and Robert Downey Jr., the movie's "Iron Man," too.

Hohenberg was the first theorist for Bell Labs hired by my PhD advisor Platzman. One day, I made an appointment to meet Platzman at the lab, and he told me to take Hohenberg's car from New York. That was my first meeting with Hohenberg, unacquainted, sitting in his little, open-top sports car for nearly an hour, not talking much.

Soon after, I, a doctoral student at Columbia University, attended a one-day conference on statistical mechanics at the Stevens Institute of Technology in Hoboken, New Jersey, located next to the Statue of Liberty. Hohenberg spoke on stage about the latest results of the dynamic critical phenomenon, answered questions after speaking, argued with his opponent, didn't give an inch, and "killed" the other party until the latter's piece of armor was not left. Boom! I had seen T. D. Lee debate with speakers at Columbia University's weekly colloquia and debates between colleagues at Bell Labs, but it was simply incomparable to Hohenberg's "performance," not on a notch.

I guess Hohenberg recreated the debate in the town square that the ancient Greek philosopher Socrates delivered, and reminded me of *The Three Musketeers* movie (1973) in which a French swordsman singles out a crowd, sword to sword fight, with the tension and romance of the master.

In China

When Hohenberg was deputy provost at Yale, he traveled to China several times to establish cooperation programs between Yale and China.

From June 11 to 15, 2001, I collaborated with the Institute of Physics, Chinese Academy of Sciences, to hold an international conference on Pattern Formation and Self-Organization of Nonlinear Complex Systems at the Fragrant Hill Hotel, which was attended by Hohenberg and contemporary elites. Hohenberg, 67, agreed to the invitation without hesitation.

He went up to the Great Wall after the meeting and took a picture in the emperor's dragon robe. I went shopping with him at Beijing's Qianmen (前门) district and asked him what he was most interested in in the world. He said it was the stuff in the shop window.

Ten years later, my understanding is that Hohenberg meant people should remain childlike, like a child in a candy store, curious about each candy, and interested in tasting it. This is the basic principle of learning.

Pierre Hohenberg, an eloquent, with big talent and great care about the world.

Remarks

Bertrand Halperin, born in New York City in 1941, parents born in the Soviet Union; BA from Harvard University; PhD in Physics, University of California, Berkeley, 1965; worked at Bell Labs from 1966 to 1976. Since 1976, he has been a professor at Harvard University.

Honors: member of the National Academy of Sciences, 1982; Lars Onsager Prize, 2001; Wolf Prize in Physics, 2002. In 2018, he was awarded the 2019 APS Medal for Exceptional Achievement in Research for his pioneering work on condensed matter theoretical physics, especially topology in classical and quantum systems.

11.13 Lax, the generous

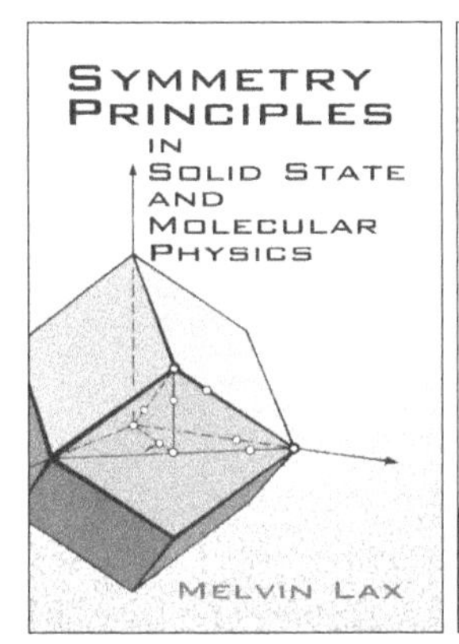

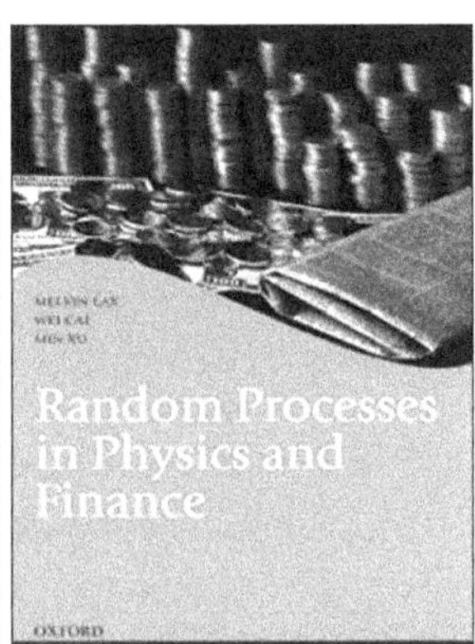

In 1972, after I finished my doctoral thesis, Phil Platzman, my doctoral supervisor at Bell Labs, asked me to go to MIT (Massachusetts Institute of Technology) for a postdoc. I said no since I wanted to stay in New York City. He didn't persuade me or ask why. In fact, the reason was that I wanted to stay in Manhattan's Chinatown and do what I wanted to do [11.11].

Bell Labs' corridors are designed to be narrow, to allow people to greet each other when they pass each other. One day, Platzman and I walked briskly down the hallway from his office and were greeted by Mel Lax, whom I had no previous contact with. Platzman stopped him and said: "You let Lui be your postdoc." Lax said: "Good." That's it, two seconds, and I got a job, which is one of the advantages of going to a prestigious school. It was later I learned that Lax had just been hired to be a Distinguished Professor at City College of the City University of New York, and there were money and vacancies.

Melvin Lax (1922-2002), born in New York City on Women's Day 1922; BS in physics, New York University, 1942; PhD in Physics, MIT, 1947; one of the founders of quantum optics theory; worked at Bell Labs from 1955 to 1971 (Director of the Department of Theoretical Physics from 1962 to 1964). He has been a distinguished professor at the City

College of New York since 1971. He was elected to the National Academy of Sciences in 1983, and was awarded the Willis E. Lamb Medal for Laser Science and Quantum Optics in 1999.

Doing Liquid Crystals

Lax and I did not talk about what to do when he hired me. It did not matter because doctoral graduates from prestigious universities do not do previous projects and can work on new topics. On my first day at work, Lax asked me to do liquid crystals.

Lax is one of the founders of quantum optics theory and has written a series of groundbreaking articles. This is his habit: He studies the areas he focuses on very thoroughly and comprehensively. He has just published a 37-page paper proposing to derive all the effects and formulas of crystal electrodynamics from the Lagrangian function, which is very successful, predicting several new phenomena, and producing several papers with experimental confirmation in the top journal *Physical Review Letters* (PRL). His Lagrangian theory is still the *only* correct and self-consistent theory in this field.

At the time, Paul Martin (1931-2016) of Harvard University published a PRL paper arguing that liquid crystals were crystals (which turned out to be wrong), so Lax thought he (actually me) could do better with his superior Lagrangian theory. The problem is that neither he nor I had ever done liquid crystals, knew nothing about the field, and did not know how to proceed.

So I started looking through a lot of liquid crystal literature, two years before the publication of book *The Physics of Liquid Crystal* (1974) by Pierre-Gilles de Gennes (1932-2007, Nobel Prize 1991), the first book on that topic. I still didn't have a clue.

In the summer of 1973, I went to the molecular fluids class at the Les Houches Summer School in France. De Gennes gave several lectures on (nematic) liquid crystal dynamics based on his book manuscript. Unfortunately, he taught in French, and I did not understand a word. Luckily, several chapters from De Gennes' book, written in English, were distributed before we left the school. Back in New York, I discovered a derivation error in De Gennes' book and wrote to him. He replied with a postcard, saying that someone else had already pointed out the mistake.

I was still at a lost. In the third year, one day, a breakthrough came and the problem was solved. The breakthrough came from an article in a book I stumbled upon in the City College library that used dissipative function to construct irreversible thermodynamics in continuum mechanics. I ended up using Lax's Lagrangian function for the reversible process part and the dissipation function for the irreversible process part. My theory retains the elegance of Lax's theory because everything (including the necessary space-time symmetry and material symmetry) is built into the Lagrangian *and* the dissipative function from the beginning, so that conservation laws and reciprocal relations emerge naturally and consistently from the equations of motion, rather than being hand-added from the outside.

This theory remains the *only* correct and self-consistent theory dealing with the dynamics and irreversible thermodynamics of all viscous materials (simple and complex fluids, liquid crystals, thermoviscous solids). But I was busy in Chinatown at the time and did not have time to publish the results.

I was at City College for three years. Although I did publish five papers based on the doctoral works I finished before, I did not write a single report or co-write an article for Lax. He never expressed impatience, and he kept paying me a thousand dollars a month. Not only that, but he also took care of two important things for me.

China Visit

In the three years after my PhD while working at City College, I lived in Chinatown and mainly did three things: doing mass work through the Chinatown Food Co-op [11.11], assisting the patriotic newspaper *China Daily News* (founded 1940) to resume daily publication, and publicizing New China.

In the summer of 1974, I was invited to visit China for seven weeks. The whole trip was by train, starting from Luohu (罗湖) to Guangzhou; has been to Hangzhou, Xi'an (西安), Yan'an (延安), Beijing, Dalian and other places; had stayed in caves, went up the Great Wall, and went to the beach. There was no tourism at the time. There was a full-trip escort, plus a local escort (not yet called a tour guide). In Hangzhou, there were only a few of us tourists in the huge West Lake, and Longjing (龙井) tea was indeed delicious. In Xi'an, I wanted to go to Northwest University to meet a chemistry professor because his research was relevant to me. As a result, the professor was taken to the Xi'an Mansion where I was staying. We talked for half an hour in an ultra-dark and ultra-large hall. That was the only academic person I met.

In the seven weeks in China, I had not done any physics but my salary was still paid by Lax.

After Postdoc

In the summer of 1975, my post-doctoral period ended. Lax was so good-hearted; he was afraid that I would be jobless and introduced me to a job at Los Alamos National Laboratory, which I declined (because I thought the job required US citizenship). Finally, Platzman, my PhD supervisor, came out again and arranged a job for me in Antwerp, Belgium.

I have extended my appointment date to October 1 so that I can fulfil my responsibilities as a stage manager for Chinatown's celebration of China's national day. Because of the organizational skills I developed at the Chinatown Food Co-op, I held this volunteer position for several years.

In the summer of 1976, I moved from Belgium to the Universitat des Saarlandes in West Germany. The following year, feeling that it was possible to return to China and settle down, I sat down and wrote four papers based on the postdoctoral work I completed alone at City College, each writing for a month. The first three, with me as the sole author, are published in *Zeitschrift für Physik B* in 1977, and one of them was my only liquid crystal paper before I returned to work in China, in January 1978. The fourth paper, by Lam and Lax, was published in *Physics of*

Fluids in 1978. Ten years later, Lax (and Ouyang Zhong-Can, 欧阳钟灿) and I published an article on liquid crystal nonlinear optics in *Physical Review A*. Those are the only two papers co-authored by Lax and me.

After I went to work at the Institute of Physics, Chinese Academy of Sciences, Beijing, Lax and another physics professor, Frank Martino (1937-2020), visited Beijing on behalf of City College. After visiting me at my apartment, at midnight, there was no bus (and no taxi yet then in China). I asked the students to lend the two their bicycles, and the three of us rode back to the Friendship Hotel together, and the memory was still fresh. Since then, Lax has trained several visiting scholars for China (see right figure).

Lax, like Platzman, inherited the scientific tradition pioneered by the ancient Greeks: learning for the sake of learning, adhering to physics, and being faithful to physics. For them, doing physics is not only a job or profession, but also a dialogue with nature and contributing to the progress of physics. Moreover, success does not have to be with me; it could be with the students they trained. They made sure that the science torch invented by the Greeks will be passed down from generation to generation, to keep the science enterprise going and healthy.

Mel Lax, the generous, is meticulous in physics and lenient.

Remarks

1. In prestigious universities, doctoral students and postdocs in theoretical physics are not simple labor forces of the "boss." They will not be assigned to do the work that the boss already knows how to do, because the "research" that already knows how to do is not called research, Feynman said. Cultivating doctoral students and postdoctoral fellows is to cultivate new blood in physics, so that the tree of physics will grow forever, thrive, and there will be successors.

2. How physics is made. There are four persons who influenced me in doing physics:

- Tsung-Dao Lee: Exemplified the non-conformist style of topic selection.
- Pierre Hohenberg: Exemplary style of debate.
- Philip Platzman: Let me learn to link theory to experiments.
- Melvin Lax: Let me learn to do basic theory.

3. Research experience: Keep eyes open, look around, ask why, think more, relax, and breakthroughs will come.

4. Lax's Lagrangian theory contains the interaction of electromagnetic fields with crystals, while the dissipative function theory for viscous materials I created does not contain electromagnetic fields. Extending my theory to include electromagnetic fields by imitating Lax's approach would be a fruitful new field.

11.14 Anderson, the brave

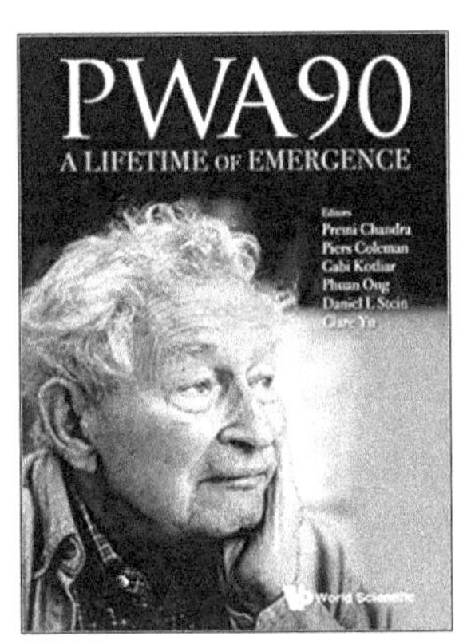

Philip Anderson (1923-2020), after graduating with a PhD from Harvard University in 1949, worked at Bell Labs from 1949 to 1984, during which he spent half a year at Bell and half a year at Cambridge University from 1967 to 1975. In 1984, at the age of 61, he became a professor at Princeton University.

Josephson Effect

After producing the Nobel-Prize work in 1958, Anderson lectured at Cambridge University in 1961-1962. Brian Josephson, a 22-year-old graduate student, listened to him and came up with what is later called the Josephson junction (or effect): Superconductivity can pass through an insulator sandwiched between two pieces of superconductor.

After Anderson returned to Bell Labs, he collaborated with the young experimentalist John Rowell and published a paper: P. W. Anderson, J. M. Rowell [1963]. Probable observation of the Josephson tunnel effect. Phys. Rev. Lett. **10**(6), 230.

The success and speed of this work come from Bell Labs' unique recipe: Theory and experiment are both done at Bell. Josephson won the Nobel Prize in 1973, four years before Anderson's Nobel Prize.

Two Phil's

American life is fast-paced and people's name has to be short. Therefore, anyone named Philip will be called Phil. Bell Labs has a lot of people, many named Philip, but everyone knows: Bell Labs has two Phil's. One Phil was Philip Anderson and the other was Philip Platzman (1935-2012), who was 12 years younger than Anderson. The latter studied under Feynman and was my doctoral supervisor. Anderson worked at Bell Labs from 1949-1984 and Platzman from 1960-2001. Both joined Bell Labs after graduating with a PhD; both stayed for nearly 40 years; and their main research careers were at Bell Labs.

My doctoral dissertation was done independently, and I only saw my PhD supervisor two or three times a year when I went to him on my own initiative [11.11]. At one time on such an occasion, he had something else to do, so he put me in Anderson's office (who wasn't there that day). The office was larger than usual, and in particular, on the wall bookshelves, there was a horizontal row of notebooks of the same size, brown skin, each about 15 x 20 centimeters in size, one centimeter thick. The notebooks lined up one and a half meters wide, which were Anderson's notebooks over the years. It was before the PC or iPad, and notes must be written with pen and paper, which was actually more convenient and safer.

Of course, I didn't peek at the notes; otherwise I might have won a Nobel Prize.

The China Report

It should be the late 1970s. When the Institute of Theoretical Physics (ITP) of the Chinese Academy of Sciences (CAS) was established in 1978, it was housed in a small place borrowed from the Zhongguancun Primary School. Anderson came to visit China and wanted to visit the ITP. The Foreign Affairs Office of ITP refused (because the place is unmentionable). Anderson insisted, saw it, and got to talk to the ITP members.

After returning to the United States, Anderson gave a seminar called "Physics in China." After showing many photographs of landscapes and people from his China visit, he ended it by saying "China has no physics." This was reported to the Institute of Physics (IoP), CAS, by an overseas Chinese physicist who had attended the seminar that day. I was working at the IoP at the time, and I soon learned about it. The report should still be in the library of the IoP. Anderson did not visit the IoP and I did not meet him.

Master Went Down the Mountain

Bell Labs was established by AT&T in 1925, where Masters in condensed-matter physics gathered. There are many innovations, and the results are brilliant. Bell Labs was like a tall mountain where newly minted PhDs wanted to go to work. However, when AT&T was forcibly broken up by the government in 1984, Bell Labs began to decline.

So, when Anderson left for Princeton University in 1984, it was like a Master who went down the mountain because the mountain was destroyed. Condensed-matter physics has not progressed much in the past few decades, and I always feel that it has something to do with the destruction of the Bell mountain.

More Is Different

After physics made obvious contributions in ending World War II, it was sought after by the US government with funding increased drastically every year. It was the golden age of high-energy physics, resulting in some elementary-particle physicists being hot-headed. They believe that particle physics is the most important in physics: After elementary particles are figured out all physical problems will be solved through reductionism. This is pure ignorance—ignorance of high-energy physicists, not the problem with reductionism. To counter this misconception, in 1972, Anderson wrote an article in *Science*, "More is different," which has been passed down to this day. This article is reproduced in full in my 1998 book *Nonlinear Physics for Beginners*.

In fact, we all know what this article says, only some particle physicists do not understand. Saying it through Anderson's writing of course is better and more efficient.

"Many is different" means that the physical properties/characteristics of many-body systems (including condensed matter) will be different from those of single-body or few-body systems. The former will have some unforeseen physical properties at the microscopic or macroscopic level—a phenomenon called *Emergence*. For example, the concept of "flow" of water at the macroscopic level does not exist at the molecular level (water molecules only see collisions between molecules and nothing about flow). The counterpart of reductionism is holism, which is based on the *unprovable* assumption that there are systems that cannot be studied through the properties of components.

Many holism proponents misunderstand reductionism and give it a bad name. Reductionism is to understand the whole of the upper layer of description through the properties of the components *and* their interactions. For example, water molecules + molecular interactions → water. That is, $1 + 1 + x > 2$, not $1 + 1 > 2$ as claimed by some holism people.

More and Different

In 1993, Anderson commented on James Gleick's book *Genius* in the journal *Science*, saying that Feynman "was probably the only man since Einstein who can be called a genius." This review and other similar articles are available in Anderson's book *More and Different* (2011). The book covers personal and Bell Labs history, philosophy and sociology, science strategy, science wars, politics and science, futurology, complexity, and more, with a cover illustration from daughter Susan.

Anderson, like Feynman, spoke bluntly and profoundly. All those involved in science (especially science history, science philosophy, and popular science people) should buy a copy of *More and Different* and read it.

One of the articles in the book is “Is it possible for modern America to invent quantum mechanics?”, imagining that Heisenberg, Schrödinger, and Dirac were living in the United States. The conclusion is that quantum mechanics will not be made, for surprisingly the same reasons of problems they would encounter if they were in China. The article is funny and worth reading.

A Daring Innovator

Anderson’s theories are always accompanied by experiments, both before and after experiments. He is good at developing new theories, including localization of disordered systems (which won the Nobel Prize), the Hubbard model with mutual attraction, the spin glass model, the Anderson model for magnetic impurities, and the pairing model for high-temperature superconductors.

He has always been a leader. His new theories continue to challenge the knowledge that all beings possess, and provide new research opportunities for others.

Phil Anderson, the brave, dares to charge forward, dares to innovate.

11.15 China's second-generation returnees

"Overseas returnees" (called *Haigui*, 海归) refers to those Chinese who have settled (or are eligible to settle) overseas since 1949 and have voluntarily left and opted to return to new China for settlement (excluding those sent out officially). There are three generations.

Three Generations of Returnees

The first generation returned to China around 1950, mainly from Europe and the United States, with thousands of people (Qian Xue-Sen 钱学森, Shi Ru-Wei 施汝为, Xie Yu-Zhang 谢毓章, Li Yin-Yuan 李荫远...). The second generation returned to China in the late 1970s and early 1980s, mainly Hong Kong and Taiwan scholars (Cai Shi-Dong 蔡诗东; Xie Ying-Ying 谢莹莹; Lin Lei 林磊, aka Lui Lam...). The third generation was in the 1990s and later, when China's economic conditions were better, much better (Shi Yi-Gong 施一公, Rao Yi 饶毅, Yan Ning 颜宁, Chen-Ning Yang 杨振宁...).

The three generations of returnees have all returned to China to participate in building up the country. But the first two generations are assigned work by the state and have no personal career considerations when they returned. Contrarily, the third generation was enjoying a market economy and has career considerations.

The conditions of China faced by the three generations are very different. The first generation faced a poor China, *before* various political movements emerged. The second generation is also facing a poor China, but *after* various political movements have taken place and could resume. The third generation faces an era when the memory of political movements has *faded* in China, which is beginning to prosper or has prospered.

Thus, the second generation needs great courage to return to China. I returned to China in 1978, belonging to the second generation of overseas returnees, and was the first person to return to work at the Chinese Academy of Sciences (CAS) after 1976.

The Baodiao Movement

The second generation of returnees have experienced the Baodiao (保钓) movement; Baodiao is short for Protecting Diaoyutai (钓鱼台, called Diaoyudao in mainland China). The movement began in December 1970 with a mass meeting at the basement of Columbia University's Teachers College. In January 1971, demonstrations were held in New York and other cities, and in March there were 3,000 demonstrators from all over the United States gathering in Washington, DC. The movement quickly swept the world of overseas Chinese, from university professors to shopkeepers.

In September of the same year, the all-US political congress to discuss Chinese politics (全美国是大会) was held in the campus of the University of Michigan, Ann Arbor. The leftists recognized Beijing as the sole legitimate government of China and turned Baodiao into a unify-China movement; the rightists withdrew on site. Baodiao split. Yet, Baodiao has continued in Hong Kong and Taiwan. In Europe, Chinese students stayed in the unify-China movement and rallied around the magazine *European Communications* (欧洲通讯). I worked in Belgium and West Germany for two years (1975-1977) and was one of them. So far only one person in the Baodiao movement has died: a Hong Kong man who went to Diaoyu Islands by boat to declare China's sovereignty, fell into the water, and lost his life.

In September 1971, we demonstrated in front of the United Nations (UN) headquarters in New York in support of China's admission to the United Nations. In October of the same year, China successfully joined the UN, and a large number of people with good Chinese and English skills were urgently recruited to join the UN as translators. Others went three different ways:

1. A small group of people were eager to return to China and participate in the rejuvenation of the motherland. (Some even went through Hong Kong to the border of Shenzhen and were turned back.)
2. A few went to Hong Kong in the late 1980s to establish the Hong Kong University of Science and Technology (Chia-Wei Woo 吴家玮 , Qian Zhi-Rong 钱致榕, Yu Zhen-Zhu 余珍珠...).
3. Most people stay in their spots, serve the mother country with professional knowledge, and promote people-to-people exchanges between China and the United States.

In the first few years of 1978, due to the lack of conditions in China (no new houses), except for a few people (Chen Ruo-Xi 陈若曦 and her husband…), no one was allowed to return to China. I applied in 1974; waited three years, took the letter of introduction written by Chih-Kung Jen (任之恭, aka Ren Zhi-Gong) to Qian San-Qiang (钱三强) and returned to China in 1977 to join the National Day celebration, and was told by Qian that my application had been approved. Finally, in January 1978, I showed up at the Institute of Physics (IoP) of CAS to work (in fact, participate in the revolution, in my opinion).

Second-generation Talents

Most of the second-generation returnees were about 30 years old, more in science than the humanities. In 1981, we wanted to set up a Taiwan and Hong Kong Alumni Association, but the authority disagreed. It was changed to Taiwan Alumni Association (台湾同学会), and everyone joined in, including the Hong Kong and Macau alumnius. According to the statistics of the 2001 member directory of the alumni association, there are less than 100 people, including the following science members:

- Deputy **Director**: Liang Xiao-Guang 梁晓光 (Institute of Photochemistry, CAS).

- **Academician**: Cai Shi-Dong (IoP, CAS), Li Chun-Xuan 李椿萱 (Beihang University), Li Jia-Ming 李家明 (IoP/Tsinghua/Shanghai Jiaotong University), Zhao Yu-Fen 赵玉芬 (Institute of Chemistry, CAS/Tsinghua/Xiamen University), Su Ji-Lan 苏纪兰 (Institute of Oceanology No. 2).
- **Institute of Physics**, CAS: Cai Shi-Dong, Lin Lei, Huang Zhou-Mou 黄周谋, Zhang Zhao-Qing 张昭庆, Feng Guo-Guang 冯国光, Yang Si-Ze 杨思泽, Fu Pan-Ming 傅盘铭, Hou Mei-Ying 厚美英.
- **Institute of Systems Science**, CAS: Wang Shan-Shan 王珊珊.

Discussion

The first generation of returnees did the foundation work (creating research institutes and cultivating talents). By 1985, they gradually retreated from the science-leadership stage, and the successors of each institute had a big gap in education from their predecessors. For example, the directors of the Institute of Physics: American PhD Shi Ru-Wei (1901-1983) handed over to Soviet PhD Guan Wei-Yan 管惟炎 (1928-2003) in 1983, who went to the University of Science and Technology of China in 1985 to become the president; the succeeding director was Yang Guo-Zhen 杨国桢, who had a (equivalent) MS degree from Peking University.

The second generation could have filled the generation lost in China's ten years of Cultural Revolution and played a role in inheriting the future, but the effect was not great. There are two reasons:

1. Not enough people came back (100 was much less than a few thousands in the first generation, and a batch left in the late 1980s).
2. The people who came back were too young.

Imagine that if Chen-Ning Yang, the Nobel laureate in physics, had not been a third-generation returnee but a second-generation, and had returned in 1978 at the age of 56, he might have stopped the practice of counting papers that began in China in 1986, resulting in no major innovations in science for more than 30 years. And China's present situation of being constrained everywhere might not have appeared.

11.16 Bump, switch, cross, jump

Most physicists do physics all their life, very loyal, like never having an affair or divorce in a marriage. But there are exceptions, also like in marriage: some physicists occasionally *bump* into another discipline, *switch* fields, *cross* two disciplines back and forth, or *jump* to another discipline. Jumping is a high-risk action; jumping outrageously could have serious consequences. Examples are as follows:

- **Bump** into biology: Erwin Schrödinger (1887-1961), Nobel Prize in Physics in 1933, published *What Is Life?* in 1944.
- **Switch** to biology: Max Delbrück (1906-1981), 1969 Nobel Prize in Physiology or Medicine.
- **Cross** to biology: Xiaowei Zhuang (庄小威), MacArthur (Genius) Award in 2003, Max Delbrück Award in 2010.
- **Switch** to humanities: Lui Lam (林磊), created Histophysics in 2002, founded Scimat in 2007; did research in history, art, and philosophy.
- **Jump** to medical practice: He Jian-Kui (贺建奎), no medical training, gene-edited baby in 2018, sentenced to three years in prison.

Additionally, there are those who either transfer into or out of physics after a bachelor's degree. Examples:

- Transfer **into**: John Bardeen (1908-1991), BS in Electrical Engineering, PhD in Physics, twice Nobel Prize in Physics—1956 and 1972.

- Transfer **out**: Elon Musk, BS in physics, founded Zip2 in 1995, founded X.com in 1999 (became part of PayPal), founded SpaceX in 2002, joined and led Tesla in 2004.

Overall, physics is a good basic training, suitable for all future careers.

All creative professions, including art, have their own stories of bump, switch, cross, and jump.

11.17 Forced to be humble

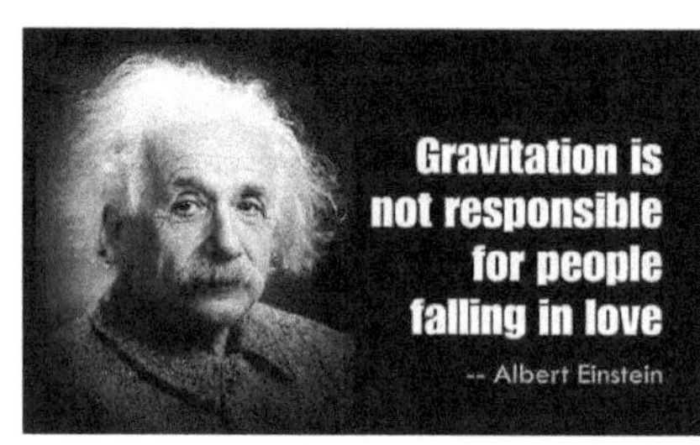

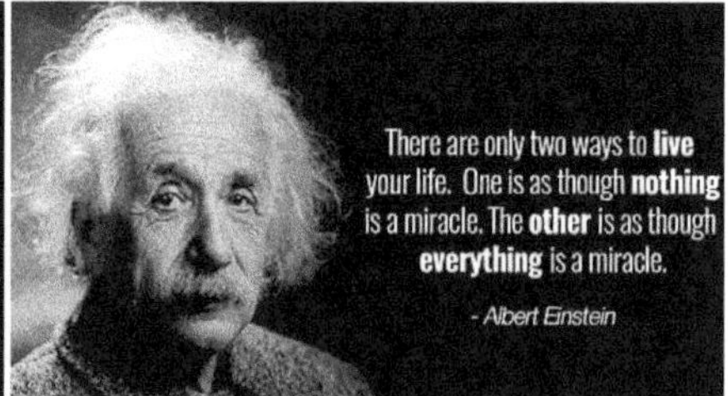

Nobel Prize in Physics laureate Lev Landau (1908-1968) rated the competency of physics peers: Einstein 1 (top), others 2 or below. Just because there is a super competent Einstein in the physics world, other physicists are forced to be humble. Only some mainland China's graduate students and folk scientists dare to boast openly that they are better than Einstein.

Einstein is not insurmountable, but only when one really surpasses him that one should come out and say it. Before that, one could only tell it to girlfriend/boyfriend in private. Tell dad that and he will reply: You're crazy! Tell mom and she will reply: Finish washing the dishes first!

In other areas, because there are no super competent people, there is no forced humility. For example: Alexander the Great or Genghis Khan fought prominently but no warrior would be forced to be humble; Monkey King in *Journey to the West* can go to heaven and earth at will and no monster is forced to be humble.

World peace is difficult because no one is forced to be humble, except in the physics world.

11.18 For the benefit of humanity

In 1952, five young scientists visited Albert Einstein. Einstein told them that *Sci-tech* (science and technology [10.6]) can only benefit humankind if it is combined with human values.

Sci-tech is *neutral*: Sci-tech itself does not benefit or harm humanity. The application of science and technology requires decision makers. It is the *decision makers* who can benefit or harm humanity. For example: Technology product kitchen knives, can cut vegetables or can kill, depending on who uses it.

12

Popsci Book

Popsci: bright and dark

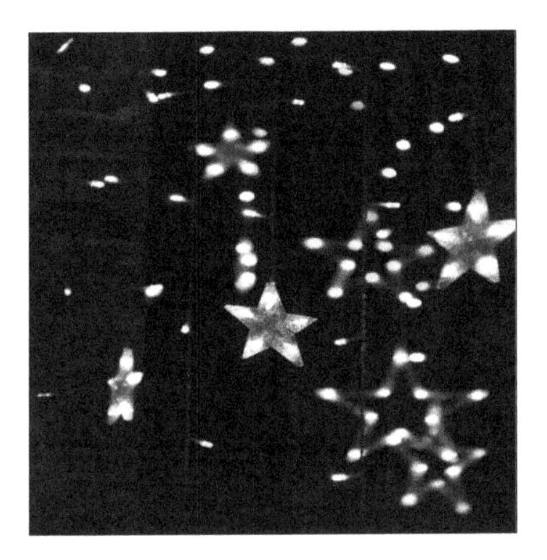

Popsci is short for popular science. *Popsci* has been proposed in 2008 to be the formal name of the discipline called science popularization or science communication (see my book *Science Matters*).

A successful stage performance requires excellent work on stage and behind the scenes. Before, there have to be first-class production work: script writing and editing, directing, rehearsals, set and prop production, etc. The performance on stage is visible and can be seen clearly by the audience—the "bright" part. Behind the scenes and the production work are invisible to the audience—the "dark" part.

Similarly, popsci works consist of two parts: bright and dark. The bright part is the popsci results seen by the public, which may appear in books, newspapers or platform articles, performances, movies, or other media. The dark part is the behind-the-scenes pre-works of the bright part. Without good *dark* popsci, there will be no good bright popsci.

In fact, everything is dark before bright, and the dark part is the gestation period. For example, a baby has to live ten months in the mother's womb before birth—it is a dark life, and only after birth it becomes a bright life.

12.2 Dark more important than bright

The crucial person who determines whether a movie is good or bad is the director (according to the French New Wave filmmakers). The actors can be replaced, but if the director is replaced, it is not the same movie.

Crouching Tiger, Hidden Dragon (2000) directed by Ang Lee (李安) is about the world of martial arts, bright (open) love, and dark (secret) love. In the sequel *Crouching Tiger, Hidden Dragon: Sword of Destiny* (2016), the actress is still Michelle Yeoh (杨紫琼; Oscar best actress, 2023) but the director changed to Woo-Ping Yuan (袁和平) and the movie becomes a purely martial-arts film—not the same thing.

Similarly, it is the professional level and taste of the *editor-in-chief*—the major figure of the dark part—who determines the quality of a popsci platform—the bright part, not the individual authors. Good authors are easy to find but good editors-in-chief are hard to find, unless the platform is just a small playground for the editor-in-chief.

In contrast, popsci books involve only one or two authors. It is the author's skills and taste that determines the quality of the book.

12.3 Ovshinsky: folk scientist successful

A "folk scientist" is a scientist without professional training. Folk scientists can be divided into two categories: *unreliable* (such as claiming to have invented perpetual motion machines or turned water into gasoline) and *reliable* (discovering new stars in the sky).

The most prominent among the latter is Stanford Ovshinsky (1922-2012), who is less known among the public. Ovshinsky, an American, graduated from high school only, with more than 400 patents in his name and known as the "contemporary Edison." In 1968, he published an article on disordered materials in the prestigious journal *Physical Review Letters*. (That year, Columbia University students in New York made a revolution, occupied the president's office; the whole school was suspended, and the departments held a discussion on reform. I was a doctoral student there and still remember this article.) There were not many people who knew the significance of that article but Nobel Prize laureate Nevill Mott (1906-1996) praised it on the front page of *The New York Times*. Ovshinsky's inventions include lithium batteries, rewritable CD/DVD, hydrogen batteries, and phase conversion memory. Awesome, right?

So, should the government allocate public funds to support folk scientists? No! Because the success rate is too low, but the wealthy with private funds can.

12.4 Science myths

Myth refers to the misunderstanding through false information or cognition. Many are provided by historians, philosophers, and communicators of science who are not critical in thinking and careful in doing research.

There are two books published by Harvard University on debunking the myths of science.

The first book is Ronald Numbers' *Galileo Goes to Jail and Other Myths about Science and Religion* (2009). In this book, Myth 27 mentions that US courts twice excluded creationism from science classes in public schools. Yet, in my opinion, although the religious community lost, the opinion offered by the scientific witnesses in court about what science is wrong. And both the witnesses and the judge did not know where it is wrong.

The correct answer would be, following the 1867 or 2007 definition of the word Science: Science is the study of nature without introducing God/supernatural considerations. Accordingly, creationism can be discussed in other classes (e.g., philosophy) but not in science classes [8.1]. It is that simple.

The second book is *Newton's Apple and Other Myths about Science* by Ronald Numbers and Kostas Kampouradis (2015). Myth 27 points out that Karl Popper's falsification criterion of science, accepted by many physicists, is fundamentally wrong and pseudoscience cannot be demarcated (see also [9.7, 10.2]). Myth 26 states that there is no such thing as the scientific method.

It is recommended that all those who care about science (especially popsci people) read these two books as soon as possible. Both books are not thick or expensive, and there are electronic versions. The book *Newton's Apple* has been translated into Chinese in 2017.

12.5 Folk medicine witch

A witch is often understood as a person who makes a pact with the devil in exchange for the magic of performing evil deeds. However, many so-called witches were simply folk (natural) healers or so-called "wise women" whose choice of profession was misunderstood.

Fear and persecution of witches took root in Europe in the mid-1400s. In the 160 years between 1500 and 1660, as many as 80,000 suspected witches were executed in Europe, about 80% of whom were women.

In recent years, many women attempt to "re-appropriate the label of witch as a powerful signifier of feminine power rather than as a dangerous insult."

Witches, Midwives & Nurses was first published in 1973. The second edition was published in 2010 to cover today's healthcare crisis. The medical crisis described in the book is a short but informative history of the problems that have plagued medical institutions with roots that can be traced all the way back to the witch hunt. According to a review, this book is highly inspiring; it will change how you think about modern medicine, and the female healers who have been eliminated by modern medicine.

12.6 Weird things

My first popsci presentation was given in Mexico City in 1994, at the Universidad Nacional Autónoma de México, entitled “Nonlinear physics is for everybody!” (left figure). Five years later, in December 1999, I started a public science lecture series “God, Science, Scientist” at San Jose State University, where I teach. The word God is to attract the audience to come.

The first speaker I approached was Charles Townes (1915-2015), the Nobel laureate who invented the laser and was a devout Christian who often talked about science and religion. I did not know him personally, but a phone call to him and he said yes. Two days later, he said that he had checked his work calendar and could not come because of time conflict. (A year later, he became the third speaker in my series.)

In a hurry, I contacted Michael Shermer, founder and editor-in-chief of *Skeptic* magazine, who publishes, lectures, and writes books for a living. I did not know him personally either. After the call was connected, he said he wanted me to pay $2,000. I said, “No. I will give you two choices: One is to turn me down, and the other is that you promise to accept $500 (including air tickets).” I added: “If you agree, you will be my friend, and I will invite you to Beijing." Shermer was smart. He had never been to China and he picked the latter right away.

In November 2000, the first international popsci conference in China, sponsored by the China Association for Science and Technology, was held in Beijing. The organizers are Li Da-Guang (李大光) of the China Research Institute of Science Popularization and Yang Xu-Jie (杨虚杰) of the *Sciencetimes*, a daily published by the Chinese Academy of Sciences. As a member of the International Advisory Board, I invited Shermer and three other California popsci experts to the conference. In the middle of the meeting, Li Lan-Qing (李岚清), member of the Politburo Standing Committee, received the main speakers of the conference, Shermer and I included, which was my first visit to Zhongnanhai.

In Beijing, Shermer went up the Great Wall, drank coffee at Starbucks in the Forbidden City, rode tricycle in the narrow alley—hutong (胡同), and went to the Peking Man Site Museum in Zhoukoudian (周口店) but found it closed. At that time, he had just started as a monthly columnist for the *Scientific American* magazine, and so after returning to the United States, he wrote two articles on his trip to Beijing.

In order to improve the scientific literacy and anti-cult ability of the Chinese people, in 2001, I had a book by Shermer (middle figure) and five books by the well-known magician James Randi (1928-2020, winner of the MacArthur "genius" award) translated into Chinese and published, respectively, by the Hunan Education Press and the Hainan Press. Neither of them made a fortune from this. I know. I was their Chinese copyright agent.

The translator of Shermer's book *Why People Believe Weird Things* is Lu Ming-Jun (卢明君; right figure), who holds a master's degree in English and American literature from the School of Foreign Languages of Nanjing University (later a PhD in English literature from the University of Toronto). I was the proofreader of this Chinese book. Working with Lu in proofreading was a pleasant experience.

12.7 Popsci wonderland

Popsci is a wonderland that allows you to meet and gain new friends from all walks of life. It might even take you to Paris. Here is how it happened to me.

Seattle

In 2000, Michael Shermer invited me to write a popsci article in the quarterly journal *Skeptic*, which he founded and edited. My popsci article "How nature self-organizes: Active walks in complex systems" appears in Volume 8, Issue 3, 2000 [16]. Near the end there is a section entitled "Modeling History." Sesh Velamoor, vice president of the Foundation For the Future in Bellevue, Washington, saw this article and invited me to the Foundation's annual workshop "Humanities 3000: Symposium No. 3" as one of five keynote speakers. ("Humanities 3000" refers to predictions of the human situation in 3000 AD.) The symposium was held in Seattle, August 12-14, 2001. There were only 23 invited "participants" (formal title, plus a few "observers"). In addition to me, there was Edward O. Wilson of Harvard University (author of *Consilience*) and Richard Dawkins of Oxford University (author of *The Selfish Gene*), but I was the only physicist and the only Chinese among the participants. I was on the same four-person discussion group as Dawkins and Shermer.

At the symposium there were experts and celebrities in various fields. The Q&A session after each talk was full of fierce debates. Unlike other conferences, the entire meeting was videotaped, and the proceedings (published in 2003) put each sentence uttered into words.

Paris

Doug Vakoch of the Search for Extraterrestrial Intelligence Institute (SETI) in Bay Area, California, was among the observers in the Seattle symposium. He appreciated the sharpness of my debates with the other speakers and invited me to speak at the workshop on how to communicate with aliens he was organizing for the next year, in Paris.

And so, in March 2002, I found myself in Paris in this small workshop, all by invitation, devoted to discussing what science-and-art messages should be sent to aliens. My presentation is "A science-and-art interstellar message: The self-similar Sierpinski gasket." Vakoch liked it so much that he included my paper in his conference report, published in the MIT journal *Leonardo* [20]. After this meeting, I had one more identity—an alien expert.

Tamsui

Clement Chang (张建邦, 1929-2018), president of Tamkang University from 1964-1986, was another observer in the Seattle workshop. Chang considered me a futurology "expert" after hearing my talk and invited me to give the Tamkang Chair Lectures at Tamsui. Tamkang University is a private university and, among other things, to distinguish it from public universities, it has established a Graduate Institute of Futures Studies.

As a result, on December 9-11, 2003, as an "internationally renowned physicist and futurist," I gave three lectures in Tamsui. The lectures are published as *This Pale Blue Dot* (2004), which was my *first* popsci book [20].

12.8 Martial arts and physics

Wuxia (武侠) literally means martial arts-cavalry. The wuxia world is an invention of the novelists. It usually has a small numbers of super martial arts masters and a few lesser ones. It often involves a martial art manual that is craved by everybody since whoever gets it could follow the book's instructions and become a super master. And a lot of fighting ensure. In this imagined world, honoring one's promise is most important and helping the weak is highly praised. Extreme love stories are common in wuxia novels, too. The wuxia world is exemplified splendidly in two superb movies: *Crouching Tiger, Hidden Dragon* (2000) directed by Ang Lee (李安) and *The Assassin* (2015) directed by Hsiao-Hsien Hou (侯孝贤) .

Accidentally or not, the wuxia world mimics closely the academic world, especially the world of doing physics. There are many parallels:

- Learn from masters → become a graduate student
- Original swordsmanship → theoretical/experimental innovation
- The *Nine Yin Manual* (九阴真经) → Theory of Everything (desired by everyone)
- Master competition → seminar debate

- Hua Mountain sword fight (华山论剑) → international top conference (such as the Solvay Conference of quantum mechanics)
- *Xianu* (侠女) → female scientist with personality
- *Hongyan* (红颜) → female close colleague

So, physicists have a unique experience when reading martial arts novels, thinking that it is the physics world (inherited from ancient Greeks) they write about.

Master-apprentice love exist in both worlds (e.g., Yang Guo 杨过 is in love with his mentor Xiaolongnü 小龙女 in the novel *Divine Condor Heroes* 神雕侠侣). But sexual assault of female (or male) students in academia does not happen in the wuxia world, at least not in wuxia novels.

In the academic world, mentors and students faking papers together happened. There is nothing like this in the wuxia world because fake kung fu will get themselves killed, unlike those who fake papers could be rewarded and promoted in the real world. After all, the two worlds are a little different: The wuxia world is purer, with clearer moral standards, and more fascinating than the academic world.

Therefore, to popularize the fascinating world of science, increase the public's understanding and support for science, or attract more young people to do science, one can start with martial arts novels, and give examples with contrasts in the two worlds: the wuxia world and the academic worlds. It would be absolutely fascinating.

Of course, popsci should be authentic. Don't forget to introduce the bloody acts and tears of academia such as deliberately ignoring the contributions of collaborators and waiting for the Nobel Prize every year (when the Swedish phone did not come the heart dripped blood). Of course, there are also and more examples like these in the wuxia world and they are equally captivating.

12.9 Science of martial arts

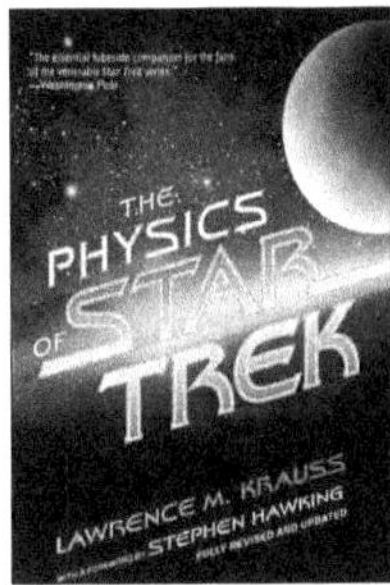

Outside China, relying on popular movies to do popsci books is a proven, effective and money-making successful approach. The authors vary from physics professors, popsci writers to illustrators.

In contrast, although we have many hot-selling wuxia novels we have not seen popsci books that write on the science of martial arts. Isn't it a pity?

For example, one can talk about physics and the mechanics of human body by first mentioning walk-fast on water surface, rooftops and walls described in the wuxia novels. Similarly, one can introduce brain science by discussing how probable one can use one hand to fight the other hand, a skill told in the novels; physiology according to the *Nine Yin Manual* (九阴真经); and emotional science according to Xiaolongnü's (小龙女) concept of love in *Divine Condor Heroes* (神雕侠侣) .

The popsci books on the science of martial arts would be lively and fun to read. Start early and be the first one to write it.

12.10 Science appreciation

Image of liquid crystal sandwiched between cross polarizers from an optical microscope.

Rational thinking and details are enjoyable for only a small number of the public. Popsci should turn their attention to science appreciation instead of disseminating scientific knowledge, borrowing the successful tactic from the art profession. This is especially true in the age of the Internet/ChatGPT when information is readily available.

Art Appreciation

Art popularization is mainly based on art *appreciation*. Of course, there are art classes to teach children or adults to draw, make pottery, etc. But the purpose is only to arouse their interest in art and improve the public's appreciation of art; it is not to train future artists. Moreover, occasionally, there will be public lectures and videos on the techniques of producing artworks but they will not be too detailed.

On the contrary, few people specialize in *science* appreciation. Popsci always wants to instill scientific principles and rational thinking into the public. Why?

The reasons come from two misunderstandings by the popsci workers:

1. People will enjoy rational thinking.
2. People naturally will appreciate science.

Two Misconceptions

Explanations are in order. 1. **Pleasure**. A person's pleasure comes from the release of "happy hormone" (see Remarks) in the brain. Rational thinking, like writing software with logic, works and is useful but does not bring pleasure, except for the moment when breakthrough ideas pop up suddenly—the Eureka moment. However, *Eureka* makes you happy because the frustration you have held back for too long suddenly disappears, not the rational thinking or any thinking before it.

2. **Science**. Not all science can be and is appreciated. The public's appreciation of science, apart from appreciating that scientists can actually discover so many mysteries of nature, comes from first-hand experience of scientific results (such as refrigerators or air conditioners). Only a small number of people can appreciate the mysteries of scientific principles per se. Furthermore, scientific achievements often have both positive and negative sides in applications. If one popularizes only the positive side, the popularizer will lose credibility when the public encounter the negative side.

For example, in addition to being good for businesses, why do ordinary people change their mobile phones every year or two? When PCs or mobile phones came out a few decades ago, the selling point was that productivity would increase and people would have more time to go fishing or visit art museums. The result? The result is that the boss can find you 24 hours a day.

Conclusion

In addition to arousing everyone's interest in scientific knowledge, popsci for the public can learn from the strategy of art popularization and engage in more science appreciation.

There is a fundamental difference between science and art. Except for a few public art, art can only be seen in art museums or galleries while scientific achievements are everywhere (from glass windows to air conditioners to mobile phones). That is why science museums are more about old science than contemporary science. Accordingly,

1. One way to popularize science appreciation is to engage the public in "bitter recalls." That is, by comparing the situation before and after the emergence of a certain scientific discovery or technology, the public can recognize the meaning and contribution of science and will end up as science supporters.
2. The second way is to engage in more art exhibitions with scientific themes (see figure).

Remarks

1. Dopamine, oxytocin, endorphins, serotonin are well-known "happy hormones" that promote positive feelings of pleasure, happiness and even love. Hormones and neurotransmitters are involved in many essential processes, such as heart rate and digestion, but also in mood and sensation.

Dopamine: pleasure; plays a motivating role in the brain's reward system. *Oxytocin*: bond, love, trust. *Endorphins*: pain relief, runner's excitement, relaxation. *Serotonin*: mood stabilizer, well-being, sense of well-being. Note: Up to 90% of serotonin in the body is produced by the intestines, so there is no reason to cleanse the intestines. Also: the release of happy hormone can come from acts of kindness (helping the elderly cross the street, volunteering), buying a new dress, or winning the lottery.

2. Feynman did not say much wrong. But when he said that scientists appreciate the beauty of a flower more than artists because scientists know more the biological details of flowers, he was *wrong*. Think about it: When one enjoys the sunset but thinks about the sun's photons scattered through the atmosphere to your eyes, through the eyes to stimulate the electrical waves in the brain, and then divide into two ways in the brain... that will definitely reduce your enjoyment of watching the sunset. The *basic* reason is that the brain's perceptual cognition and rational cognition are governed by different parts of the brain, and when both parts are excited the sensitivity of either part is reduced.

That is why classical music uses only instruments without singing and ballet with dancing without speaking, which is to deliberately turn off certain parts of your brain in order the heighten the parts left behind. This is also the reason why multimedia art is not necessarily "the more media, the better." Some people with partial brain damage suddenly find that they have more skills like foreign language skills, and the principle is the same.

3. Many popsci people have two blind spots. (1) The constant fear of not being accurate enough when disseminating scientific knowledge. In fact, not all scientific knowledge can be accurately communicated to the public. For example, velocity or acceleration is a vector quantity and cannot be defined without calculus; the usual statement that "velocity is how far per second" is wrong—it is just the average speed, not the velocity.

Another example, to explain quantum mechanics to the public, we can only use analogy, i.e., comparing the microscopic world described by quantum mechanics (QM) with the macroscopic world that humans can experience. This is impossible to succeed because the microscopic world is different (scientists don't know why). It will only make QM more mysterious; e.g., the so-called "particle-wave duality" is absurd since particle and wave are two phenomena/concepts existing only in the macroscopic world, not the microscopic world.

Popsci is not impossible, but its purpose should only be to arouse the public's *interest* in science and direct them to read a thin textbook of QM if they really want to learn more, or, better, encourage them to take a physics class. Otherwise, if the public end up with more respect for physics or physicists that would be enough.

(2) The more details, the better. In general, people actually hate details. When we hear the announcement of the Nobel Prize, we just want to know who took it, what affiliation, and why. If one wants to know more one will go to the Nobel Prize official website or professional society to read. Today, when almost any detail can be found on the Internet or ChatGPT popsci should pay more attention to science appreciation.

13
Personal

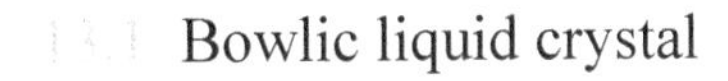

Bowlic liquid crystal

There is a scene in the movie *Red Cliff* (2008): Zhou Yu's beautiful wife Chi-Ling Lin (林志玲) assists a horse giving birth, and as soon as the pony comes out of the womb, it can stand up and walk a few steps by itself, remember? Similarly, upon graduation with a PhD in physics from prestigious colleges like the Columbia University, the student can stand up on her own like this pony. Not only that, but also like the Jen (玉娇龙) of *Crouching Tiger, Hidden Dragon* (2000), she is able to go out and fight in the real world immediately and stand her own grounds alone.

In other words, after graduation, do not continue to do the thesis topic— do not "gnaw (PhD) supervisor." There are two reasons for this:

1. The thesis topic is usually the project of the doctoral supervisor, not your own. Continue doing it is like help pushing the sedan for the supervisor, and a young researcher who aspires to be a master does not do this for anybody.
2. Doing something else early and developing your own experience can help innovation, because many innovations succeed by borrowing something from different fields.

New York

In 1972, after I finished my doctoral thesis, my Bell Labs' thesis advisor, Philip Platzman (1935-2012), asked me to go to MIT as a postdoc. I said no. I told him I wanted to stay in New York City (because I was living in Manhattan's Chinatown at the time and was doing community work). He immediately arranged for me to go to City College of the City University of New York to work with Melvin Lax (1922-2002). Lax, after misled by a wrong Harvard article, asked me to do liquid crystals (LCs). The problem is, neither he nor I have ever done LC and neither we know how to do it.

I worked with Lax for three years and in the last year solved the problem alone. But it was not until 1977 in West Germany that I sit down and wrote four papers on this work done at City College, one of which was the only LC paper I had published before returning to work in China, in January 1978 [11.13].

Beijing

In the fall of 1978, Chia-Wei Woo (吴家玮, later the founding president of the Hong Kong University of Science and Technology) came to visit for few months at the Institute of Physics (IoP), Chinese Academy of Sciences (CAS). In order to do LC theory with him, the Institute asked me, the only LC expert there, to train a few colleagues. That was how my six years of LC research at the IoP began.

Bangalore

Liquid crystals were discovered by the Austrian in 1888 (the 14th year of Emperor Guangxu of the Qing dynasty) and the molecules are rod-shaped (called *rodics*). Later, the Indians synthesized (hexagonal) disc-shaped LCs (called *discotics*) in 1977, which is the second type of LCs. At the end of 1979, I was invited to the Raman Research Institute in Bangalore, India, to attend a LC conference (accompanied by one person from the IoP and another one from Peking University), which was the first time that mainland scholars participated in an international LC conference. Needless to say, the disc-shaped liquid crystal was the protagonist of this meeting. It was also the first LC conference I attended, and I did not know anyone there. At six o'clock in the evening of December 4, I sat alone in a hall waiting to watch the Indian dance performance arranged by the conference. Bored, looking around, accidentally looking up, I saw the decoration on the ceiling consisted of hexagonal boxes strung together. With a wit, I asked myself: Can't these three-dimensional hexagonal boxes also become LCs? This is the source of the bowl-shaped LCs—which I called *Bowlics*—I later proposed in 1982 [A2].

Beijing

After returning to the IoP, I went to Tsinghua University to find a chemist and asked him to synthesize the bowlics—China's own invention. He turned me down; he wanted to make discotics because it was popular. Later, in 1982, I wrote a review of LCs in the journal *Wuli* (Physics) and formally proposed the bowlics at the end of the paper. I then translated important parts of the article into English and sent it to major foreign colleagues. The following year, I mentioned the bowlics in an English paper about the development of liquid crystals in China. In 1985, two European groups successfully synthesized the bowlics, and Lin Lei (my name in pinyin) was recognized as the inventor of bowlics.

At present, the younger generation does not know the research conditions of 40 years ago, so I would like to add here. In the first 30 years of New China, following the Soviet model, scientific research was dominated by the Academy of Sciences while universities focused on teaching. In 1978, as far as basic research was concerned, the working conditions of the IoP of the CAS should be the best in the country. But the lasers of the whole institute were unstable and domestically produced, and the only photocopy machine was also domestically produced, too, which broke down every day. We could not buy white paper, so I used translucent Chinese manuscript paper for calculations and writing; the paper's front is printed with light green squares and I write on the back. No air conditioning in summer. In winter, everyone sits in the cold library wearing thick cotton clothes and pants, reading or hand-copying copies of slightly outdated foreign periodicals (because only two original copies of each periodical were imported for the whole country).

After eating breakfast in the morning, I was hungry by 11 o'clock (because of insufficient calories). Lunch time, I took my own iron bowl to the canteen to eat—no tofu, no lean meat. After lunch, a nap (covered with the quilt I bought to the workplace) was important; otherwise, not enough energy to work in the afternoon.

Conclusion

Under such poor research conditions, we had still made innovative work that led the world. The key is to pick the right topic to study. Do not follow the crowd; do not follow the trend; do not publish papers for the sake of publishing papers. A good chef can cook good dishes according to the ingredients in hand. See the opportunity to act, don't blame the ingredients for being incomplete or inferior, and never sit and wait.

A sophisticated instrument just allows you to do some experiments that require that instrument, but it does not mean that you will produce good works. When there are good instruments and air conditioning it is not surprising that good works are produced, but when innovative works are still absent there must be other reasons. Also, good works should be published first in domestic journals (as is the case of bowlics).

In general, it is the management system that goes wrong, and the consequences are self-tossing and mutual suffering. Can change! Change quickly!

Remarks

I further predicted bowlic polymers in 1987, which was first synthesized by a Chinese doctoral student studying in the United States in 1999. The term Bowlic (or bowl-shaped liquid crystal) created by me has been officially recognized by IUPAC and has been included in the *Handbook of Liquid Crystals*. The editorial of the 2017 No. 4 issue of *Liquid Crystals Today*, the official journal of the International Liquid Crystal Society, introduced the bowlics. The story of the bowlics is a story of successful innovation in China [A2].

Bowlic room-temperature superconductor

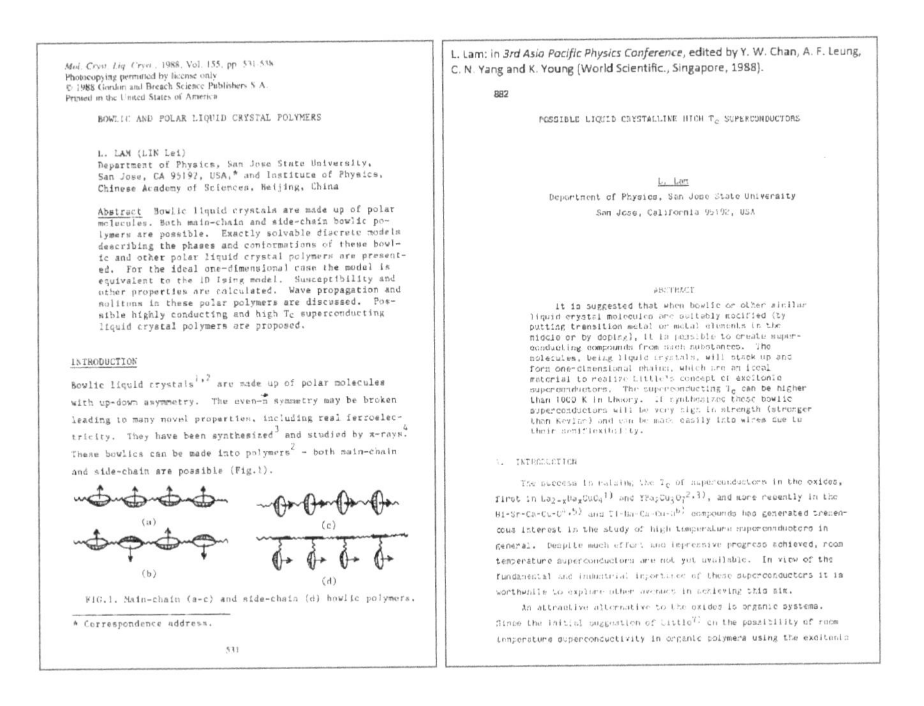

Mol. Cryst. Liq. Cryst., 1988, Vol. 155, pp. 531-538
Photocopying permitted by license only

BOWLIC AND POLAR LIQUID CRYSTAL POLYMERS

L. LAM (LIN Lei)
Department of Physics, San Jose State University, San Jose, CA 95192, USA,* and Institute of Physics, Chinese Academy of Sciences, Beijing, China

Abstract Bowlic liquid crystals are made up of polar molecules. Both main-chain and side-chain bowlic polymers are possible. Exactly solvable discrete models describing the phases and conformations of these bowlic and other polar liquid crystal polymers are presented. For the ideal one-dimensional case the model is equivalent to the 1D Ising model. Susceptibility and other properties are calculated. Wave propagation and solitons in these polar polymers are discussed. Possible highly conducting and high Tc superconducting liquid crystal polymers are proposed.

INTRODUCTION

Bowlic liquid crystals[1,2] are made up of polar molecules with up-down asymmetry. The even-$\vec{n}$ symmetry may be broken leading to many novel properties, including real ferroelectricity. They have been synthesized[3] and studied by x-rays.[4] These bowlics can be made into polymers[2] - both main-chain and side-chain are possible (Fig.1).

FIG.1. Main-chain (a-c) and side-chain (d) bowlic polymers.

* Correspondence address.

531

L. Lam: in *3rd Asia Pacific Physics Conference*, edited by Y. W. Chan, A. F. Leung, C. N. Yang and K. Young (World Scientific., Singapore, 1988).

882

POSSIBLE LIQUID CRYSTALLINE HIGH T_c SUPERCONDUCTORS

L. Lam
Department of Physics, San Jose State University
San Jose, California 95192, USA

ABSTRACT

It is suggested that when bowlic or other similar liquid crystal molecules are suitably modified (by putting transition metal or metal elements in the middle or by doping), it is possible to create superconducting compounds from such substances. The molecules, being liquid crystals, will stack up and form one-dimensional chains, which are an ideal material to realize Little's concept of excitonic superconductors. The superconducting T_c can be higher than 1000 K in theory. If synthesized these bowlic superconductors will be very high in strength (stronger than Kevlar) and can be made easily into wires due to their semiflexibility.

1. INTRODUCTION

The success in raising the T_c of superconductors in the oxides, first in $La_{2-x}Ba_xCuO_4$[1] and $YBa_2Cu_3O_7$[2,3], and more recently in the Bi-Sr-Ca-Cu-O[[illegible]] and [illegible][[illegible]] compounds has generated tremendous interest in the study of high temperature superconductors in general. Despite much effort and impressive progress achieved, room temperature superconductors are not yet available. In view of the fundamental and industrial importance of these superconductors it is worthwhile to explore other avenues in achieving this aim.

An attractive alternative to the oxides is organic systems. Since the initial suggestion of Little[7] on the possibility of room temperature superconductivity in organic polymers using the excitonic

Superconductivity refers to the phenomenon that certain materials have zero resistivity. Cryogenic superconductivity was discovered in 1911 by the Dutchman Heike Kamerlingh Onnes (1853-1926), and its microscopic mechanism, called the BCS theory, was proposed in 1957 by three Americans: John Bardeen (1908-1991), Leon Cooper, and John Schrieffer (1931-2019). In 1986, German Johannes Bednorz and Swiss Alex Müller discovered high-temperature superconductivity.

All of these people have won the Nobel Prize in physics, and the first prize of the National Natural Science Prize in China has been awarded twice to the Institute of Physics, CAS, for high-temperature superconductivity work (1989 and 2013), which shows how important superconductivity is in basic research. However, room-temperature superconductors under *atmospheric* pressure are the Holy Grail in this field. If successful, it will greatly reduce the losses during power transmission, help solve the energy crisis, and make faster computers, better electronic parts and devices, which will definitely change the world. The problem is that the path to success is unknown.

Bowlic Superconductors

Personally, I believe that the way out of room-temperature superconductors is in bowlic liquid crystals. The bowlics was invented and named by Lin Lei (aka Lui Lam) in 1982 at the Institute of Physics of the Chinese Academy of Sciences. They have long been experimentally verified abroad and recognized by peers and international academic institutions.[1, 2]

In 1988, bowlic polymers and bowlic *room-temperature* superconductor were proposed by Lam,[3, 4] which uses the *excitonic* mechanism that has been implemented in one-dimensional organic superconductors. If successfully synthesized, these bowlic superconductors will be very strong (stronger than Kevlar) and, due to their semi-flexibility, can be easily made into wires.

Physicists, chemists, and materials scientists are strongly encouraged to begin serious research into the synthesis and properties of this new superconductor, as it is one of the most promising approaches to room-temperature superconductors compared to existing high-temperature superconductors. After success, it will obviously be of great significance to scientific progress and industrial development, and will immediately allow the participating scientists to win the Nobel Prize.

In addition, bowlics is one of the three types of liquid crystal in the world, and China's self-innovation in the 1980s, with the involvement of many Chinese scientists. The room-temperature superconducting bowlics, once made, will complete this unique program and make this Chinese innovation story perfect [A2].

References

1. Lin Lei (Lam, L.) [1982]. Liquid crystal phases and "dimensionality" of molecules. Wuli (Physics) **11**(3): 171-178.

2. Wang, L., Huang, D., Lam, L., Cheng, D. [2017]. Bowlics: history, advances and applications. Liquid Crystal Today **26**(4): 85-111.

3. Lam, L. [1988]. Bowlic and polar liquid crystal polymers. Molecular Crystals and Liquid Crystals **155**: 531-538.

4. Lam, L. [1988]. Possible liquid crystalline high T_c super-conductors. *3rd Asia Pacific Physics Conference*, Chan, Y. W., Leung, A. F., Yang, C. N. & Young, K. (eds.). Singapore: World Scientific.

13.3 Liquid crystal soliton

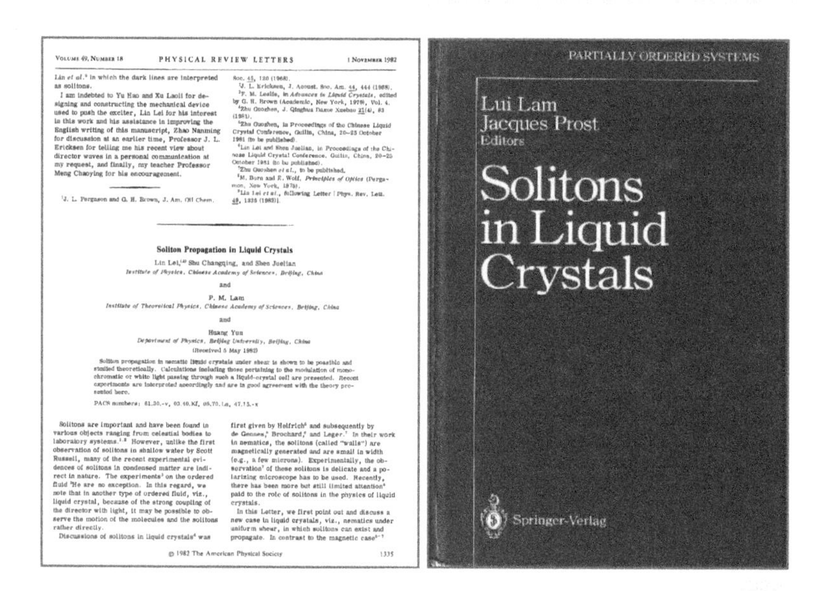

VOLUME 49, NUMBER 18 PHYSICAL REVIEW LETTERS 1 NOVEMBER 1982

Lin *et al.*[9] in which the dark lines are interpreted as solitons.

I am indebted to Yu Hao and Xu Laoli for designing and constructing the mechanical device used to push the exciter, Lin Lei for his interest in this work and his assistance in improving the English writing of this manuscript, Zhao Nanming for discussion at an earlier time, Professor J. L. Ericksen for telling me his recent view about director waves in a personal communication at my request, and finally, my teacher Professor Meng Chaoying for his encouragement.

[1] J. L. Fergason and G. H. Brown, J. Am. Oil Chem. Soc. 45, 120 (1968).
[2] J. L. Ericksen, J. Acoust. Soc. Am. 44, 444 (1968).
[3] F. M. Leslie, in *Advances in Liquid Crystals*, edited by G. H. Brown (Academic, New York, 1979), Vol. 4.
[4] Zhu Guozhen, J. Qinghua Daxue Xuebao 21(4), 83 (1981).
[5] Zhu Guozhen, in Proceedings of the Chinese Liquid Crystal Conference, Guilin, China, 20–25 October 1981 (to be published).
[6] Lin Lei and Shen Juelian, in Proceedings of the Chinese Liquid Crystal Conference, Guilin, China, 20–25 October 1981 (to be published).
[7] Zhu Guozhen *et al.*, to be published.
[8] M. Born and E. Wolf, *Principles of Optics* (Pergamon, New York, 1975).
[9] Lin Lei *et al.*, following Letter [Phys. Rev. Lett. 49, 1335 (1982)].

Soliton Propagation in Liquid Crystals

Lin Lei,[(a)] Shu Changqing, and Shen Juelian
Institute of Physics, Chinese Academy of Sciences, Beijing, China
and
P. M. Lam
Institute of Theoretical Physics, Chinese Academy of Sciences, Beijing, China
and
Huang Yun
Department of Physics, Beijing University, Beijing, China

(Received 5 May 1982)

Soliton propagation in nematic liquid crystals under shear is shown to be possible and studied theoretically. Calculations including those pertaining to the modulation of monochromatic or white light passing through such a liquid-crystal cell are presented. Recent experiments are interpreted accordingly and are in good agreement with the theory presented here.

PACS numbers: 61.30.-v, 03.40.Kf, 05.70.Ln, 47.15.-x

Solitons are important and have been found in various objects ranging from celestial bodies to laboratory systems.[1,2] However, unlike the first observation of solitons in shallow water by Scott Russell, many of the recent experimental evidences of solitons in condensed matter are indirect in nature. The experiments[3] on the ordered fluid ^{3}He are no exception. In this regard, we note that in another type of ordered fluid, viz., liquid crystal, because of the strong coupling of the director with light, it may be possible to observe the motion of the molecules and the solitons rather directly.

Discussions of solitons in liquid crystals[4] was first given by Helfrich[5] and subsequently by de Gennes,[5] Brochard,[6] and Leger.[7] In their work in nematics, the solitons (called "walls") are magnetically generated and are small in width (e.g., a few microns). Experimentally, the observation[7] of these solitons is delicate and a polarizing microscope has to be used. Recently, there has been more but still limited attention[8] paid to the role of solitons in the physics of liquid crystals.

In this Letter, we first point out and discuss a new case in liquid crystals, viz., nematics under uniform shear, in which solitons can exist and propagate. In contrast to the magnetic case[5-7]

The first *Physical Review Letters* (PRL) paper by mainland-only author(s) was authored by me, so is the second (one of the two) one. Both articles are about liquid crystal physics, which I did when I was at the Institute of Physics (IoP), Chinese Academy of Sciences (CAS).

The former is a single author, about liquid crystal (LC) phase transitions, published in 1979. The latter concerns *solitons* in (shearing) liquid crystals, by five authors, from *three* research units, published in 1982 (left figure). Three units were used to impress the editors and reviewers of PRL. This article was the second PRL paper from the IoP, but the first PRL paper by the Institute of Theoretical Physics, CAS, and Peking University. But no one knew the significance of PRL at that time, unlike the later ones when each paper could receive a prize of 10,000 yuan. (This is the first paper in the LC literature to have the word Soliton appears on the title.)

The soliton project began in the summer of 1981. I was the only doctoral supervisor in liquid crystal *physics* at that time, belonging to the first generation of doctoral supervisors which is an official title in China. I put one doctoral and one master's degree student in the project. The way I supervised PhD student was to discuss with him at home for half a day every day, and then let him solve the problem by himself. I basically did the work of a ship captain at the helm, for nearly a year. Over the next ten years, we led the global research on *propagating* solitons in LCs, published 23 papers in 12 domestic and international journals, and finally published the book *Solitons in Liquid Crystals* (1992, right figure) .

In 1982, when the PRL soliton paper was published on November 1, I was visiting the Laboratory of Solid-State Physics in Orsay, southern Paris (where Nobel laureate Pierre de Gennes had previously led the whole French LC research). I gave the LC soliton seminar before returning to Beijing, via India and Hong Kong. From December 6 to 10, the 9th International LC Conference was held in India, and after arriving in Hong Kong in mid-December, Kenneth Young (杨纲凯) of the Chinese University of Hong Kong (CUHK) invited me to give the same seminar. Before the seminar I had a one-to-one discussion with Chen-Ning Yang (杨振宁) in Yang's office (he happened to be staying at CUHK). I told Yang that there are also good research in China (at the Institute of Physics, CAS), such as LC solitons. After that, Yang attended my seminar; he did not ask questions but gave information about a physics topic.

In 2021, in an article about Yang, someone[1] mentioned: "In 1983, he proposed to pay attention to the progress of liquid crystal theory." Yang's proposal is most likely related to my conversation with him at CUHK in mid-December 1982. Unfortunately, Yang's words were obviously not widely circulated; I did not hear it at the time; and of course, not many people paid attention to liquid crystal theory. Yang's words on LC theory was like a small wave in the history of Chinese physics, which has appeared and disappeared in an instant. In hindsight, it is a bit pity.

References

1. Ge Mo-Lin (葛墨林), "Yang Zhen-Ning: Great Physicist, Outstanding Patriot," Intellectual (知识分子), 2021.03.09.

13.4 Active walk

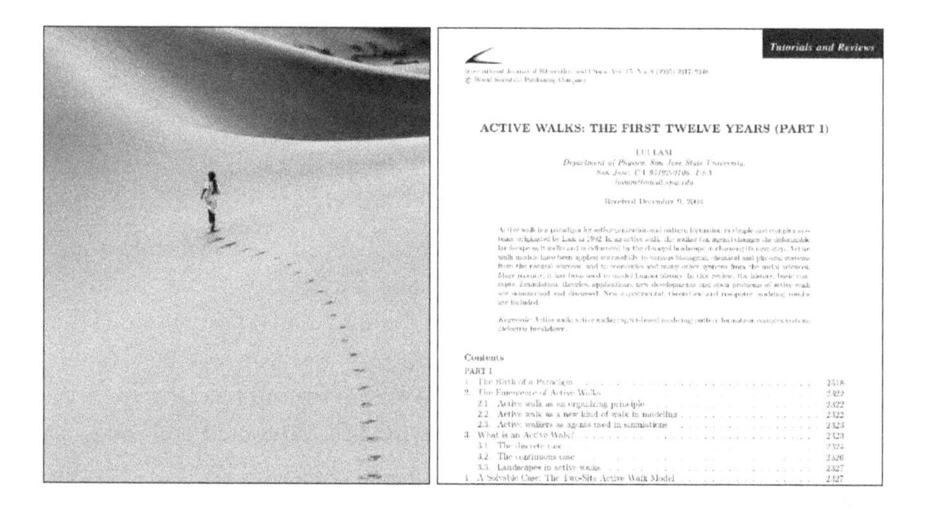
Tutorials and Reviews

ACTIVE WALKS: THE FIRST TWELVE YEARS (PART I)

LUI LAM

Contents

PART I

1990 was my very busy year. In January, I organized a Winter School in Nonlinear Physics at San Jose State University. In March, in Anaheim, California, I chaired the American Physical Society Symposium on Instabilities and Propagating Patterns in Soft Matter Physics, which I originally proposed. In June, I was the Director of the NATO Advanced Research Workshop on Nonlinear Dynamical Structures in Simple and Complex Liquids, in Los Alamos. This was followed in July by a Nonlinear and Chaotic Phenomena conference in Edmonton, Canada, a few days before the international liquid crystal conference in Vancouver.

It was a few days before this conference in Edmonton that we rushed through a simple experiment in our Nonlinear Physics Laboratory at San Jose State University, in Room 55 in the basement of the Natural Science building. The experiment was so simple that it could be finished in less than one second. What we did was take a liquid crystal cell—a thin layer of liquid crystal between two conductive, transparent glass plates (the same one you would find in any digital watch or calculator), put in liquid crystal or oil, and apply a high enough voltage across the cell. After a flash of light, the experiment was finished and we found a complicated filamentary pattern left on the inner surfaces of the coated glass plates. I wrote up the paper right before I rushed to the airport, and later presented the results in Vancouver. On July 27, 1990, the last day of the conference, establishment of the International Liquid Crystal Society was announced [13.5].

The physical and chemical processes giving rise to these filaments are rather complicated, and are still being investigated. Essentially, the filaments are the locations that a series of chemical reactions induced by dielectric breakdown takes place. After this experiment, without knowledge of the physical mechanisms, we set out to do a computer modeling of this filamentary pattern. We soon realized that a growing filament could be identified as the track of a walker. To grow the filament, we only have to tell the walker how to walk, and specify how the walker changes the local environment as it walks. When I wrote up the paper for my book *Modeling Complex Phenomena* (Springer, 1992), I named the walkers "active walkers" and the model the "active walker model." This is the first paper on *Active Walk*. In 1994, I called the process an "active walk" (AW) and started to refer to these models as "active walk models."

And quite immediately, I realized that AW is not just good for modeling filamentary growth, but is in fact a general paradigm applicable to many other complex systems. Because of its novelty, in less than a year, I was invited to publish the first review of AW in the journal *Computers in Physics*. The 1992 Fractals and Disorder conference in Hamburg, Germany, even printed our AW results on the cover of the proceedings.

Since then, AW has been applied by many people worldwide to various simple and complex systems from the natural and social sciences as well as the humanities. Examples include pattern formation in physical, biological and chemical systems such as surface-reaction filaments and retinal neurons, formation of fractal surfaces, food foraging of ants, worm moving, bacteria movements and pattern forming, spontaneous formation of human trails, localization-delocalization transitions, granular matter, oil recovery, economic systems, positive-feedback systems, and human history (helping to form the new discipline *Histophysics* in 2002 [14]).

At the same time, in the last decade or so, agent-based models gained momentum and became a favorite in tackling problems from social systems. In these models, a large number of agents are employed; each agent keeps altering an environment and vice versa. These agents are in fact active walkers, and the models are AW models.

Among them, Dirk Helbing of Germany used AW to simulate the formation of pedestrian trajectories (*Nature*, 1997), and developed it into *crowd dynamics*. One application is crowd control, which helps solve the problem of crowd trampling during large gatherings and can save human lives.

Before the invention of AW, walks in modeling various systems such as economic and physical simulation were *passive* walks; i.e., the walker did not change the environment. The simplest example is random walk, which was used by Einstein in 1905 to explain Brownian motion (indirectly proving the existence of atoms). Conversely, the AW's walker is an *active* walker: It changes the environment, and the changed environment in turn influences the walker's next step. This environment can be physical or virtual. The former example is the distribution of odor concentrations emitted by ants when walking; the latter is the "bookkeeping" record used by search engines or machine learning in artificial intelligence (using positive feedback principle).

The invention of AW is a typical example of generalization from a special case (simulated results of an electrical breakdown experiment in a liquid crystal cell) to a universal paradigm. Innovation is actually not that difficult. The secret is keep your eyes open, seize opportunities, and spend time on topics that others have not done.

Remarks

For a comprehensive review on AW, see Lui Lam's two-part articles, "Active Walks: The first twelve years," International Journal of Bifurcation and Chaos **15**, 2317-2348 (2005); **16**, 239-268 (2006).

13.5 Liquid crystal society

中国液晶学会
（第一届1980～1984）

会员录

证号	姓名	[illegible]	年令	职称	入会时间	学会职务	工作单位
0082	[illegible]安	男	42	助研	80	理事	物理所
0083	林[illegible]	〃	39	付研	〃	付理事长 兼秘书长	中科院物理所
0084	刘家尚	〃	42	助教	〃		北京林学院物理组
0085	俞昌清	〃	41	研究生	〃		中科院物理所
0086	王心宜	〃	35	〃	〃		中科院物理所
0087	谈曼琪	女	51	助研	〃	理事	中科院生物物理所
0088	胡坤生	男	43	〃	〃		〃
0089	张树霖	〃	48	讲师	〃	理事	北京大学物理系
0090	宋增福	〃	56	付教授	〃	理事	〃

• 10 •

LIQUID CRYSTALS *Today*

Vol. 1 No. 1
January 1991

Newsletter of the International Liquid Crystal Society

A Message from the President

S CHANDRASEKHAR

WELCOME TO READERS

Comments and Contributions should be sent to

COPY DATE for next issue: 31st MARCH 1991

In This Issue:

Officers and Board Members of the ILCS
New Books
News from around the Liquid Crystal World
Liquid Crystal Display Applications, J Constable
Society News
Meeting Reports
Forthcoming Meetings

LOGO COMPETITION

I have helped to found *two* liquid crystal (LC) societies: one in China (1980) and one international (1990). Two completely different experiences.

In early January 1978, I arrived to work at the Institute of Physics of the Chinese Academy of Sciences (CAS) as a second-generation returnee, those who returned from abroad voluntarily after 1949 [11.15]. At that time, there were only dozens of LC researchers in China. Some worked in applications, some in chemistry and biology, and basically only two people doing LC physics: Xie Yu-Zhang (谢毓章, 1915-2011) and me. But it seems the only LC physics paper by any Chinese was the one I published before returning to China [11.13]. The working environment faced by researchers at that time was:

1. The monthly salary was generally only 60 yuan per month, and it costed 30 yuan (plus coupons for cotton) to buy a pillow.
2. Work six days a week, except for Spring Festival and National Day, and a few other short holidays.
3. To order a bottle of milk, you must be at the associate-professor rank or above. Additionally, one needed a letter of introduction issued by the working unit and ride a bicycle to the dairy factory to place the order (no taxi or private cars). Staying in a hotel or buying a train ticket also required a letter of introduction from the unit, and there was no private trips. In other words, letters of introduction (meaning approval) for everything.

Therefore, under the condition of three no's (no holidays, no money, and no excuses), the only way to go to Guilin and other scenic places for sightseeing is to go there for a conference, at public expense.

Chinese Liquid Crystal Society

An academic society of course will meet once or several times a year. Establishing a new society, in addition to academic reasons, is also very beneficial for the individuals (as explained above). Liquid crystal is multidisciplinary but the idea of forming a Chinese LC society came from Tsinghua University. After several deliberations, it was decided to attach it to the Chinese Physical Society (the venue is in the Institute of Physics, CAS).

In the summer of 1980, I led a delegation to Kyoto, Japan, to participate in the 8th International LC Conference, which was the first time that mainland scholars participated in this biennial conference series. After returning to China, my colleagues and I established the Chinese Liquid Crystal Society (officially known as the Chinese Physical Society LC Branch) in Beijing on July 18. Xie Yu-Zhang from Tsinghua was the president and I was the vice president and secretary general. I drafted the Society's bylaws.

International Liquid Crystal Society

On December 24, 1983, I left Beijing to return to work in New York. The following year, I was elected to the power center in the LC profession: the Planning and Steering Committee of the International LC Conference Series. The solid state laboratory of the CNRS (equivalent to the Chinese Academy of Sciences) in Orsay, in the southern suburbs of Paris, was full of eminent LC scientists (where Nobel laureate De Gennes' liquid crystal work was done) and was a place I visited frequently.

I met Mireille Delay (1951-1987) there in 1982 and talked for five minutes in her lab. In July 1987, I went to Bordeaux, the country of fine wine, to attend a LC polymer conference. I stopped in Orsay on my way and asked about Mireille, only to learn from Georges Durand that she had died. Shocked, I felt that there should be an international liquid crystal society and a magazine like *Physics Today*, so that colleagues can know each other's news. Years later, I learned from her colleague that Mirielle committed suicide just four months before I arrived, leaving behind two children. Many female researchers are "superwomen," playing the roles of wife, mother and researcher at the same time every day. Needless to say, it is stressful, physically and mentally exhausting. Society and peers should care them more and provide necessary supports.

Over the next three years, I initiated, single-handedly planned the International Liquid Crystal Society (ILCS). I traveled around the globe to garner the support and cooperation of peers in the LC community, and drafted the Society's bylaws. Finally on the last day of the 13th International LC Conference held in Vancouver, July 27, 1990, the ILCS was established.

Conclusion

Unlike within China, there is complete freedom to build a learned society. There is no need to get approval, affiliate it with some bigger organizations, or register it with any government. In fact, after the establishment of the ILCS, none of my colleagues in my university even knew about it.

The ILCS may have set two world records:

1. It is the only international society founded by a Chinese.
2. It is the only international society founded by one person.

In fact, there may be another record: This is the only international society founded because of a woman. A romantic story. (For the record, Mireille and I have only seen each other for five minutes in our lifetime.)

Remarks

It is Alfred Nobel (1833-1896) who established a prize, the Nobel peace prize, for a woman [3.2].

For more details about the ILCS, see my two articles:

1. Lam, L. [2014]. The founding of the International Liquid Crystal Society. *All About Science: Philosophy, History, Sociology & Communication*, Burguete, M. & Lam, L. (eds.). Singapore: World Scientific.
2. Lam, L. [2017]. Prehistory of International Liquid Crystal Society, 1978-1990: A personal account. Molecular Crystals and Liquid Crystals **647**: 351-372.

13.6 Art and killing time

Of all the research papers I put on ResearchGate, the most read is "Arts: a science matters," more than my bowlic review or articles on science. Quite a happy surprise. The paper is about the origin and nature of art, published as Chapter 1 in *Arts: A Science Matter* (2011). Here is how that paper came about.

In the second international scimat conference, held in Portugal in 2009, we decided to focus on art and science [17]. Before that, I knew little about art even though, like others, I am fond of them. That has been my practice of gaining knowledge by forcing myself to enter a new field.

The Dawn

My talk consists of some random thoughts on art that I gathered few months before the conference. But after the conference, as the chief editor of the proceedings, as the practice, I had to write a comprehensive introduction to the subject of art and science, starting with the common nature of art which includes visual arts, literature, film, performance arts, music, architecture, new media arts and so on. After reading some articles and books on the subject, including *Philosophy of Art* (1999), *The Methodologies of Art* (1986) and *What Is Art For?* (1988), which I found all wrong and scratching my head for a few days, I stumble on the idea of sitting in front of the TV and watch the program called *ARTS*.

The program is meant to fill up TV times by randomly showing old video clips of ballets, classical music singings or orchestra performances, and occasionally some paintings. After returning to the program a few times and found it relaxing, I asked myself: What have they been in common? My answer: They kill my time, gently. And that is it!

Killing Time

On May 8, 2010, Saturday evening, I was in a Chinese restaurant by myself. While waiting for the food to be served I had nothing else to do but think. It suddenly became clear to me that what differentiates human from other animals, and, in fact, between any two biological systems, is how they spend their spare time. That is, the way that they "kill time."

All biological systems (bacteria, fish, dogs, chimps, humans, etc.) have to do two things without exception: **eat** and **reproduce**—as observed by many people including the ancient Chinese philosopher Mencius (c.372-289 BC). Other activities such as sleep is to serve these two functions.

Sleep is to preserve energy (for snakes, a whole winter), to help digest food (only sometimes, since you can work and digest food at the same time), but essentially is to regenerate energy so you can go on to eat and reproduce. Reproduce means sexual activities for some animals, or asexual like fish laying down eggs and sperms.

The sleep cycle depends on where the biological system resides. For those residing on Earth's surface, it is regulated by the rise and setting of the sun, and hence usually a 24-hour cycle. For fish living deep in the ocean, that cycle is different (or no cycle, but the oceanographers must know about that already).

Work, like sleep, is to support the two functions of eat and reproduce. Millions of years ago, work consisted of hunting and gathering fruits. Later, 10,000 years ago, agriculture and the like. Modern day, 9 to 5, five days a week, say.

In all cases, beyond sleep and work, there are plenty spare time that is not needed to fulfil the eat and reproduce functions. And how do the biological systems use this spare time?

In the Discovery channel, you see that lions are sitting there mostly or wandering slowly occasionally. Chimpanzees may sit there cleaning themselves or teasing each other, it seems. Fish, resting or moving to other areas with new food sources, perhaps.

Human

Among all animals, *only* humans do art to kill time. The practice of art has accelerated the breadth and depth of human thinking and imagination, which gives human the advantage in evolution over other animals.

All these acts like drawing and dance that we call art could, and very likely, happen one to two million years ago since human already developed the ability to do this two million years ago, which is called *mimesis*—to mime, imitate, gesture, and rehearsal of skill [7.1, 7.2].

The emergence of art is *not* for sexual selection because millions of years ago, that was basically a free-sex community. And art is not about *aesthetics* alone, since nature provides plenty, and better, aesthetic experiences—like the sunset.

Art helps us to survive!

PART III

ARTICLES

14 History of Histophysics

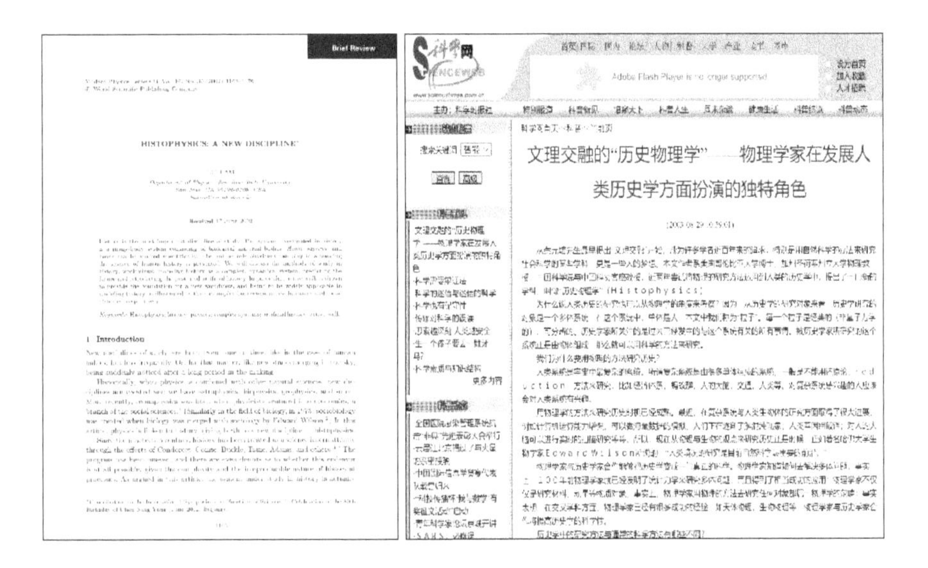

Brief Review

HISTOPHYSICS: A NEW DISCIPLINE

1 Introduction

文理交融的"历史物理学"——物理学家在发展人类历史学方面扮演的独特角色

Histophysics is a new discipline proposed by Lui Lam in 2002 with a paper published in *Modern Physics Letters B* (left figure). Histophysics means physics of history, which uses physical methods to tackle human history problems. From 2002 to 2010, three journal papers and two reviews are published. This article details how the idea of Histophysics came about and how Histophysics was developed and published. In short, this is the Histophysics story, told here for the first time.

1 Separation and Integration of the Two Cultures

The two cultures refer to the science culture and the humanities culture [18]. In the era of Mozi (c.476-c.290 BC) and Aristotle (384-322 BC), science and humanities were not separated, and there was no question of two cultures. During the Renaissance following the Middle Ages, Leonardo da Vinci (1452-1519) was probably the last person to be fluent in the two cultures; for him, the two cultures was not a problem either. In 1918, Cai Yuan-Pei (1868-1940), the president of Peking University and the founder and first president of the Chinese Academy of Sciences, initiated *General Education* requirement for students, ahead of the rest of the world. That is, before graduating, humanities students should take some science courses, and vice versa. This incident indicates that the separation of the two cultures had already appeared at that time. Later in 1956, Charles Percy Snow (1905-1980) formally raised the issue of the serious gap between the two cultures in society [Snow 1998; Lam 2008a].

Recently, someone pointed out: "The culture of the future will be a culture based on humanities-science synthesis, and a culture of continuous synthesis of different cultures. Future scholastic masters will be produced in the field of cross-integration of humanities and science, and of different cultures." [Hu 2005]

Histophysics—using physical methods to study human history—is exactly such a discipline that integrates the humanities and science [Lam 2002]. This article gives the background and developmental history of this new discipline and uses it as an example to illustrate the process of academic and disciplinary innovation. The experience we have gained from this is relevant to the current development of science in China, and may also inspire the answer to the so-called Needham Question.

2 The Founding Process of Histophysics

1998

In 1982, I invented a new type of liquid crystals—*Bowlics* [Lin 1982; Lam 1994]. In 1992 I proposed a new paradigm for dealing with complex systems—*Active Walk* [Lam et al. 1992; Lam 2005, 2006]. Liquid crystals is a branch of condensed matter physics; complex systems are broader than liquid crystals and include many systems in the natural and social sciences.

In 1998, my book *Nonlinear Physics for Beginners* was published [Lam 1998] (Fig. 1), in which Chapter 7 on complex systems talks about active walk, with particular reference to a new phenomenon discovered by my Nonlinear Physics Group in 1995: *Intrinsic abnormal growth* (Lam et al. 1995). This phenomenon shows that in probabilistic complex systems (such as human **history**) composed of chance and necessity, if the control parameters are appropriate, the effect of chance will increase, resulting in abnormal system evolution. The "intrinsic" character explains the unpredictability of some historical phenomena, the abnormal growth of some plants, the irreproducibility of some medical experiments (not due to falsification), and rejects Steven J. Gould's (1941-2002) assertion that evolutionary **history** is *necessarily* unrepeatable in his book *Wonderful Life* [Gould 1989]. Our conclusion on this issue is different: If the evolution of life on Earth can be "repeated," it is *possible* that humans will still appear on Earth [16].

Fig. 1. My two books on nonlinear physics. *Left*: *Introduction to Nonlinear Physics* (Springer, 1997). *Right*: *Introduction to Nonlinear Physics* (World Scientific, 1998).

After that, I wanted to do a broader and more important work, to combine Active Walk with a branch of the humanities or social science to create a *new* discipline. However, I did not know anything about these disciplines, and at that time I had considered literature, marketing, and political science. Then, by chance in 2000, I picked **history**.

2000

In **May** 2000, I invited Michael Shermer—a historian, and founder and editor-in-chief of *Skeptic* (Fig. 2)—to my college to be the speaker for my newly established “God, Science, Scientist” public lecture series. I asked him what the main journal of history is, and his answer was *History and Theory*.[1]

In 2000, Shermer invited me to write a popular science article in the quarterly journal *Skeptic*. The article is titled “How nature self-organizes: Active walks in complex systems,” near the end of which is the section: Modeling History [Lam 2000]. This is the first time I have written directly on **history**. The first paragraph of this section:

> When historians write narrative histories, they use such phrases as “leave a mark in history” or “follow the giant’s footsteps.” But to model these scientifically, one needs a deformable landscape so that marks and footsteps can be left on it. Moreover, one needs a walker to do these things. As noted by the Chinese, through the action of individuals bad things can sometimes be turned into good things, and vice versa. The walker should thus be an *active* walker. The active walk model, then, fits naturally into historical analyses. With prudent application, it could be the first step in achieving a quantitative theory of history.

This idea was fully reflected in the founding article of Histophysics two years later [Lam 2002].

Fig. 2. *Left*: Cover of *Skeptic* magazine, Volume 8, Issue 3, 2000. Inside there is the author’s article [Lam 2000]. *Right*: Michael Shermer (2003; photo/Lui Lam).

2001

In 2001, I found the journal *History and Theory* in the college library. But after reading a few issues, I still did not know much about the current state of the history profession.

On **August** 12-16, 2001, I gave a keynote presentation at a futurology conference in Seattle, "Modeling history and predicting the future: The Active Walk approach" [Lam 2003a] (Fig. 3).[2] It is an extension of the "Modeling History" section of the 2000 *Skeptic* article. The future can be predicted because the active walk model can simulate the history of the past, so when time is moved forward, it is about the history of the future [16].

Fig. 3. The "Future of Humanity: Seminar 3" hosted by Foundation For the Future, Seattle, Aug. 12-16, 2001. (Source: futurefoundation.org/programs/ hum_sem3 /html)

2002

On **January** 9, 2002, my luck came. I entered the University of Toronto bookstore in Canada and was shocked but delighted to see the General History bookshelf standing there, which is full of history books (Fig. 4). This happened because there is a well-known history professor at the university. I bought Richard Evans' *In Defence of History* [Evans 1997]. After reading it, I realized that history was in crisis of being oppressed by postmodernism, and the crisis represented an opportunity for me to enter the discipline as an outsider.

Fig. 4. Bookstore at the University of Toronto, Canada. The four bookshelves in the second row are all general-history books. (2002; photo/ Lui Lam)

On **April** 4, 2002, I found a key book in the Stanford University bookstore: *History and Historians: A Historiographical Introduction* (2000) by Mark Gilderhus. The book was thin (140 pages) and at once taught me the traditional methods of historiography. My conclusion is that, in general, historians do not have an accurate understanding of the nature of science, and the *scientificity* (scientific level) of historical research is obviously not high, and that is where physicists can contribute and make breakthroughs. (I also bought the 2000 American edition of Evans' book: *In Defense of History*. Note that Defense is spelled differently than the 1997 version of Defence.)

On **April** 25 of the same year, I gave my first report on **Histophysics** in my physics department, entitled "Histophysics: A new discipline,"[3] which was well received.[4] The talk was reported by the student-run newspaper *Spartan Daily*. I started drafting an article based on the talk in San Jose in May and finished it in Beijing in June.

On **June** 18 of the same year, I gave the same talk "Histophysics: A new discipline" at the International Symposium on Frontier Science, held by Tsinghua University in Beijing for Chen-Ning Yang's 80th birthday.[5] The transcript of the talk was submitted in September and published in the proceedings by World Scientific of Singapore in 2003 [Lam 2003b].

In **July** of the same year, I found two good books in the University of Toronto bookstore: *An Introduction to the Philosophy of History* (1998) by Michael Stanford and *Dynasties of the World* (2002) by John Morby. The former tells me about various aspects of history (including the relationship between history and philosophy/social science as well as the themes, causes, and interpretations of history). The latter, published by Oxford University Press, has data on the years of emperors and dynasties in all countries of the world [Morby 2002]. And, of course, data is necessary for quantitative analysis—the basic skill of physicists. But I put this "dry" book on hold and did not touch it for almost two years.

2004

On **March** 13, 2004, a few days before I went to Montreal, Canada, to attend the March Meeting (March 22-26, 2004) organized by the American Physical Society, I opened the book *Dynasties of the World*, traced two curves based on the data on the Chinese dynasties, and found two *quantitative* laws about Chinese history and a *quantitative* prediction [Lam 2006]. These results were presented at the Montreal meeting.

Finally, one of these laws developed into the broader *Bilinear Effect* [Lam et al. 2010] [16].

2010

I and collaborators created a three-layer network model that mimics the structure of the ancient Chinese state; numerical results show that the Zipt plot [5.3] of the dynasty lifespans consists of two straight lines [Lam et al. 2010]. However, the mechanism and necessary conditions for the emergence of the bilinear effect remain unknown.

In short, I did Histophysics for ten years, from 2000 to 2010. In the meantime, since 2007, my research focus has shifted to *Scimat*—a multidiscipline that I pioneered in 2007 [17].

3 The Publication History of Histophysics

2002

My two academic innovations, bowlic liquid crystals [Lin 1982] and active walk [Lam et al. 1992], were published in 1982 and 1992, respectively, ten years apart. For the sake of symmetry perfection, I hope that the first article in Histophysics will be published in 2002. Therefore, in **June** 2002, I wrote an article entitled "Physics in history" and submitted it to the journal *Wuli* (Physics), the official publication of the Chinese Physical Society. In **November**, two referee reports returned, one saying it could be published, the other saying that the article involved social science and should be cautious. The following year, in February 2003, the editor-in-chief of Wuli decided to reject the manuscript. Wuli thus missed the opportunity to publish the first article in a new discipline.

In **December** 2002, World Scientific, without informing me, published my article "Histophysics: A new discipline" in their *Modern Physics Letters B* journal [Lam 2002].

2003

On **January** 14, 2003, while I was lamenting that the publication symmetry had been broken and that the first article in Histophysics could not appear in 2002, I suddenly received an email from Alexandre Wang from France, asking me to send her/him a copy of the Histophysics article. I scratched my head and looked it up on the Internet, only to know that there was such an article published [Lam 2002]. Thankfully, the publication symmetry is not broken.

On **August** 29, 2003, I published in *Sciencetimes* (a newspaper published by the Chinese Academy of Sciences) a short Chinese article entitled "The humanities-science synthesized Histophysics: The unique role of physicists in developing human history" [Lin 2003a], with my email address at the end of the article. It was reprinted by several websites in China, and some readers contacted me. And I met with several of them and developed a cooperative research relationship.

In **December** of the same year, my long Chinese article "The humanities-science synthesized Histophysics" appeared in the book *Emerging Interdisciplinary Disciplines* published by Tsinghua University Press in Beijing [Lin 2003b].

In 2003, my article "Modeling history and predicting the future: The Active Walk approach" was published [Lam 2003a]. That was my presentation at the Foundation For the Future's workshop in Seattle in 2001.

2004

In **December** 2004, "How to model history and predict the future" [Lam 2004a] appeared in my first popular science book *This Pale Blue Dot: Science, History, God*, published by Tamkang University [Lam 2004b]. This book is a collection of papers from the Tamkang Chair Lectures (淡江讲座) I gave at Tamkang University on December 9-11, 2003 [15, 20].[6]

2006

In **February** 2006, quantitative results on Chinese dynasty history were first published in Section 5.2 of my review article on Active Walk [Lam 2006]: Two quantitative laws and a quantitative prediction in Chinese history. The review was published in the International Journal of Bifurcation and Chaos, a World Scientific journal.

2008

The first book in the Science Matters series, *Science Matters: Humanities as Complex Systems*, which I founded and edited, was published by World Scientific. Chapter 13 is "Human history: A science matter" [Lam 2008b].

2010

The paper "Bilinear effect in complex systems," with me as the first author, was published in *EPL* (Europhysics Letters) [Lam et al. 2010]. One of the effect's three examples shown there is the distribution of Chinese dynasty lifespans, which I first reported in my 2006 review of active walks [Lam 2006]. In addition, in the paper, the term **Histophysics** appears in reference 29.

2013

The book *Renke: Humanities as Complex Systems* was published by China Renmin University Press, Beijing. *Renke* (人科) is Scimat's official Chinese name. Chapter 13 is my article "Research in human **history**: An example of scimat" [Lin 2013]. The book is a Chinese translation of the English book *Science Matters*, published in 2008 by World Scientific.

In summary, with regard to Histophysics,[7,8] I published 3 journal papers [Lam 2002, 2006; Lam et al. 2010] as well as 2 English reviews [Lam 2003a, 2008b], 1 Chinese review [Lin 2013], and 2 major Chinese popularization articles [Lin 2003a, 2003b].

4 Discussion and Conclusion

1. The creation of the new discipline of Histophysics, using the triple-jump method in track and field, took two years. The three-step jumps: (1) *Skeptic* article of 2000; (2) Seattle presentation 2001; (3) San Jose seminar 2002. (In those two years I also did granular physics research and organized two international conferences.)

Many times, my work habit is to set the time and topic of a talk first, and then spend one or more months forcing myself to come up with some content for the talk, rather than making a presentation after I have the content. This requires quick thinking and academic skills. The creation of Histophysics took only two years; the key is *not* reading those books. Engaging in innovation without reading books is the secret of innovation that only a few people who have successfully innovated know. For more discussion, see "Two types of reading" in the book *Research and Innovation* [Lam 2022a].

2. The material systems of nature vary in size but have continuity. Looking down at the small scale, there are molecules, atoms, nucleons and so on. Looking up at the large scale, there are molecules, condensed matter, cells, biological tissues, human and so on. Therefore, there is coherence between all disciplines, the so-called "knowing one, know them all." This is the reason that one can cross disciplines in doing research [11.16].

3. Some people think that science includes both the natural and social sciences but not the humanities. This view is wrong, because natural science is the study of all material systems in nature, including the biological system of human, and the humanities are also about the study of human, so humanities, like social science, should be part of natural science. History, on the other hand, studies the history of humans, i.e., what has happened to individuals and societies. So, history is part of the humanities or social science, and hence is part of natural science. The conclusion is that history can be studied in physics [5.1].

4. Scientific innovation or the birth of a new discipline requires three conditions:

(1) Sometimes it only takes a far-sighted and determined person who can use all the resources. For example, Einstein, who established the general theory of relativity.

(2) There is an academic environment without the pressure of publishing papers. For example, Einstein was an employee of the patent office at the time and had no pressure to publish papers; I am a tenured professor, with the freedom not to apply for research grants and not to publish papers—doing research and publishing

papers is purely a personal interest. Unfortunately, tenure no longer exists in China.

(3) The person must be courageous and dare to challenge authority and popular opinion of the time.

For scholars to innovate and cross disciplines, the basic conditions are an "iron rice bowl" (guaranteed income) with a tenure system (no assessment, no form filling) and a "one-part salary" system (no need to apply for projects or doing administrative works).[9]

5. The so-called Needham Question asks: Why did modern science not arise in China? [Wang 2003]. Part of the answer is that the third condition above—those who dare to challenge authority—did not exist or were not numerous in old China, because of Confucianism and the suffocating feudal social system. For more discussion, see [9.5].

6. Finally, in China, there is a lack of good bookstores like that at Stanford and Toronto universities. It is these bookstores that allow outsiders in a discipline to know the ins and outs of the discipline and the latest progress in a short period of time. This lack may be one of the reasons why Chinese scholars tend to stick to a discipline, while academic innovation is the opposite, which quite often involves crossing disciplinary boundaries [Lam 2022a].

Notes

1. In December 1999, I created a public science lecture series "God, Science, Scientist" at San Jose State University, where I teach. The first speaker in this series was Michael Shermer, who delivered a lecture on "How people believe: Seeking God in the age of science." In November of the same year, he and I went to Beijing to attend the International Conference on Science Communications [12.6]. After returning to the United States, Shermer wrote several articles on his visit to China as a columnist in the monthly magazine *Scientific American* [Shermer 2001a, 2001b].

2. In 2000, Sesh Velamoor, vice president of the Foundation For the Future in Bellevue, Washington, noticed my *Skeptic* article, particularly the section on "Modeling History" [Lam 2000]. He invited me to give a keynote presentation at the Foundation's Humanity 3000 Program, "The Future of Humanity: Workshop 3," Seattle, August 12-14, 2001 [12.7].

3. The English name Histophysics for physics of history, is a new word I coined in 2002. (I have coined many other new words, such as Bowlic, Active walk, Bilinear effect, and Scimat.) At that time, my consideration was that Westerners often drew inspiration from Greek mythology or Latin when making up new words while most peoples in the world are not familiar with these ancient Western cultures, but were more familiar with English. Therefore, I merged the two English words History and Physics into Histophysics—the privilege of pioneers in naming their new-born "baby."

4. Actually, my colleagues in the physics department knew that Histophysics was a new thing but did not understand why it was important, because they did not understand History. Colleagues in the history department reacted emotionally, believing that physicists had invaded their turf, and even more because they did not understand Physics. They mistakenly think that physics is only about nonhuman systems, and they do not know that there are three levels of research in any discipline: empirical, phenomenological, and bottom up [4.3]. Starting with historical figures is the bottom-up level, which is the traditional approach of historians, while methods such as Zipf plot and active walk belong to the empirical and phenomenological levels. In general, the scientific training of historians is lacking [16].

5. Chen-Ning Yang's 80th birthday international symposium was held at Tsinghua University in Beijing from June 17 to 19, 2002, with a large group of distinguished people, including 14 Nobel laureates and one mathematician who won the Fields Medal. Before going to Tsinghua, I prepared dozens of preprints of "Histophysics: A new discipline" and gave them out when I met the right people, one of whom was Murray Gell-Mann (1929-2019), the doctoral mentor of Philip Platzman (1935-2012)—my doctoral mentor at Bell Labs. Gell-Mann's parents immigrated from Austria to New York after World War I, and his father was a language teacher, so Gell-Mann had a deep understanding of various languages and a very good memory. He took my preprint, looked at the title and the author's name Lui Lam. His first sentence was: You are from Hong Kong. I said: Yes. The second sentence is: Histo- stands for tissue. This is also true; for example, Histology stands for (cellular) tissues in pathology. But I responded super quickly, and I said: How about Histogram? Gell-Mann was speechless for a moment, and walked away silently with my preprint. I do not know if he later read the preprint.

6. Clement Chang (1929-2018) founded a futurology institute at Tamkang University in Tamshui, when he was president there. He attended the Foundation For the Future's "Future of Humanity: Seminar 3" workshop in Seattle in August 2001, listened to my talk on predicting the future, thought I was an expert in futurology, and invited me to give Tamkang Chair Lectures at his school [12.7].

7. On January 8, 2011, I accidentally discovered a new online journal *Cliodynamics*, run by the University of California. (Clio is the goddess of

history in Greek mythology). After checking the Internet, I learned that Peter Turchin had published a book of the same name in 2003, so I sent him an email telling him that I had proposed the term and concept of Histophysics in 2002, a year before him, and pointed out that according to the content, Cliodynamics should be part of Histophysics, and Histophysics is part of Scimat. Turchin replied to my email five days later, not disagreeing with my claim, saying that he did not know about my work on Histophysics, but would read the literature I had sent him.

On April 4, 2014, I emailed Turchin and told him that I wanted to organize a session on "The Science of Human History" at the AAAS Annual Meeting in San Jose, February 12-16, 2015. And I proposed that he gives a review on Cliodynamics and I would talk about Histophysics. He replied two days later, saying thank you for the invitation but did not have time to attend. In the end, the session did not work out.

8. On July 19, 2022, someone posted a link to a paper on the physics of history on the WeChat group. The paper is entitled "Cliophysics: A scientific analysis of recurrent historical events" with 9 co-authors [Aruka1 et al. 2022]. This paper is published in the same journal EPL as my article 12 years ago on Histophysics [Lam et al. 2010]. It does not mention the word Histophysics or Cliodynamics or any works related to them. I emailed the nine authors the same day to tell them about it.

Only two authors responded to my emails. Bertrand Roehner of France replied one day later (July 20), saying that his Chinese co-author at Beijing Normal University mentioned my name and work to him after the article was published, which was too late to mention it in their paper but will mention my Histophysics works in their future papers.

Another author, Peter Richmond of Ireland, replied two days later (July 21) that the three words Histophysics, Cliodynamics, and Cliophysics, will compete in the academic world for peers to decide. He also promised to mention my work in future papers.

In this Cliophysics paper, 4 of the 9 authors are Chinese (Chen Xiaosong, Di Zengru, Wang Qing-hai, Yang Yang); 3 of them work in mainland China (Chen, Di, Yang are all at Beijing Normal University); 4 of them are in physics (Kim, Roehner, Richmond, Wang). The unusual thing is that all these 9 authors failed to notice the many Histophysics articles I published in physics journals and books, in English and Chinese, from 2002 to 2010 (see Sec. 3). At the same time, somehow, the long academic tradition of "who publishes first will name it" is broken by latecomers who show up 20 years later.

9. China presently uses the "three-part salary" system in universities/research institutes: basic salary, administrative salary, and performance salary. That is, one's salary will be drastically reduced if you are not leader of a

group/department/college and if your *yearly* assessment performance is not good enough.

References

Arukal, Y., Baaquie, B., Chen Xiaosong, Di Zengru, Kim, B., Richmond, P., Roehner, B. M., Wang Qing-hai & Yang Yang [2022]. Cliophysics: A scientific analysis of recurrent historical events. EPL **138**: 22004.

Evans, R. [1997]. *In Defence of History*. London: Granta Books.

Gould, S. J. [1989]. *Wonderful Life: The Burgess Shale and the Nature of History*. New York: Norton.

Hu Xian-Zhang (胡显章) [2005]. Preface to Tsinghua Humanities book series. *Misconduct of Scientists*, Shigeaki Yamazaki (山崎茂明). Beijing: Tsinghua University Press.

Lam, L. [1994]. Bowlics. *Liquid Crystalline and Mesomorphic Liquid Crystals*, Shibaev, V. P. & Lam, L. (eds.). New York: Springer.

Lam, L. [1998]. *Nonlinear Physics for Beginners: Fractals, Chaos, Solitons, Pattern Formation, Cellular Automata and Complex Sys*tems. Singapore: World Scientific.

Lam, L. [2000]. How nature self-organizes: Active walks in complex systems. Skeptic **8**(3): 71-77.

Lam, L. [2002]. Histophysics: A new discipline. Modern Physics Letters B **16**: 1163-1176.

Lam, L. [2003a]. Modeling history and predicting the future: The Active Walk approach. *Humanity 3000, Future of Humanity: Seminar 3 Proceedings*. Bellevue, Washington: Foundation For the Future. pp 109-117.

Lam, L. [2003b]. Histophysics: A new discipline. *Proceedings of the International Symposium on Frontier of Science, 2002, Beijing: In Celebration of the 80th Birthday of C. N. Yang*, Nieh, H. T. (ed.). Singapore: World Scientific. pp 456-471.

Lam, L. [2004a]. How to model history and predict the future. *This Pale Blue Dot: Science, History, God*, Lam, L. Tamshui: Tamkang University Press. pp 15-34.

Lam, L. [2004b]. *This Pale Blue Dot: Science, History, God*. Tamshui: Tamkang University Press.

Lam, L. [2005]. Active Walks: The first twelve years (Part I). International Journal of Bifurcation and Chaos **15**: 2317-2348.

Lam, L. [2006]. Active Walks: The first twelve years (Part II). International Journal of Bifurcation and Chaos **16**: 239-268.

Lam, L. [2008a]. Science Matters: A unified perspective. *Science Matters: Humanities as Complex Systems*, Burguete, M. & Lam, L. (eds.). Singapore: World Scientific. pp 1-38.

Lam, L. [2008b]. Human history: A Science Matter. *Science Matters: Humanities as Complex Systems*, Burguete, M. & Lam, L. (eds.) Singapore: World Scientific. pp 234-254.

Lam, L. (林磊) [2022a]. *Research and Innovation.* San Jose: Yingshi Workshop.

Lam, L. [2022b]. *Science and Scientist.* San Jose: Yingshi Workshop.

Lam, L, Freimuth, R. D., Pon, M. K., Kayser, D. R., Fredrick, J. T. & Pochy, R. D. [1992]. *Filamentary Patterns and Rough Surfaces. Pattern Formation in Complex Dissipative Systems*, Kai, S. (ed.). Singapore: World Scientific.

Lam, L., Veinott, M. C. & Pochy, R. D. [1995]. Abnormal Spatio-Temporal Growths. *Spatiotemporal Patterns in Nonequilibrium Complex Systems*, Cladis, P. E. & Palffy-Muhoray, P. (eds.). Redwood City: Addison-Wesley.

Lam, L., Bellavia, David C., Han Xiao-Pu, Liu Chih-Hui A., Shu Chang-Qing, Wei Zhengjin, Zhou Tao & Zhu Jichen [2010]. Bilinear effect in complex systems. EPL **91**: 68004.

Lin Lei (Lam, L.) [1982]. Liquid crystal phases and “dimensionality” of molecules. Wuli (Physics) **11**(3): 171-178.

Lin Lei [2003a]. The humanities-science synthesized Histophysics: The unique role of physicists in developing human history. Sciencetimes, Aug. 29, 2003.

Lin Lei [2003b]. The humanities-science synthesized Histophysics. *Emerging Interdisciplinary Disciplines*, Liu Guo-Kui (ed.). Beijing: Tsinghua University Press.

Lin Lei [2013]. Research in human history: An example of scimat. *Renke*: *Humanities as Complex Systems*, Burguete, M. & Lam, L. (eds.). Beijing: China Renmin University Press. pp 232-251.

Morby, J. E. [2002]. *Dynasties of the World.* Oxford: Oxford University Press.

Wang Qian-Guo-Zhong (王钱国忠) (ed.) [2003]. *The Bridge between East and West in Scientific Culture: A Study of Joseph Needham*. Beijing: Science Press.

Shermer, M. [2001a]. Starbucks in the Forbidden City. Scientific American, July 1, 2001. 10.1038/scientificamerican0701-34.

Shermer, M. [2001b]. I Was Wrong. Scientific American, Oct. 1, 2001. 10.1038/scientificamerican1001-30.

Snow, C. P. [1998]. *The Two Cultures.* Cambridge, UK: Cambridge University Press.

Original: Lam, L. [2022]. History of Histophysics. *China Complex*, Lam, L. San Jose: Yingshi Workshop. English translation here.

15 How Nature Self-organized: Active Walks in Complex Systems

Caffe Trieste, San Francisco (photo Lui Lam 2022)

Ants do it. Birds do it. Humans do it.

1 Self-organization

When ants go out and look for food, they do not know where the food is initially. Somehow, they find it, carry it home, and recruit more ants from the nest to join them. Some kinds of trails are formed spontaneously, as shown in Figure 1. There is no central planning, no central command. The ants just self-organize and get the job done efficiently. How do they do it?

On telephone wires or building ledges we often see birds sitting side by side, chatting, or so it seems. Do they have a chat group? Or even a chat group leader? No. They just *self-organize*.

In early human history people self-organized to form groups, and groups self-organized to form societies. More recently, prompted by an external event—the World War II—countries self-organized to form the United

Nations. On a smaller scale, in the city of San Jose (where I teach), everyone (or almost everyone) gets his or her hair cut, in one way or another. The number and locations of the barbershops are not decided by any organization or committee. It just happens, seemingly by a miracle. But it is not a miracle; it is *self-organization.*

Self-organization does not happen in a vacuum. It depends on the nature of the individuals, the interaction between the participants, and the environment. All three of these components mutually influence each other. We have a good understanding of self-organization from physics—for example, thermal convection in a glass of beer.

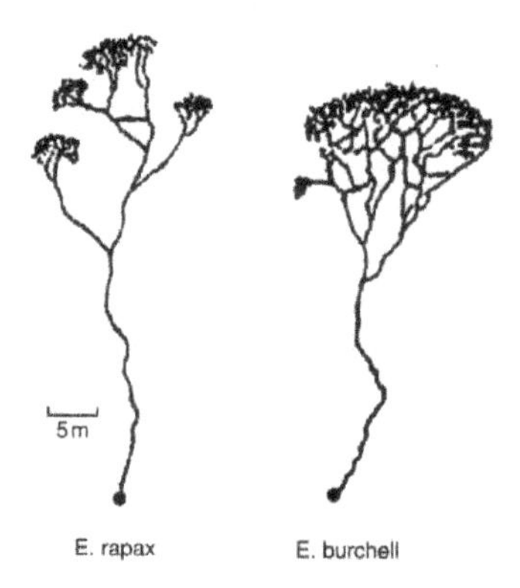

Fig. 1. Some real ant trails.

However, for complex systems like ant and human societies, we know much less. The aim of scientists is to find a unified description of self-organization that covers all complex systems. Exciting headway has been made in the last eight years in describing complex self-organization through a theory called *Active Walk.*

2 Universality

In the long pursuit of an understanding of the universe, and especially of human society, the emphasis has always been on the universal characteristics or behavior that are shared by all systems. As noted by Erwin Schrödinger in the preface of his 1967 classic *What Is Life*? this emphasis is reflected in the choice of the word "university" in naming our institutes of highest learning. In this regard, one may also note that the highest degree of learning conferred by a university is the Doctor of Philosophy (Ph.D.), irrespective of the subjects studied. Incidentally, philosophy is a Greek word meaning the "love of wisdom"; it is "wisdom" as the object, not any particular subject.

It is interesting to recall that Aristotle (384-322 BC), a Greek scholar, did not focus his attention on just one or two branches of knowledge but studied and contributed significantly to all fields, including biology, psychology, physics, and literary theory, as well as inventing formal logic as a system of thought. The fragmentation of learning and the compartmentalization of knowledge into different disciplines such as physics, chemistry, and economics is a rather recent phenomenon, occurring only in the last few centuries. It is suspected that this trend is due more to administrative convenience than to the nature of science itself.

Recent attempts to recoup a unified approach to knowledge with a view to finding a common description of both natural and social systems, include cybernetics and general systems theory from the 1940s to the 1960s, Ilya Prigogine's *dissipative structures* and Hermann Haken's *synergetics* in the 1970s, and the blossoming study of complex systems since the 1980s.

3 Complex Systems

A complex system is one that consists of a large number of simple elements, or "intelligent" agents, interacting with each other and the environment. The elements/agents may or may not evolve in time, and the behavior of the system cannot be learned by the reduction method, meaning that knowledge of the parts is not enough to predict the behavior of the whole system. But such a definition is not without problems. For example, a system may appear complex only because we do not understand it yet (by the reduction method or not). Once understood, it becomes a simple system.

Moreover, whether a system is complex or not may depend on the particular aspect of it that one happens to study. For example, if want to know about the inner structure and formation mechanism of a piece of rock, the rock could be a complex system. But if we want only to know how the rock will move when given a kick, then the classical Newtonian dynamics will do, and the rock is considered simple.

A precise definition of a complex system is thus difficult to devise, as is frequently the case in the early stages of a new research field. However, this difficulty has not hindered the study of complex systems much. In practice, it is safe to say that almost the subjects covered in the various departments of a university—except for those in the conventional curriculums of physics, chemistry, and engineering departments—are in the realm of complex systems. The topics studied span a wide spectrum, including human languages, the origin of life, DNA and information, evolutionary biology, economics, psychology, ecology, earthquake geology, immunology, cellular automata, neural networks, and the self-organization of nonequilibrium systems. In other words, the study of complex systems is a study of the real world. The recent surge in the study of complex systems could be attributed to three developments.

First, in the early 1980s the science of *Chaos* was better understood, and time series obtained in almost every discipline—from a dripping faucet to a heartbeat to the stock market—were subjected to the same analyses as inspired by chaos theory. The importance of this development was that chaos theory seemed to offer scientists a handle (or an excuse if one was needed), to tackle problems from almost any field. A psychological barrier was broken; no complex system was too complex to be touched.

Second, a crucial influence that helps to propel and sustain complex systems as a viable research discipline is the prevalence of personal *computers* and the availability of powerful computational tools, such as parallel processors. While theoretical study of complex systems is usually quite difficult and sometimes appears impossible, one can always resort to some form of computer simulation (or computer experiment, as some like to call it.)

Finally, in the last 20 years the study of basic physics by going to smaller and smaller scales as practiced in the discipline of particle physics has hit an *impasse*. The forefront of particle physics is superstring theory—called the Theory of Everything in some quarters—which is very complicated mathematically and could not be tested by experiments for the time being. The impasse encouraged many physicists to turn their attention elsewhere, looking at systems of increasing scales and squarely facing the issue of complexity characterizing these large systems. It should be noted that biologists, engineers, paleontologists and many others have studied complex systems daily. What makes the present period of the study of complex systems distinct and exciting is the involvement of a large number of physicists who, working with computer scientists and others, bring with them new tools and concepts that give new hope in unifying the field.

4 Paradigms

If there were a universal law governing all complex systems, what would it look like? How would we recognize it? In my opinion, there are two signs of a successful universal law applicable to all complex systems:

1. The law should be simple enough to be stated in one or two reasonably short sentences.
2. The law should conform to our daily experience.

The first one follows from the wide applicability of the law. The second one is due to the fact that each one of us is a complex system by herself or himself. (As an example, the second law of thermodynamics that states "heat cannot flow on its own from cold to hot bodies" easily satisfies these two requirements.)

Unfortunately, such a simple, universal law for complex systems has not been found, yet. Instead, in the last 20 years or so there emerged *four* general paradigms, each of which was found to be applicable to a large number of complex systems, if not all of them. The first is **Fractals**, founded by Benoit Mandelbrot in the early 1980s. Many complex systems exhibit the self-similar characteristic that a small part of itself, when enlarged in scale, resembles the whole; these are called fractals. Examples of fractals include the random walk, blood vessels, trees, rivers, clouds, stock market fluctuations, and so forth. Ubiquitous as they are, not all complex systems are fractals; for example, the ant trails in Fig. 1 are not fractals.

The second paradigm is **Chaos**. Chaos was studied by the French mathematician Henri Poincaré at the turn of the last century, but was resurrected in the late 1970s through the works of Edward Lorenz and Mitchell Feigenbaum. In the realm of science, chaos is a technical word characterizing the behavior of some deterministic nonlinear systems that are sensitively dependent on initial conditions. This usage of the word chaos obviously differs from that adopted in our daily lives, in which chaos is synonymous with a state of utter confusion. When the distinction between these two usages is ignored—consciously by a few scientists, unconsciously by most lay people and some science writers—a state of utter confusion does arise! And we are left with a bundle of dubious books like *The Tao of Chaos* and *Seven Life Lessons of Chaos*. (It is hard to claim that life is a deterministic system, or that life always depends sensitively on what you do in your early life, even though it may occasionally happen that way.)

As noted, it is obvious that chaos and fractals are two very different things. The connection between the two lies in the fact that for dissipative chaotic systems, in the mathematical *phase space* there exists one or more structures called *strange attractors*; and it happens that these strange attractors are fractals. (An example of a dissipative system is a simple pendulum immersed in water—energy kept drained away by friction and the pendulum will eventually come to a stop. This is not a chaotic system, however; but a pendulum with the hanging point in oscillation is.) This confusion between chaos and fractals led Martin Gardner to remark that chaos theory "is fashionable and interesting, but…it is mostly fractal geometry...more of a temporary fad like catastrophe theory" (see *Skeptic*, Vol. 5, No. 2, 1997, p 61). The truth is that chaos theory has been tested and confirmed in many real systems, from dripping faucets, to lasers, to the planet Pluto. Chaos theory is regarded as a great revolution in 20th-century physics, along with relativity and quantum mechanics. (See, for example, my 1997 book

Introduction to Nonlinear Physics and Abraham Pais' 2000 book *The Genius of Physics: A Portrait Gallery*.)

Another unnecessary confusion arises from the occasionally interchangeable use of the words "nonlinear dynamics" and "chaos theory." Chaotic systems are nonlinear systems, but not vice versa. In other words, not every nonlinear system (such as a simple pendulum) is chaotic. More recently, the word "chaoplexity" has crept into the literature. This is unfortunate and must be clarified. While a chaotic system does appear complex, it does not follow that every complex system is chaotic in origin. Chaos is just one of many routes that a system can become complex. A swarm of ants is very complex, but it is not chaotic. Each one of us is a very complex system—perhaps the most complex system in the universe—but most of us, most of the time, are not chaotic.

The third paradigm of complex systems is **Self-organized Criticality**, first proposed in 1987 by Per Bak, Chao Tang, and Kurt Wiesenfeld. Using the model of a sandpile, it asserts that large dynamical systems tend to drive themselves to a critical state with no characteristic spatial and temporal scales. While self-organized criticality, in the incarnation of power laws, shows up frequently in many computer models, it is less established in controllable, testable real systems, in contrast to the cases of fractals and chaos. Nevertheless, Vice President Al Gore apparently forgot what he learned as a kid playing on the beach, when he claimed that he found self-organized criticality "irresistible as a metaphor," and that it helped him to understand change in his own life.

The fourth paradigm is **Active Walk**, initiated by myself and my students in 1992. We proposed that the elements/agents in a complex system communicate indirectly with each other through their interaction with the deformable landscape they share. Each element is an active walker in the sense that it changes the landscape when it moves on its surface, and is influenced by the changed landscape in choosing its next step (Fig. 2). Complex behavior of the system could result from very simple rules governing the walker's interaction with the landscape.

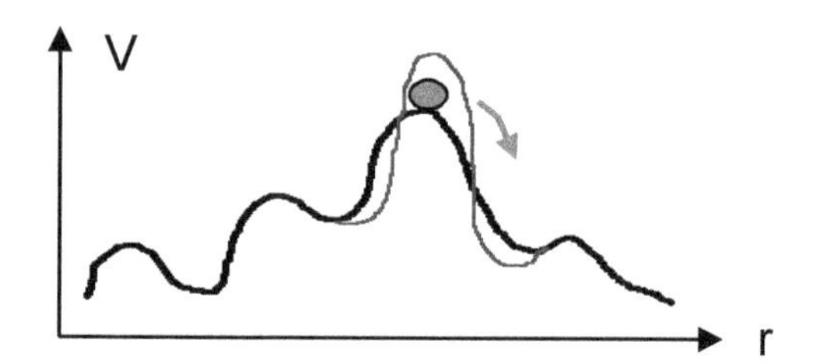

Fig. 2. Sketch of an active walk. The walker, represented by the solid dot, changes the landscape (from the solid line to the broken line) around itself according to a landscaping rule.

For example, an ant is a living active walker. It releases a chemical as it walks, and it moves toward regions of high chemical concentration. In this case, the landscape is the spatial distribution of the chemical concentration, the so-called chemical concentration "field" or "potential." A swarm of such ants can efficiently self-organize to do a number of nontrivial chores, such as foraging for food, without any central direction from the queen ant. Active walks have been applied successfully to a variety of problems in both the natural and social sciences.

5 Active Walk

In 1990 I took an active walk myself, in the metaphorical sense. In January I organized a Winter School in Nonlinear Physics at San Jose State University. In March I chaired the American Physical Society Symposium on Instabilities and Propagating Patterns in Soft Matter Physics in Anaheim, California. In June I was the Director of a NATO Advanced Research Workshop on Nonlinear Dynamical Structures in Simple and Complex Liquids, in Los Alamos. This was followed in July by a meeting in Vancouver, British Columbia, Canada, where I established the International Liquid Crystal Society during the 13th International Liquid Crystal Conference.

It was a few days before this conference in Vancouver that we rushed through a simple experiment in our Nonlinear Physics Laboratory at San Jose State University, in the basement of the Science Building. The experiment was so simple that it could be finished in less than one second. What we did was take a liquid crystal cell—a thin layer of liquid crystal between two conductive, transparent glass plates (the same one you would find in any digital watch or calculator)—replaced the liquid crystal by oil, and applied a high enough voltage across the cell. After a flash of light, the experiment was finished, and we found a complicated filamentary pattern left on the inner surfaces of the coated glass plates. An example is shown in Fig. 3, left. We presented these results in Vancouver.

The physical and chemical processes giving rise to these filaments are rather complicated, and are still being investigated. Essentially, the filaments are the locations that a series of dielectric breakdown—like lightning in the sky—takes place in the cell. After this experiment, without knowledge of the physical mechanisms, we set out to do a computer modeling of this filamentary pattern. We soon realized that a growing filament could be identified as the track of a walker. To grow the filament, we only have to tell the walker how to walk, and specify how the walker changed the local environment as it walks. One of our computer results is presented in Fig. 3, right, which agrees fairly well with the experiment. When I wrote up the paper for my 1992 book, *Modeling Complex Phenomena*, I named it the "Active Walker Model," and subsequently called the process an "Active Walk."

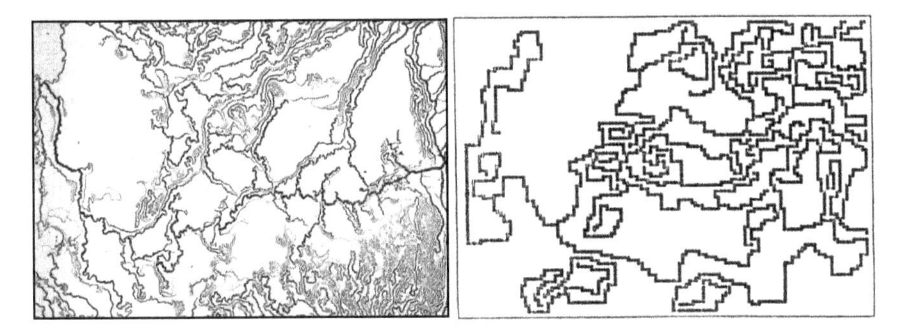

Fig. 3. *Left*: An experimental filamentary pattern. *Right*: An active walk computer simulation.

The use of a walker in modeling physical and other phenomenon runs a long history. The most well-known is a random walker, which has been used in mimicking the motion of a completely drunk person, the Brownian motion of a particle suspended in a liquid, or the fluctuations in a financial market. A random walker does not change anything in its environment and is what we would call a *passive* walker. In contrast, an active walker changes the landscape as it walks and is an *active* walker.

The description of an active walk thus involves two interacting components: the location of the walker as a function of time, and the deformable landscape as a function of time and space. The dynamics of an active walk are determined by three constituent rules:

1. The landscaping rule, which specifies how the walker changes the landscape as it walks.
2. The stepping rule, which tells how the walker chooses its next step.
3. The landscape's self-evolving rule, which specifies any change of the landscape due to factors unrelated to the walker, such as, in the case of ants, chemical evaporation or blowing wind. These rules may evolve in time for "intelligent" active walkers. The details of these three rules should depend on the system under study.

The track of the active walker forms a filamentary pattern, while the landscape usually becomes a rough fractal surface after some time. Of course, any number of active walkers may coexist, and they communicate with each other indirectly through their individual interaction with the shared landscape.

The landscape could be one of two types. The first type are physically existing surfaces; the second type are abstract mathematical artifacts. Examples of physical landscapes include the following cases:

1. a woman walking on a sand dune; the sand dune surface is the landscape.
2. Percolation in soft materials; when water flows through a porous, deformable medium, the shape and distribution of the pores would

be changed, and the porous medium itself is the (three-dimensional) landscape.

3. For ants, as already noted, the chemical distribution is the landscape. Frank Schweitzer, Kenneth Lao, and Fereydoon Family have successfully constructed an active walk model of ants. Ant trails like those in Fig. 1 are recovered. Furthermore, simulation of ants in the presence of randomly located food sources shows that the ants essentially attack and exhaust the food sources one at a time, a behavior observed in real ants.

4. Bacteria or fishes tend to move to regions with higher nutrients; as they move, they consume and reduce the nutrient concentration—the nutrient distribution thus constitutes the deformable landscape.

Abstract landscapes can be illustrated by two examples:

1. Urban growth can be modeled by the aggregation of active walkers. Initially, a "value" can be assigned to every piece of vacant land lot according to its location. For example, a lot on the flatland will have a higher value than one located on the hill; a river nearby can increase the value of the lot. The value could be taken to be the probability that the lot will be developed, by having a house or factory built on it. Once this happens, the value of the lands nearby will be increased and a new house will be added somewhere. The process is repeated. In this case, the spatial distribution of the value is the abstract landscape; the house is an active "walker," which does not walk. Such a model is very realistic but has not yet been tried.

2. The second example is the fitness landscape employed in evolution biology. Every species is in coevolution with other species. The presence of species A (Fig. 4a) affects the fitness landscape of species B (Fig. 4b), which in turn changes the fitness landscape of A (Fig. 4c); the changed landscape of A then determines how A will move. If we want to describe this evolutionary process by a simple model involving A alone, we will go directly from Fig. 4a to Fig. 4c, with Fig. 4b hidden. It then seems that A deforms its own fitness landscape at every step of its movement; that is, A acts like an active walker.

A representative simulation from the active walk model is shown in Fig. 5, which compares very well with three real results from physical and biological systems. More amazingly, we produced our simulation without the experimental results in mind and before one of them was created in the laboratory. It shows that active walk is really in the bag of tricks of Mother Nature when she wants to produce complicated patterns with minimal effort.

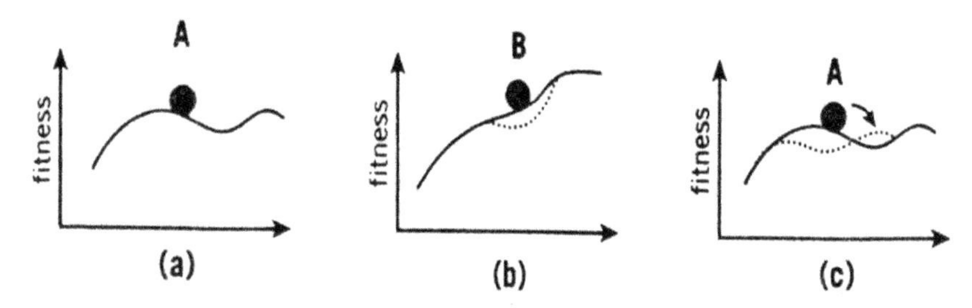

Fig. 4. Sketch of coevolution of two species A and B (see text).

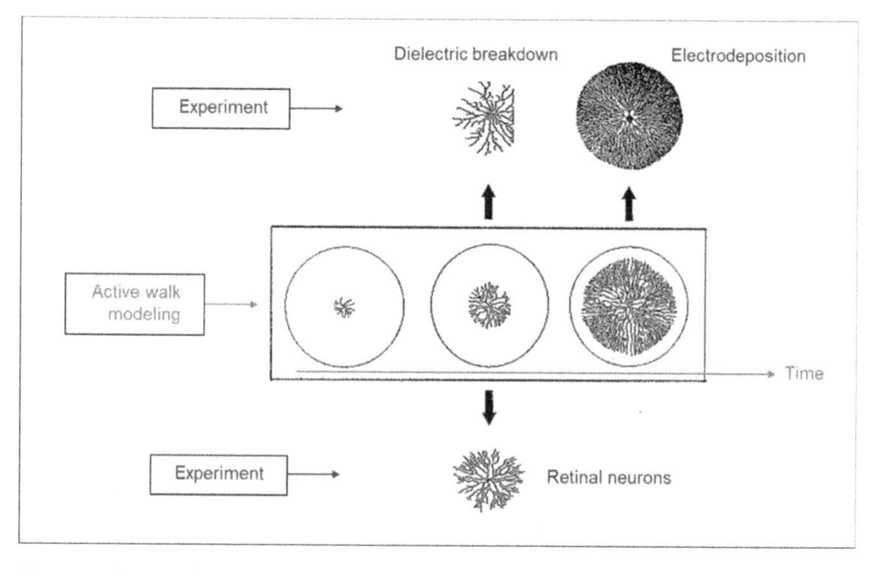

Fig. 5. Comparison of active-walk computer simulation with experimental results. *Middle*: Time development of a computer generated pattern; time increases from left to right. *Top left*: Chemical reaction pattern induced by a series of dielectric breakdown in a thin layer of oil. *Top right*: An experimental electrodeposit pattern. *Bottom*: A retinal neuron.

6 Applications in Economics

Increasing returns refers to self-reinforcing mechanisms in the economy, whereby chance events can tilt the competitive balance of technological products. For example, the Beta video system lost out to the VHS system, despite the slightly technical superiority of the Beta version, because more people happened to buy the VHS version in the beginning. (A student told me that this was because videotapes of X-rated movies were recorded in the VHS format at that time-this claim remains to be checked.) Increasing returns can be easily modeled by having an active walker jumping among the sites on a fitness landscape of the product types (Fig. 6).

Fitness is a measure of how desirable the product is. The probability that a product would be bought by the next customer could be taken to be an increasing function of its fitness; that is, the higher the fitness, the more likely the product would be bought. The uncertainty, in the form of the probability, is due to the customer being only partially rational or having only partial information about the products, or both. A typical numerical result is shown in Fig. 7. We assume that the two products are equally good in the beginning. When information is completely lacking, customers choose the products randomly and both products will survive and coexist.

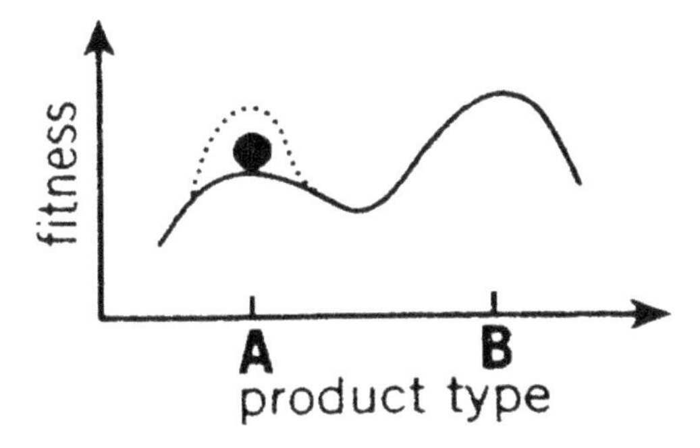

Fig. 6. The active-walk description of competition between two different products A and B. The solid dot represents the active walker. When the dot is on site A, it means product A is bought by a new customer. The fitness of A is then raised by a constant amount. The same goes for B.

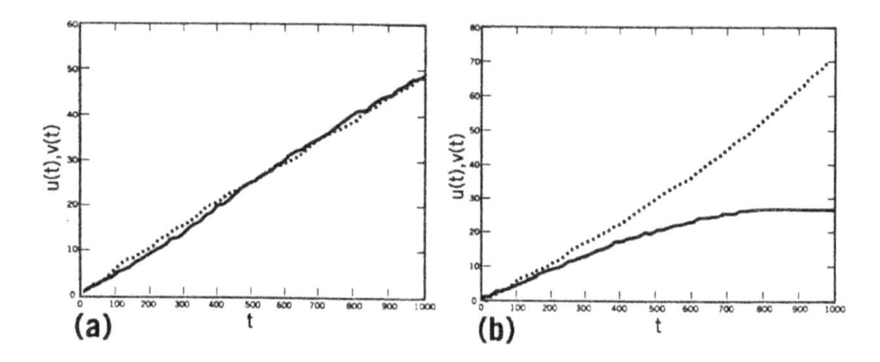

Fig. 7. Numerical results for two competing products. Here u(t) and v(t) are the fitness of products A and B at time t, respectively; each is proportional to the total number of that product sold. (a) The two products coexist in the market, when customers' buying is completely random. (b) One product wins out, by chance (see text).

But when some information is available or the customers are only partially rational, after a period of coexistence, one of the products will clearly win out (Fig. 7b)-like in the case of the PC and the Apple computers.

Which product will win is purely by chance and cannot be predicted. Recently, Daniel Friedman and Joel Yellin at the University of California, Santa Cruz, have applied the active walk model to two problems in economics, namely, population games and the dynamics of rank dependent consumption. Robert Savit and coworkers at the University of Michigan, Ann Arbor, succeeded in using active walk to model adaptive competition and market efficiency. These works also have implications for biological systems.

7 Modeling History

When historians write narrative histories they use such phrases as "leave a mark in history" or "follow the giant's footsteps." But to model these scientifically, one needs a deformable landscape so that marks and footsteps can be left on it. Moreover, one needs a walker to do these things. As noted by the Chinese, through the action of individuals bad things can sometimes be turned into good things, and vice versa; the walker should be an active walker. The active walk model, then, fits naturally into historical analyses. With prudent application, it could be the first step in achieving a quantitative theory of history.

For example, the "model of contingent-necessity" of Michael Shermer (see the final chapter of his 2000 book *How We Believe*), which essentially says that contingencies are important in the early stages of a historical sequence but less so in the later stages, could be understood as follows. In the beginning, the walkers are close to each other at the center of the landscape plane. There is less space for the walkers to navigate and, in a crowded situation, each walker is easily affected by the action of others through the changing landscape they share. In other words, each step counts. Later in the sequence, when the walkers are more separated from each other and away from the center, each step counts less.

In another example, through the study of active walk models we find that the relative importance of chance—versus necessity—could depend on where in the parameter space that the system belongs (Fig. 8). If our world happened to sit in the sensitive zone, then history could indeed be very different if life's tape is replayed, as advocated by Stephen Gould in his 1989 book *Wonderful Life*; otherwise, history could be repeated, more or less. The problem is to know where our world sits.

Mingjun Lu, a translation scholar formerly at the Nanjing University (and later at University of Toronto), observes that there is more than one way to translate a sentence from one language to another. However, the choice is narrowed and affected by the particular translation of the previous sentence. Lu suggests that the translation process can be modeled by active walk, a very innovative idea in the theory of translation. In fact, any decision-making process—translation and management in particular—is a "history-making" process and thus could be modeled by active walks.

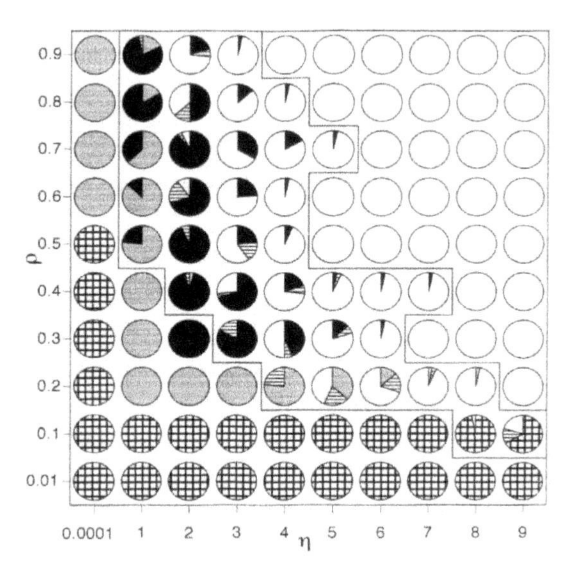

Fig. 8. A summary of the outcomes from a probabilistic active walk model. The two control parameters in the model are recorded on the horizontal and vertical axes. The pie chart at each point represents the percentage of each type of five outcomes obtained from 30 runs of the algorithm. Within the "sensitive zone" in the middle region—where chance plays a dominant role—the outcome is unpredictable.

8 What Does All This Mean?

There are more successful active-walk applications: Migration of workers and economic agglomeration, collective opinion formation, and pedestrian traffic studied by Frank Schweitzer in Germany; cold production in heavy oil by Jian-Yang Yuan in Canada; tumor growth by Thomas Deisboeck of the Harvard Medical School; and surface filamentary pattern formation by Ru-Pin Pan in Taiwan.

Active walk is a (classical) field theory of complex systems. The field—the landscape—could be either physical or mathematical. In physics, we know that elementary particles interact with each other through a field they share. It is the same between elements and agents in a complex system.

Space limitation prevents us from spelling out all the "Life Lessons of Active Walk" here. But let me tell you the first one: Each of us is an active walker. By changing the surrounding landscape with each action, every person has the potential to make history!

Further Reading

Lam, L. [1998]. *Nonlinear Physics for Beginners: Fractals, Chaos, Solitons, Pattern Formation, Cellular Automata and Complex Systems*. Singapore: World Scientific.

Original: Lam, L. [2000]. How nature self-organizes: Active walks in complex systems. Skeptic **8**(3): 71-77.

16 Histophysics: How to Model History and Predict the Future

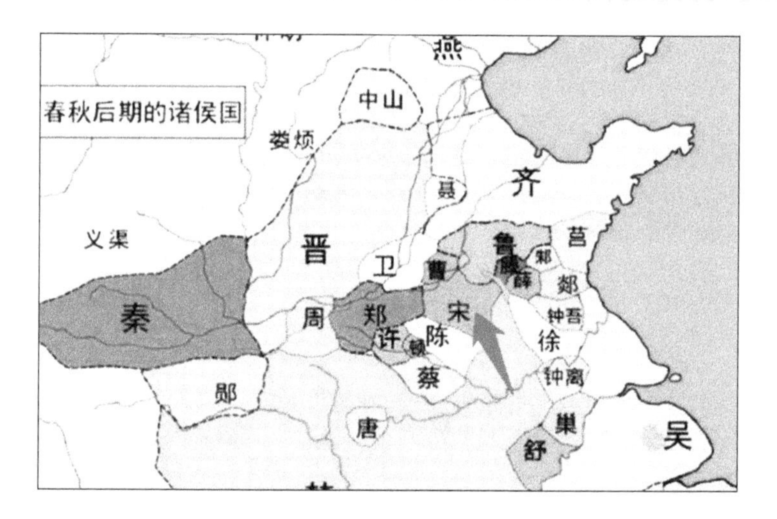

History is the study of events that occurred in the past. Human is a (biological) material system consisted of *Homo sapiens*, and natural science studies all material systems, so history should be a part of natural science. However, so far, the average person does not think so. Why? To answer this question, we must explore the nature of history. So, what exactly is history?

1 Introduction

History is the study of past events and is not repeatable, unlike many cases in natural science that can do controlled experiments. Most historians therefore believe that history cannot be a science, but some important scholars in the past, including Robin Collingwood (1889-1943) [5.1], believe that history, like physics, can also be studied in a scientific way and some laws can be derived [Breisach 1994; Stanford 1998]. However, this recognition is not yet universally accepted, and we think this is because historians have not received enough scientific training.

We will point out the unique role that physicists can play in developing the history of humankind and discuss in detail several aspects: 1. Historical research methods. 2. Worldview. 3. History as a complex dynamical system. 4. How to predict the future and retroact the past. 5. Artificial history. We will illustrate that "active walk" can be used to construct new worldviews and can be widely used to model history, and illustrate it with five examples from the history of sociology, economics, and evolution.

In fact, since Cai Yuan-Pei proposed "humanities-science synthesis" in 1918, especially the use of science methods to study liberal arts, the integration of arts and sciences has become an important goal of scholarship [4.1]. In the past two years, the author of this article has applied the research methods of physics to the study of human history and proposed a new discipline called *Histophysics* (physics of history) [Lam 2002].

2 What Is History About?

The object of historical research is a many-body system. In this system, each component is an individual, which we call "particle" here. Every particle is classical (not quantum mechanical) and distinguishable. Due to differences in size, age, and race, this many-body system is also a heterogeneous system composed of different species.

The interest of the historians is about anything happened in the *past* that is related to these agents (which exist presently or in the past). The system being investigated by the historians is thus a system of material bodies, and thus can be studied scientifically. But how?

For this purpose, let us examine the constituents more closely. For comparison, a well-studied many-body system in the physical sciences—water—will be used (Fig. 1).

1. In water, the particles are water molecules. In history, the particles are biological bodies.
2. A water molecule is composed of two hydrogen atoms and an oxygen atom; each atom is composed of a nucleus and electrons;

each nucleus is composed of neutrons and protons; neutrons and protons are composed of quarks. Correspondingly, in history, each biological body is composed of organs (brain, heart, stomach, etc.), blood and other liquids; each organ is made up of cells (a total of about 5 trillion cells in a body); each cell is made up of (mostly organic) molecules, such as DNA; each molecule is made up of atoms—and from this point on, there is no distinction between the two cases of water and history.

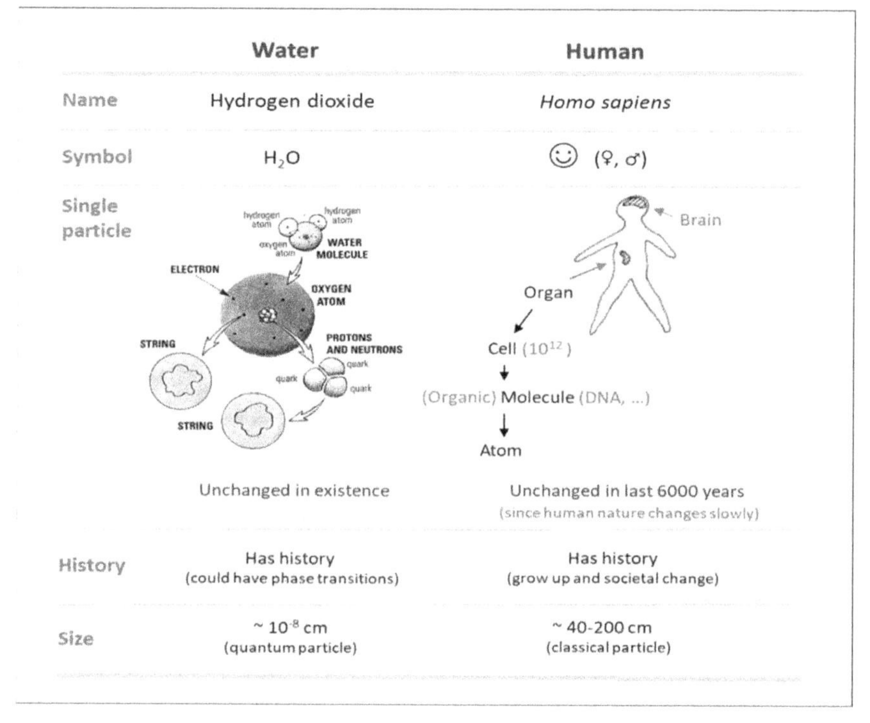

Fig. 1. Comparison between water molecule and human body.

3. In both these two systems, there are layers and layers of structures, and many *internal states* in each particle.

4. The existence of atoms in water and in historical systems has a long history. The nature of the water molecule has not changed. What about human? The history of human evolution is very long,

but human nature has basically remained unchanged. Because of this, it is possible for historians to speculate on human thoughts thousands of years ago.

5. The size of a water molecule is very small, about 10^{-8} cm, which is a quantum particle, while the size of a human body is about 40-200 cm, which can be regarded as a classical particle.

3 Motivation

Our study of history from the perspective of physics is based on the following considerations:

1. **Human system is the most complex system in the universe**. The so-called complex system is a system composed of many monomers, which generally cannot be studied by the method of reduction; examples are economic system, ant colony, human brain, transportation, human beings, etc. [2.3]. Anyone interested in complex systems should be interested in the human system.

2. **The time is ripe for the study of history using physical methods**. Recently, great progress has been made in the study of complex systems and human organisms. For example, the enhanced computing power of computers can do modeling with massive data; people are gradually understanding the phenomenon of chaos, deciphering the human genome, and doing real-time scanning on the human brain, etc. So now is the time to study history from a physical and biological point of view. As the famous Harvard University biologist Edward O. Wilson (1919-2021) once said: The study of the human condition is currently the most important frontier of natural science.

3. **Historians need to draw on the power of other disciplines to turn history into a real science**. Historians in past and present have received inadequate scientific training. For example, in universities, neither undergraduates nor graduate students of history receive systematic education in advanced mathematics and computer programming.

4. **Physicists working with historians can turn history into a real science**. Physicists know how to solve many-body problems. In fact,

physicists invented statistical mechanics to study many-body problems 100 years ago, and there are many successful applications. Physics is not merely the study of materials, energy, etc.; the use of physical methods to study any object belongs to the category of physics. History shows that physicists have had a lot of successful experience in interdisciplinary fields, such as astrophysics and biophysics. Collaboration between physicists and historians will raise the scientificity of history as a discipline.

4 Methods of Study in History

In any science discipline, the scientific method consists of the following steps [23]:

1. Starting from raw data or existing theory, a new hypothesis is proposed.
2. Consequences from the new hypothesis are checked with data.
3. In agreement, the hypothesis is confirmed and turns into a new theory—new laws may be discovered; otherwise, go back to step 1.
4. New findings from the theory are published in peer-reviewed research journals. In exceptional cases,
5. Popular science books are written, by the scientists themselves or, mostly, by others.

In the profession of history, these steps are fragmentized. The proposal of a new hypothesis, step 1 above, is called *constructionism*. The discovery of new laws, step 3, is called *reconstructionism*. And many research results in history, expressed in the narrative form, are published as popular books, step 5, without going through step 4—peer-reviewed journals. It is not surprising then, that history is not yet a scientific discipline.

Some representative research methods in history are noted here.

1. The "internal states" of the brain are emphasized by Robin Collingwood when he insists that historians must reenact the thoughts of historian figures in understanding history. This approach is called

"method acting" in the movie industry; it is very difficult and can only be mastered by a few gifted persons.

2. History—like archaeology, paleontology, and a large part of astronomy—is about things happened in the past. The importance of social science and long-time view in history study is advocated by the French Annales School.

3. The objectivity in, and the possibility of reconstructing past "reality" through history narratives are ruled out by Hayden White (1928-2018), Jacques Derrida (1930-2004) and other deconstructionists because, they argue, the meaning of any writing is undecidable. These and other postmodern attacks lead to a crisis in the history profession in the 1990s [Evans 1997].

However, as we shall demonstrate below, research in history can benefit from the inclusion of some scientific approaches developed in physics.

5 Worldview

Worldviews, through the action of powerful political leaders and governments, have tremendous consequence in applied history. The study of worldviews is important to historians.

Self-organized criticality, after its introduction in physics, has been invoked to be the metaphor for history. The claim is that history works like sandpiles: When sand is added continuously from the top, the sandpile builds up by itself to a critical state, crumbles, builds up, crumbles, and so on. While there is no denial that someone's life may work like this, we know for sure not everyone's life and not everything in history goes through this repetitive and depressing route.

History, resulting from a combination of contingency and necessity, is a stochastic process with many possibilities. Some aspects of it may appear periodic, sometimes; other aspects, moving in a spiral or chaotically. Yes, it may even occasionally build up and crumble like in a sandpile. But it may also go through weird paths in time.

Active walk, a paradigm introduced by Lam in 1992 to handle complex systems [Lam 2005, 2006], provides exactly the needed foundation for such a worldview. In an *active* walk, a particle—the walker—changes a deformable potential—the landscape—as it walks; its next step is influenced by the changed landscape. (In contrast, in a random walk—a *passive* walk, the particle changes nothing of its environment.) For example, ants are living active walkers. When an ant moves, it releases chemicals of a certain type and hence changes the spatial distribution of the chemical concentration. Its next step is moving towards positions of higher chemical concentration. In this case, the chemical distribution is the deformable landscape.

In the last two years, the close connection between active walk and history is suggested by the author. The connection comes naturally. When historians write narrative history, they use such words as "leave a mark in history" or "follow in a giant's footsteps." It follows that to model these scientifically, one needs a deformable landscape so that marks and footsteps can be left on it. Moreover, one needs a walker to do these things. As noted by the Chinese, through the action of individuals, bad things can sometimes be turned into good things, and vice versa; the walker should be an active walker. The active walk model, then, fits naturally into historical analyses.

6 Modeling History as a Complex Dynamical System

The first step in the scientific study of any subject is the collection of empirical data. This step is followed in both physics and history. The next step is to summarize the data, which leads usually to some empirical laws. In history, there are fewer attempts at empirical laws, the existence of which is even doubted by some historians. An example of such laws is Michael Shermer's "model of contingent-necessity," which states essentially that history results from the combination of contingency and necessity, with the former being more important in the early stages of a historical sequence of events.

In fact, there are *three* research levels in any discipline [4.3]:

1. **Empirical**. For example, the ideal gas equation: PV = nRT.

2. **Phenomenological**. For example, based on a few simple symmetry principles, physicists derived a phenomenological equation for fluids—the Navier-Stokes equations, which are still widely used today.
3. **Bottom up**. For example, when water is known to compose of molecules people can use powerful computers to do computational simulations to study the collective behavior of water molecules.

When doing computer simulations, physicists made two *simplifications*:

1. Reduce the number of molecules to several hundred, which is much smaller than the real number of 10^{23} per cubic centimeter.
2. Simplify the interaction between molecules. In the study of physics, many real details can be ignored according to the research goal.

It is because of the simplifications and approximations that physics has been able to advance by leaps and bounds in the past few hundred years.

In addition, one can add an "**artificial** level." For example, when using the "lattice gas automata" method to study water, the water molecules are greatly simplified and become artificial molecules. The results obtained from this research method are the same as those obtained from the two other levels: empirical and phenomenological.

However, these three or four research levels did not all exist in previous historical studies. Traditional historical research uses the bottom-up approach: to understand history by knowing the historical figures. A possible reason is that historians believe that to fully understand history requires an understanding of the interactions between people, and that the more details about historical events, the better. In other words, they want to enter the third research level immediately, so it is very difficult.

Yet, we have gained two valuable experiences from physics research:

1. It is not necessary to know all the details.
2. Research can move forward by simplifications while retaining the main key factors.

7 Three Examples of Stochastic Processes

The process of history is a combination of chance and necessity. To help understand the general nature of this kind of stochastic processes, here are three physical examples:

1. Random walk on a horizontal **hard surface** (Fig. 2, left). Random walk is the simplest *passive* walk—the walker does not change the environment. The walk is probabilistic and the trajectories of the walking particle cannot be predicted. But we can ask: What is the morphology of the walker's trajectory? What is the dimension of the trajectory pattern, and is it a fractal? What is the relationship between the distance R (from the start point to the end point) and time t? Physicists have asked these questions and found answers. Similarly, history is a probabilistic system, and historians must ask the right questions. For example, you cannot ask in which year a certain dynasty or someone's life will end but only the probability that it will end in a particular year.

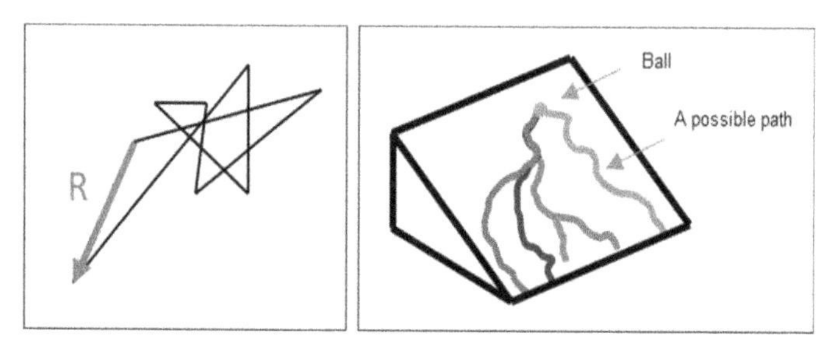

Fig. 2. Two stochastic processes. *Left*: Random walk on a horizontal hard surface —a passive walk. $R \propto t^{1/2}$. *Right*: Ball sliding down a soft slope—an active walk.

2. Ball sliding down a **hard slope** with sticks arranged in a triangular lattice (Fig. 3). The probability of sliding to the left or right gap between the stick is 1/2. After the first ball, drop the second ball from the same spot at the top of the slope. The trajectory of each ball is unpredictable, but the probability of the ball falling at a certain position on the bottom line of the slope follows the Gaussian distribution. Each ball is a *passive* walker. This example shows that in a probabilistic system such as history, although every step of an individual dynasty or individual is

unpredictable, if the questions are asked correctly, some quantitative trends or laws can be obtained.

3. Multiple balls sliding down in sequence on a **soft slope**—a deformable surface (Fig. 2, right). In this system, each ball is an *active* walker. The trace left by each ball affects the trajectory of the next ball—a stochastic process that depends on history and is potentially predictable. The process of human history is like this case—an active walk process.

If the ball in the last example is replaced by a current, like the Yangtze River entering the sea. Before entering the sea, the Yangtze River flows through a delta composed of loose soil. We cannot accurately predict the trajectory of the Yangtze River in the delta and where it enters the sea, but we can be sure that the Yangtze River will definitely flow into the sea. The "long river of history" metaphor is like the image of the Yangtze River entering the sea, with its inevitability and contingency.

Likewise, the role of a great man in history is like a big boulder moving through the Yangtze River. Although it may change the exact trajectory of the long river of history and the location where it enters the sea, it cannot change the inevitable trend of history. In other words, great heroes may change the details of history to a certain extent, but they cannot change the general direction of history.

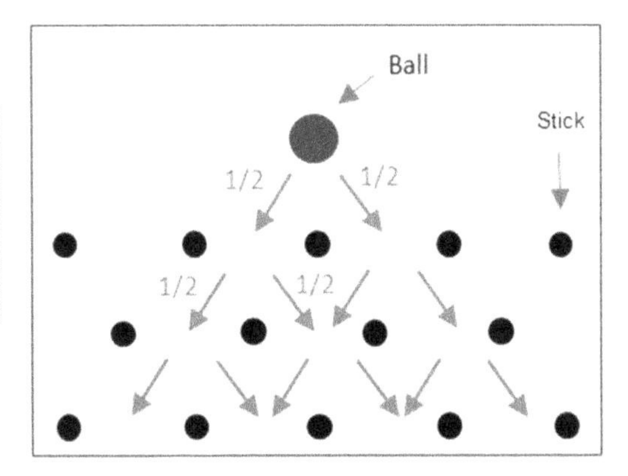

Fig. 3. Ball sliding down a hard slope with sticks—a passive walk. The actual path of the ball cannot be predicted, but the probability of the ball landing at the bottom positions can be predicted: a Gaussian distribution.

8 Six Examples of Histophysics

In historical research, our focus should not only be on the historical trajectory itself; there are many other issues that can be studied besides the trajectory. Six examples are presented here.

1. **Statistical distribution of war casualty**. Decades ago, British scholar Leslie Richardson made a statistical analysis on the number of war deaths in Europe, and found that in general, the number of large-scale wars was relatively small, and the number of small-scale wars was the largest [Richardson 1941]. This result can be represented by a power-law curve (i.e., the variable y is proportional to a certain power of the independent variable x) (Fig. 4, left). Surprisingly, there are many such complex systems in nature, such as earthquakes (Fig. 4, right) and the distribution of urban population. These results show that:

(1) Although the outbreak and scale of human wars are related to the personal behavior of some historical figures in terms of chance, they cannot change the characteristics of the power-law distribution of war deaths as a whole.

(2) The development of human history has common features with many other complex systems.

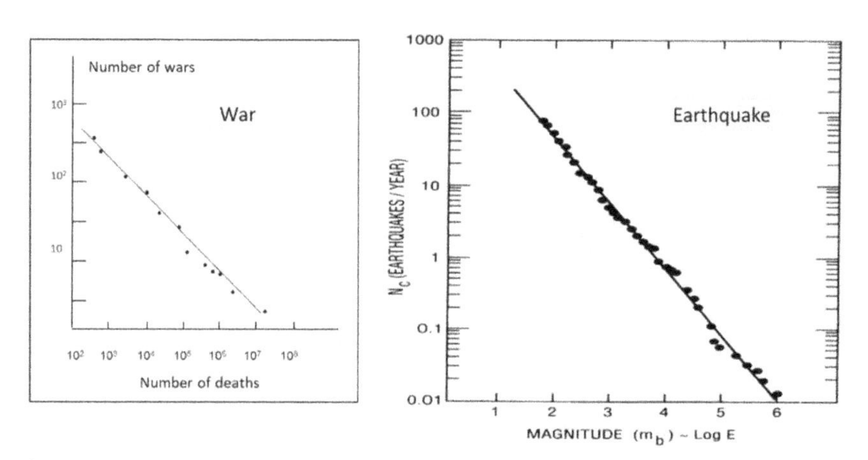

Fig. 4. Examples of two power-law distributions in complex systems. The two graphs are represented in log-log scale: A straight line indicates power-law distribution. *Left*: Distribution of war deaths. *Right*: Earthquake intensity distribution.

2. **Modeling history in economics**. Why did a product that was initially at a disadvantage gain market share and finally dominate the market? In Florence, Italy, there is a church tower clock built in 1443; the clock hands rotate counterclockwise (Fig. 5, left) [Brian 1990]. It can be guessed that two kinds of clocks existed then: One kind runs counterclockwise, another clockwise. In the northern hemisphere, before clocks were invented, sundials were used to keep time, and the shadow of the sundial in the northern hemisphere moved clockwise. In this sense, the counterclockwise clock at that time was a product at a disadvantage and was gradually eliminated. This is an example of an inferior product disappearing, which seems natural. Oddly enough, however, inferior products could end up winning the market too, such as the existence of the QWERTY keyboard [David 1986].

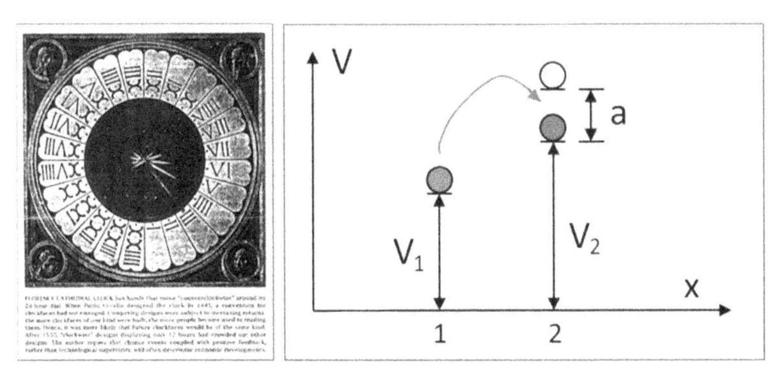

Fig. 5. *Left*: Tower clock (built in 1443) that runs counterclockwise in Florence, Italy. *Right*: A one-dimensional probabilistic active walk model with 2 positions. A particle (walker) jumps between two positions probabilistically according to a given function and changes the fitness V of the two positions.

To understand this, an active walk model with a single particle jumping between two sites is constructed [Lam 2005]. Each site, representing one product, is given a *fitness* height V_i ($i = 1, 2$). When the particle lands on a site, that corresponding product is bought by a customer, and the height of the site is increased by a fixed amount a. The probability that site i is chosen by the particle is proportional to $f(V_i)$, a given monotonic increasing function of V_i. (Note that the particle may stay in the same site.) A particular choice is $f(V_i) = \exp(\beta V_i)$, where the "inverse temperature" β varies from zero to infinity. The parameter β represents the combined effect of the rationality of the customers and how effective information is passed among the customers (through personal contacts or advertisements). A zero β corresponds to the two products being chosen randomly; an infinite β corresponds to the same product chosen all the time. This two-site active walk model can be mapped to a one-dimensional position (x)-dependent probabilistic walk and is solvable (Fig. 5, right).

Let p (x, t) be the probability of finding the particle at position x and time t. The solution p (x, t) of the probabilistic-walk equation is determined by the parameter β and the initial value x_0 of x ($x \equiv V_1 - V_2$). $x_0 > 0$ means product 1 has the first-mover advantage; $x_0 = 0$ means that products 1 and 2 have the same initial advantage; $x_0 < 0$ means product 2 has the first-mover advantage.

Figure 6 shows a typical numerical solution of the equation, using initial conditions: $p(x_0, 0) = 1$ and $p(x, 0) = 0$ for $x \neq x_0$. Shortly after the start of time, an asymmetric peak appears and evolves into two peaks. As time progresses, the two peaks move at constant velocity along the x-axis: one to the right and one to the left. Around x = 0, the positions of these two peaks are symmetrical. The area of the right peak represents the probability of product 1 winning in the market, and the left peak represents the probability of product 2 winning. Due to the probabilistic nature of the model, except for particular parameter values and initial conditions, it is unpredictable which product will actually win in the market.

Our analytic and numerical results show that:

1. For zero β and $x_0 = 0$, the two products coexist in the market.
2. For $\beta \to \infty$, the product first picked by a customer always wins out if $x_0 = 0$, and the one with initial advantage wins out if $x_0 \neq 0$.
3. For $x_0 = 0$ and finite β, each product has equal chance of winning, but which product actually wins out is unpredictable.

4. For finite x_0 and β, the product with an initial advantage has a higher chance of winning, but the other product has a non-zero chance of catching up and winning, too. And there is an optimal time that this chance of catching up becomes a maximum, implying that the initially disadvantageous product should stay in the market and not give up before this optimal time.

Of course, the initially disadvantageous product can change the rule of the game by increasing its own a, for example, by improving its quality or starting an advertisement campaign, or both. With sufficient real data, our model can be fitted to describe the competition between real products.

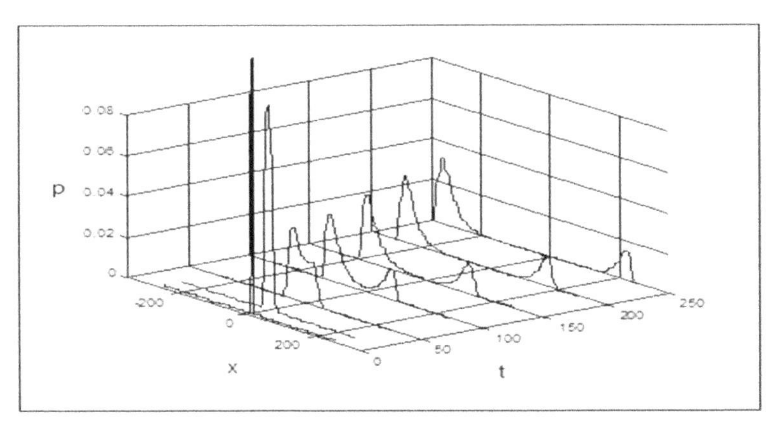

Fig. 6. Numerical solution of the function p (x, t), the probability of finding the one-dimensional position-dependent probabilistic walker at location x and time t. Here a = 1; x_0 = -2 and β = 0.1.

3. **Modeling evolutionary history**. In 1989, the famous paleontologist and science writer Stephen J. Gould (1941-2002) published the book *Wonderful Life* [Gould 1989]. From the fossil record found outside of Vancouver, Canada, it seems that some "advanced" organisms (with many legs, say) that should survive were wiped out suddenly. From this *single* data set, Gould concludes that contingency is extremely important, i.e., not the fittest will survive, contrary to what evolutionary theory of Charles Darwin (1809-1882) asserts. He asked: If life can be replayed from the beginning like a video, and the evolution of history can be repeated, will human beings still exist on Earth? I call this "The Gould

Question." His answer: **No**. (Of course, if the video tape is replayed, what you see will be the same as before. Gould's term "replaying life's tape" refers to the fact that if Earth evolves again after its appearance, will life appear on Earth as before?)

This idea has sparked a lot of debate, but no one has come up with further scientific proof. What is more, no second data set has been found so far. Our probabilistic active walk model results (Fig. 7, left) provide valuable insight into this debate. We found that the relative importance of contingency to necessity depends on the position of the system under consideration in the *morphogram* (a term I coined) consisting of control parameters. Chance is important if the system falls within the *sensitive zone* (another term I coined) of the morphogram. If so, there is a chance that life can be repeated, although the details of history may be different. In other words, our answer to the Gould Question is not the same as his. Our answer: **Maybe**.

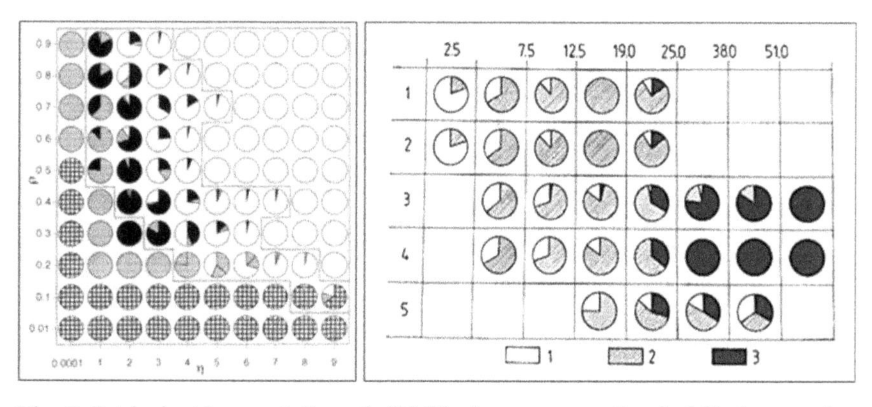

Fig. 7. Intrinsic Abnormal Growth (IAG) phenomenon of probabilistic complex systems [Lam et al. 1995]. *Left*: Morphogram resulting from a probabilistic active walk model. The two control parameters in the model are given on the horizontal and vertical axes. The pie chart at each point represents the percentage of five types of morphology obtained from 30 repeated runs (with same parameters but different sequences of random number). In the *sensitive zone* in the middle where chance plays a leading role, outcome of each run is unpredictable. *Right*: Data on the growth and reproduction of the plant Common Teasel—an example of IAG.

4. **Simulating the growth of a historical society**. A simulation of the development of a society in the Long House Valley in the Black Mesa area of northeastern Arizona, USA, was carried out by Axtell et al. [2002]. The simulation results show agreement with the quantitative historical data, which are reconstructed from paleoenvironmental research based on alluvial geomorphology, palynology, and dendroclimatology. For example, between the years AD 400-1400, the number of households has two peaks; this is reproduced in the simulation. So is the evolution of the spatial distribution of settlement (Fig. 8). In this study, heterogeneity in both agents and the landscape, hard to model mathematically, is found to be crucial.

The model starts with a landscape reconstructed from paleoenvironmental variables, which is then populated with artificial agents representing individual households. Five household attributes are specified, together with household rules guessed from historical data. The model involves 14 reasonably chosen parameters, plus eight adjustable parameters for optimization. The model is very detailed. It is interesting to see whether the model can be simplified to its bone, with fewer parameters, that can still produce the same essential results.

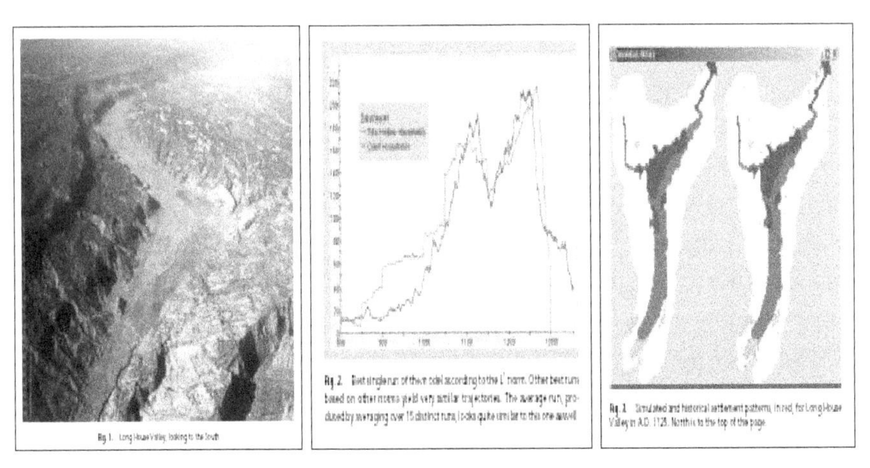

Fig. 8. A historical study of the growth and decline of a Village in Arizona, USA [Axtell et al. 2002].

5. **Modeling social history**. Will all societies end up as liberal democratic societies? Francis Fukuyama, considered one of fifty key thinkers on history, publishes in 1989 an article, End of History? [Fukuyama 1989]. He asserts that every human being needs two satisfactions, viz., economic wellbeing and "recognition," with the latter meaning respect by others. He argues that since the liberal democratic society is the only one that can satisfy its citizens on these two basic needs, consequently, given enough time, all societies will end up as liberal democratic societies. And that will be the end of history if history is understood to be directional change in societal forms. Misunderstandings of Fukuyama's thesis ensure, and debates go on in the history profession. Nothing is done scientifically to settle the issue.

In our view, the two human needs suggested by Fukuyama should be generalized to "body satisfaction" and "soul satisfaction." After all, body and soul (or spirit) comprise the whole of a human being. And we know for sure, e.g., when someone joins a revolution to change the society, the person may give up her or his life before the revolution succeeds, if at all—and recognition is not in the person's mind. The degrees of satisfaction of "body" and "soul" in each society can be quantified by an index, obtained from a survey of its citizens.

To test Fukuyama's thesis, one can represent each society as an active walker, a particle, moving in the two-dimensional space of "body" and "soul" indices (Fig. 9). At each point in this space, a fitness potential can be defined. The movement of each particle (usually, but not always, up the scales) will change the fitness landscape and influence the movement of other particles. The problem will be to find out, under what circumstances, all the particles will cluster together at the location corresponding to a liberal democratic society. It is thus a problem of clustering of active walkers in a two-dimensional deformable landscape. (The model can be generalized to include the possibility that two particles may combine into one, and some particles may split into two or three—some kind of chemical reactions—corresponding to the case in history that countries may get unified or fragmented in time.)

Such a problem has been studied before in physics in another context, and clustering of active walkers indeed occurs. The corresponding investigation in history as outlined above will bring Fukuyama's historical study one level up, to the scientific level, and serve as an example in other cases.

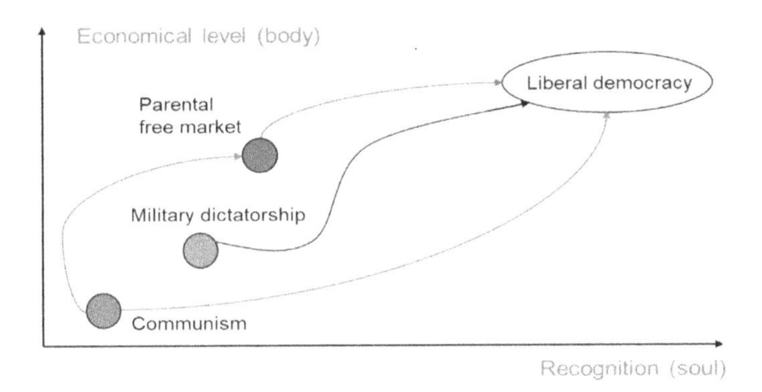

Fig. 9. Sketch of an active walk model for the evolution of political systems.

6. **The origin of Chinese historical fate**. There were 231 emperors in Chinese history from the Qin dynasty to the Qing dynasty (221 BC to 1912), and the number of years each emperor ruled, *regime* lifespan τ_R, varied from 1 year to 61 years (accuracy is 1 year), with an average of 12.5 years. If one looks at the changes in τ_R in each dynasty or arrange it over a period of more than two thousand years, one cannot see any trends. But if one looks at the histogram that includes all dynasties τ_R, one can see that the distribution of τ_R obeys a power law (Fig. 10), like the distribution of war casualty or earthquake size (Fig. 4). Thus, the rule of Chinese emperors is also a kind of complex system [Lam 2006].

The *dynasty* lifespan τ_D is defined as the sum of τ_R of all the regimes in the same dynasty. From the Qin dynasty to the Qing dynasty, if τ_D is rearranged in a decreasing sequence, and each dynasty is given a rank R, then the Tang dynasty has the largest τ_D (289 years), and R = 1; and so on. Figure 11 gives a plot of τ_D as a function of R (called a Zipf plot); the curve is divided into two straight lines—*Bilinear Effect* [Lam 2006; Lam et al 2010]. This means that if a dynasty's lifespan is less than 57 years, its possible lifespan jumps by 3.5 years each time. That is, if a dynasty can survive 3.5 years, then it can live another 3.5 years; and so on. On the other hand, if a dynasty's lifespan is greater than 57 years, its possible lifespan jumps in 25.6-year intervals. That is, if a dynasty can survive 25.6 years, then it can live another 25.6 years; and so on. Therefore, the distribution of the life expectancy of Chinese dynasties is discontinuous and "quantized." It shows 57 years is a critical threshold: if a dynasty lived for 57 years, it would take longer to end.

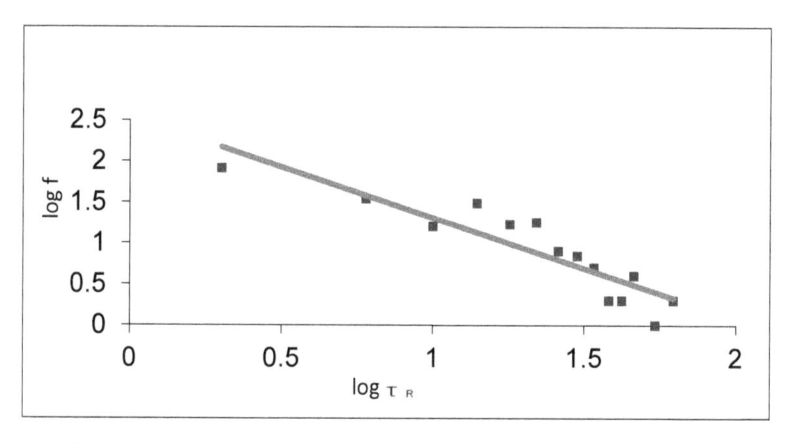

Fig. 10. Histogram of dynasty lifespan τ_R. f is the occurrence frequency of τ_R. The exponent of the power law is -1.3.

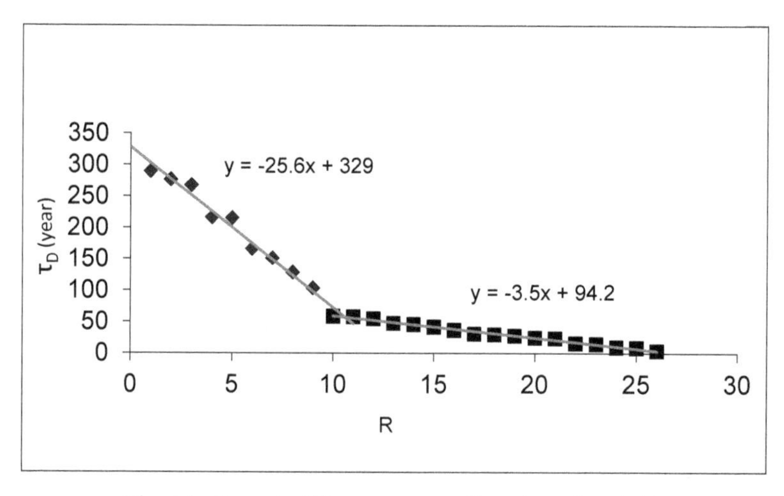

Fig. 11. Dynasty lifespan τ_D as a function of rank R.

We believe that this phenomenon is a common phenomenon in nature and in society. For example, a restaurant is most likely to collapse in the first year; if it can survive two or three years, then it may last longer. The same goes for companies in the business world. In some third world countries, due to poverty, low level of medical care, etc., some people will die at a very young age. In one country, children who live beyond the age of 13 are found to be likely to live longer.

It can be predicted from Fig. 11: If there is another dynasty after the Qing dynasty, then its lifespan has only two possibilities:

1. If its lifespan is less than or equal to 289 years, then its lifespan must fall on the two straight lines in Fig. 11.
2. If its lifespan is greater than 289 years, then it must collapse at year 329.

This is a manifestation of the "curse of history" [Lam 2006].

9 Predicting the Future and Retrodicting the Past

There are three methods in predicting the future. The first method is **time series analysis**. The future of a complex system can sometimes be predicted without knowing the nature or mechanism of the system itself. Time series analysis has been applied in predicting the stock market. (Incidentally, the 2003 Nobel Prize in economics is awarded to works in this area.)

The second method is **computer simulations** at the bottom-up level. Examples are the simulation of a village growth (Example 4 in Sec. 8), and the simulation of human-trail patterns using active walk models [Helbing et al. 1997a, 1997b]. Once the existing data are reproduced by the simulations, the future can be predicted by running the computer program further in time.

The third method is **modeling** at the phenomenological level. Sometimes, it is possible to turn the data into tracks in some suitable space and, assuming that these tracks are generated by active walkers, one can then figure out how the landscape is modified and go on to predict the future or retrodicting the past. This approach is encouraged by the success in using active walk to reproduce the filamentary patterns observed in dielectric breakdown experiments [Lam et al 1992].

It is sometimes believed that due to the complexity and contingency of history, one will never be able to recreate the trajectory of history. This is true at the bottom-up level for individuals but not true at the phenomenological level as demonstrated in the human-trail simulations. Also, any method that can explain the past is able to predict the future when time is moved forward.

10 Artificial History

Artificial life was established as a discipline by Christopher Langton in 1989. Afterwards, artificial societies were also studied. Human history cannot be repeated, but the evolution of artificial society can be repeated.

For example, to build a community/society/world video game, the computer model behind it must contain two elements: probability and laws/rules, corresponding to the chance and necessity behind history. Therefore, let historians look at the (artificial) history of video games as they evolve over time, and see what historical laws they can conclude or discover, and whether they can even find the laws behind the game—this should be a good way to train or test historians. The advantage of artificial history is that video games can be easily repeated any number of times, which is very suitable for checking the theories created by historians and judging the skills of these professionals.

Lessons learned from artificial history should help in the understanding of human history, in the same sense that artificial life has contributed to the progress in the study of real-life forms. A place to start is to use commercially available computer games such as *The Sims*, as an artificial society. The first step is to collect the time-evolving data for the characteristics of the players in this community.

Needless to say, the predicting capacity coming from the study of artificial history is relevant in many real applications, such as the writing of better computer games, movie plots, novels, and military game plays with computers.

11 Conclusion

The object of historical research is a many-body system composed of biological matter, which can be studied scientifically. And history should be part of natural science.

Due to history's intrinsic complexity and contingency, historical laws are not to be found at the bottom-up level. Unfortunately, this is exactly the level that traditional historians have been working on and, unsurprisingly, they had found no laws. Note that the existing historical laws (Figs. 4, 10 and 11) are found at the empirical level by non-traditional historians.

Historians have contributed significantly in preserving history. But history is too important to be studied by historians alone. Physicists should be able to participate more in the study of history. The complexity and contingency of the historical process cannot be used as an excuse to avoid it. We do have experience and tools from studying complex systems that can be used to study history.

It would be difficult to ask present-day historians to learn mathematics and computer programming, but collaboration between historians and physicists would be feasible and productive. Perhaps we should require university history students to take mathematics and computer language classes, and even encourage them to take physics, so that future historians will be more literate in science. Similarly, physics students taking a course on historiography will help them to do Histophysics.

References

Axtell, R. L. et al. [2002]. Population growth and collapse in a multiagent model of the Kayenta Anasazi in Long House Valley. Proc. Natl. Acad. Sci. USA **99** (Suppl. 3): 7275-7279.

Breisach, E. [1994]. *Historiography: Ancient, Medieval, and Modern.* Chicago: University of Chicago Press.

Brian, A. [1990]. Positive Feedbacks in the Economy. Scientific American, Feb. 1990, p 92.

David, P. A. [1986]. Understanding the Economics of QWERTY: the Necessity of History. *Economic History and the Modern Economist*, Parker, W. N. (ed.). New York: Blackwell.

Evans, R. [1997]. *In Defence of History*. London: Granta Books.

Fukuyama, F. [1989]. The end of history? The National Interest **16** (summer): 3-18.

Gould, S. J. [1989]. *Wonderful Life: The Burgess Shale and the Nature of History*. New York: Norton.

Helbing, D., Keltsch, J. & Molnár, P. [1997a]. Modelling the evolution of human trail systems. Nature **388**: 47-50.

Helbing, D., Schweitzer, F., Keltsch, J. & Molnár, P. [1997b]. Active walker model for the formation of human and animal trail systems. Phys. Rev. E **56**: 2527-2539.

Lam, L. [2002]. Histophysics: A new discipline. Modern Physics Letters B **16**: 1163-1176.

Lam, L. [2005]. Active walks: The first twelve years (Part I). International Journal of Bifurcation and Chaos **15**: 2317-2348.

Lam, L. [2006]. Active walks: The first twelve years (Part II). International Journal of Bifurcation and Chaos **16**: 239-268.

Lam, L, Freimuth, R. D., Pon, M. K., Kayser, D. R., Fredrick, J. T. & Pochy, R. D. [1992]. Filamentary Patterns and Rough Surfaces. *Pattern Formation in Complex Dissipative Systems*, Kai, S.(ed.). Singapore: World Scientific.

Lam, L., Veinott, M. C. & Pochy, R. D. [1995]. Abnormal Spatio-Temporal Growths. *Spatiotemporal Patterns in Nonequilibrium Complex Systems*, Cladis, P. E. & Palffy-Muhoray, P. (eds.). Redwood City: Addison-Wesley.

Lam, L., Bellavia, David C., Han Xiao-Pu, Liu Chih-Hui A., Shu Chang-Qing, Wei Zhengjin, Zhou Tao & Zhu Jichen [2010]. Bilinear effect in complex systems. EPL **91**: 68004.

Richardson, L. F. [1941]. Frequency of occurrence of wars and other fatal quarrels. Nature **148**: 598.

Stanford, M. [1998]. *Introduction to the Philosophy of History*. Malden, MA: Blackwell.

Original: Lam, L. [2004]. How to model history and predict the future. *This Pale Blue Dot*, Lam, L. Tamsui: Tamkang University Press. Example 6 in Sec. 8 added here.

The Scimat Story

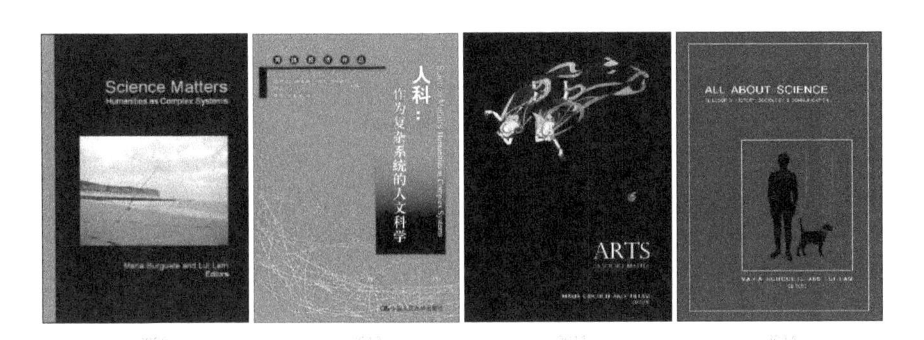

Scimat is a term I coined to represent a new multidiscipline that I proposed in 2007 [Lam 2008]. Scimat is about the scientific study of human, and can be understood as the "science of human"—the goal that the Enlightenment (1688-1789) wanted to do but did not achieve. In terms of content, scimat is a collective term for the humanities, social science, and medical science. Scimat advocates a rational discussion of any question about human—all questions, big or small—from a scientific point of view that is based on scientific knowledge about humans. It aims to raise the scientific level of the humanities and broaden the field of study of "natural science" by encouraging collaboration between scholars from the two sides. Here is the story of scimat—how it began, what had been done, and what lies ahead.

Motivation

I was interested in people when I was a child, at least with beautiful female classmates. Because I grew up with low-end people, I knew about human sufferings early on, and I thought about it from time to time. And when I grew up, my thoughts expanded into a concern for past and present world developments and the fate of humankind. Even before my physics training, I was curious in everything and loved to ask why about everything. All these led me to return to China and joined the "revolution" there, for the sake of a better tomorrow—the tomorrow of the humankind [Lam 2015].

After I invented the *Active Walk* for dealing with complex systems in 1992 [Lam et al. 1992], I began to dabble in some humanities topics (such as economic history). But it was not until 2002 when I proposed the new multidiscipline of *Histophysics* [Lam 2002] that my formal study of human problems began—marking the ultimate combination of research direction and personal interest. The reason for turning from history to *Scimat* [Lam 2008] is simple. It takes only one founding paper or a few to establish a new field of study, and that could be finished in a few years [14]. And so, by 2006 [Lam 2006], I was ready to tackle the *ultimate* question concerning the entire humanities: Why the overall research level of the humanities has been stagnant for such a long time—2,400 years, in fact, since the time of Plato (427-347 BC). Frustration with this lack of progress in the humanities has been raised by some Western scholars [Gottschall 2008].

Beijing

In July 2005, I met Maria Burguete from Portugal at the 22nd International Conference on the History of Science in Beijing. She invited me to visit. Maria holds a PhD in History from Ludwig Maximilians University in Munich, Germany, and works at the Bento da Rocha Cabral Institute in Lisbon, the capital of Portugal.

Portugal

Although I worked in Europe for two years (1975-1977) and had visited almost all countries in Western Europe, I never went to Portugal because it is located at the southwestern tip of Europe, which is not on route to other European cities, and because I was unfamiliar with Portugal. In other words, I did not find a reason to go.

This time there was a reason. So, on March 25, 2006, taking advantage of the spring break of my college and after securing the air-ticket money from the dean of the Faculty of Science, I flew to Lisbon. Maria took me around and finally we ended up at the Vila Galé Hotel (Fig. 1) in the seaside town of Ericeira, where we drank too much and wanted to do something together. After thinking about it, I concluded that we could organize a series of international conferences together. Since I have been thinking mostly about the science of human lately, I suggested this as the central theme of the conference series. The agreed division of labor: Maria manages the hardware (raising money, arranging venue), and I manage the software (conference content).

Fig. 1. Portuguese coast. *Left*: Foz da Arelho. *Right*: The Vila Galé Hotel in Ericeira.

Seoul

On May 17-19, 2006, I presented a report on "Two cultures and The Real World" at the 9th International Conference on Public Communication of Science and Technology in Seoul [Lam 2006]. The report points out two factual errors in C. P. Snow's book *The Two Cultures*, provides the historical origins of the two cultures that Snow did not address, and the remedy for bridging the gap between the two cultures. The remedy I recommend is *not* the patch-up, general-education approach that Snow suggested but rather the scimat approach, i.e., the integration of arts and science based on the fact that the two share the same roots [18].

The report's discourse on science and its conclusions—the humanities, like physics, are a branch of science—became the theoretical basis for Scimat a year later. See my tone-setting article in the first scimat book *Science Matters: Humanities as Complex Systems* [Lam 2008].

Paris

In order to make the meeting a success, I flew from Beijing to Paris on January 2, 2007, joined Maria and met Paul Caro at a restaurant. We asked him to support and attend the first scimat meeting. After three rounds of drinks, he readily agreed.

Paul, a rare-earth expert, is a Corresponding Member of the French Academy of Sciences and a Member of the French Academy of Technology, and before his retirement was a research director of the CNRS (Centre National de la Recherche Scientifique). In the 1980s, he became interested in popular science. Until 2001, he was in charge of "scientific affairs" at the Cité des Sciences et de l'Industrie in Paris. He has been a consultant for DG Research European programs in the field of education and is also a scientific advisor to the Portuguese program Ciència Viva.

Of course, I also went to the Rodin Museum, which was his former home.

Ericeira

From 28 to 30 May 2007, the First International Conference on Science Matters was held in Ericeira, Portugal. Maria and I are co-chairs (Fig. 2). I gave the conference's keynote presentation, "Science Matters: From Aristotle to you and me," and the special-subject presentation, "Histophysics: integrating history and physics."

The slogan of the conference was: Everything in nature is part of science! This simple sentence means:

1. Problems relating to human should and can be studied scientifically.
2. The study of human—the humanities, social science, and medical science—is part of natural science. The disciplinary division is only a division of labor; the three are at different stages of scientific development depending on their respective difficulty.
3. In terms of their current *scientificity* (scientific level), medical science is the highest, followed by social science, and the humanities are the lowest.

To announce and advertise a conference, one has to make a poster. But I could not think of a suitable title for the conference. In a hurry, I used the term "Science Matters," meaning that everything that the conference wants to talk about, especially the humanities, is part of science. In the first scimat book *Science Matters: Humanities as Complex Systems* (2008), *Scimat*—a new word I coined—appeared as an abbreviation for Science Matters, which was not Chinese translated until 2013. In 2013 Chinese translation of the book was published by China Renmin University Press, in which the term 人科 (*Renke*), meaning Human Science, appears as the official Chinese name of scimat and the title of this book (see figure in beginning of this article). After that, Scimat and 人科 became the official names of this new multidiscipline.

Scimat

Regarding the content of science and the position of each discipline, scimat looks at it like this:

Science = Natural Science

= Nonhuman-systems Science + Human-system Science

where Nonhuman-systems Science = "Natural Science"—the usual usage of the term; Human-system science = Humanities + Social Science + Medical science = Scimat.

International Scimat Program

The 2007 conference marked the beginning of the International Scimat Program. It has 6 points and is carried out in 6 steps:

Step 1: We launched an international scimat **conference series** which are held every two years. Each conference focuses on a specific topic. The first one in 2007 set the tone for scimat. The second one in 2009 was about art. Since our focus is on the humanities, we do not talk about "natural science" per se, although at this one we did connect art to science.

The third one in 2011 was devoted to all aspects of science: the nature, philosophy, history, sociology, and science communication of science. The sixth one in 2017 marks the 10th anniversary of scimat. In this one, we have finally revealed the ultimate goal of scimat: to build a better tomorrow. Scimat's ideal is a little grander than Plato's. Plato thinks about how to make a country better, but scimat thinks about how to make the whole world better.

I am co-chair of the first three and sixth conferences (Fig. 2) but I planned all the first to sixth ones.

Fig. 2. Posters of the International Conference on Science Matters. *Left to right, up to down*: 2007, 2009, 2011, 2017.

Step 2: We set up an International Scimat **Committee**. The committee was established on May 30, 2007, and grew from 9 to 18 members, including the 2005 Nobel Prize laureate Robin Warren and the *Scientific American* columnist Michael Shermer. I also served as the committee's coordinator and liaison. The committee aims to promote the scimat philosophy and oversee the international scimat program.

Step 3: I established and edited an English scimat **book series**, published by World Scientific in Singapore. The first three books are *Science Matters* (2008), *Arts* (2011), and *All About Science* (2014). Only one-third of the articles in these books are selected from the conference reports; the rest were written by relevant experts from around the world, solicited after the conference.

At present, in addition to my series called Science Matters, there are two other book series called Scimat, published in the UK and Portugal.

Step 4: We are in the process of establishing a number of scimat **centers** (100 eventually) around the world, which would be independent of each other but will collaborate and reinforce each other. The Center is:

1. To do fundraising to support the Center financially.
2. To organize international workshops/conferences and summer/winter schools.
3. To communicate the scimat ideas to the public.
4. To give out an Award every two years (in the donor's name perhaps) for an individual who contributes significantly in the advancement of scimat.
5. To host short-term visiting scholars who will give lectures/short courses, who will also collaborate with existing faculty members and students of any discipline, especially from the humanities.
6. To help match faculty members from the humanities and science departments, and give them release time to create new interdisciplinary courses (e.g., Physics of History, Art and Science, and Philosophy and Science).

7. To help promote the new general-education course "Humanities, Science, Scimat" for undergrads of all majors (Fig. 3).

Contents

Fig. 3. A scimat course at the 2015 International Summer School, China Renmin University. Course name: Humanities, Science, Scimat: An Interdisciplinary and Cross-cultural Experience (Number: SH1518). Lecturer: Lui Lam. *Top*: Class photo. *Bottom*: Course materials.

Note that the Center will *not* do research within itself, and so the maintenance fee is very minimal. With enough (outside) money, it can even advance scimat by funding interdisciplinary research within a university. The scimat center will be in a leading position academically in the most important multidiscipline of the 21st century.

Step 5: Establish an international scimat **society**.

Step 6: Publish an international scimat **journal**.

We are now working on step 4. Steps 5 and 6 are for the future, hopefully the near future. (For more, see: www.sjsu.edu/people/lui.lam/scimat.)

Conclusion

Scimat's motto is "Everything in nature is part of science." Scimat's key insight is that we have "One culture, two systems, three levels"—science culture, simple and complex systems, three research levels (empirical, phenomenological, and bottom-up). What we have presented here is the initial stage in the birth of a new multidiscipline—more precisely, a new paradigm—called *Scimat*. Some remarks:

1. It is similar to the case of History of Science (initiated by George Sarton early last century) and of Artificial Life (by Christopher Langton in 1986), but not quite. Scimat is much larger in scope since it incorporates the research of everything related to humans, and thus will be more far reaching in its influences. In particular,

2. It provides a unified perspective for all the disciplines in the humanities, social science and "natural science."

3. It is a rally point to raise the scientificity of the humanities, making the world a better place. (The reason is that many large-scale human tragedies can be traced to the underdevelopment of the humanities in the last 2,400 years since Plato [3.3]; see Epilog.)

4. It is the foundation behind the humanities-science synthesis [4.1], solving the so-called two-culture problem at the fundamental level.

5. It provides the basic rationale for general education and a route to make it successful.

6. It provides the broadest framework in interdisciplinary learning/ teaching, and science teaching [Matthews 2015].

Above all, Scimat is the most interesting and important multidiscipline in the 21st century.

In short, Scimat advocates the understanding of our world through science and rational thinking, whereas the humanities are recognized as part of science. Let us work together for a better humanity and make the world a peaceful place forever, for us and our children!

References

Gottschall, J. [2008]. *Literature, Science, and a New Humanities*. New York: Palgrave Macmillan.

Lam, L. [2002]. Histophysics: A new discipline. Modern Physics Letters B **16**: 1163-1176.

Lam, L. [2006]. The two cultures and the real world. The Pantaneto Forum, Issue 24 (2006).

Lam, L. [2008]. Science matters: A unified perspective. *Science Matters: Humanities as Complex Systems*, Burguete, M. & Lam, L. (eds.). Singapore: World Scientific.

Lam, L. [2015]. From physics to revolution and back. Science **348**: 1170.

Lam, L, Freimuth, R. D., Pon, M. K., Kayser, D. R., Fredrick, J. T. & Pochy, R. D. [1992]. Filamentary Patterns and Rough Surfaces. *Pattern Formation in Complex Dissipative Systems*, Kai, S.(ed.). Singapore: World Scientific.

Matthews, M. R. [2015]. *Science Teaching: The Contribution of History and Philosophy of Science*. New York: Routledge.

Original: Lam, L. [2017]. Bettering humanity: conference theme. Keynote presentation at the 6th International Science Matters Conference, *Bettering Humanity: Historic Secular Movements*, Cascais, Portugal, October 25-25, 2017.

The Two Cultures and The Real World

A brand new course for students of any major!

It is time to go beyond textbooks
and learn something about

The Real World

Phys 196 (3 units), Spring 1997
MW 4:00-5:15 pm

The course contains unified descriptions of the real world, with themes from fractals, chaos and complex systems, and applications in many social and natural systems. In addition to homeworks, the student has one of three options: (i) take a written final exam, (ii) do a report on a popular science book, or (iii) do a project on any topic selected from the daily newspaper. Topics include:

- DNA and information
- Predictions in the financial market
- Traffic problems
- Can one model Darwin?
- "The Bible" and "Gone With The Wind," What is in common?
- What does a computer scientist know about AIDS?
- Why we are here?

Prerequisite: An open mind. (No advanced math beyond algebra, computer knowledge not needed, but plenty of chance to use your computer skills if the student so desires.)

Instructor: L. Lam (Sci. 303, 924-5261, luilam@email.sjsu.edu)

The "two cultures" refer to the scientific culture and the literary culture, pointed out by C. P. Snow in the 1950s. The scientific culture derives from the study of material systems in Natural Science, while the literary culture comes from the understanding of human. However, humans are *Homo sapiens*, a (biological) material system, and is thus part of the Natural Science since the latter is the study of all material systems. Consequently, science and the humanities are unified at the fundamental level. The apparent *gap* between the two cultures comes from the different levels of scientific development, the deficiency in the school curricula, and the unfortunate misconception reinforced by current science communications. To help close this gap, a general-education course—The Real World—was introduced and taught by the author at the San Jose State University. The idea is to introduce students to the unifying principles behind the humanities and "natural science" and the world of nonlinear and complex systems.

1 The Two Cultures: Snow's Lecture

Forty-seven years ago, on May 7, 1959, Charles Percy Snow (1905-1980) gave the lecture "The two cultures and the scientific revolution" at Cambridge University [Snow 1998]. The lecture essentially contains three themes: the distinction and non-communication between the scientific culture and the literary culture in the West, the importance of the scientific revolution (defined by Snow to mean the application of the "atomic particles," presumably nuclear physics and quantum mechanics), and the urgency for the rich countries to help the poor countries. Very interesting, big themes—but nothing original, as admitted by Snow himself.[1]

The lecture generated tremendous interest and discussion around the world, which helped to earn Snow twenty honorary degrees (mostly from universities outside of England) and carve his name in history. While the other two themes are definitely worth talking about, it is the "two cultures" theme that causes the most controversy and debates. This is not at all surprising. Many in the literary circle felt slighted by Snow in his lecture and had to defend themselves or their profession.[2] And, by definition, literary people are those who can write.

The purpose of this article is not to discuss Snow's lecture but to present the deep reasons behind the apparent gap between the two cultures, which are hardly touched upon by Snow himself. (After all, the existence and importance of the gap is not in doubt.) Furthermore, our effort in helping to close this gap is described.

But before we do that, for historical sake, Snow's two errors concerning scientific matter in his lecture should be pointed out. Apparently, despite the worldwide fame of this lecture, no one has done this before.

Snow is wrong when he writes, "No, I mean the discovery at Columbia by Yang and Lee" [3] [Snow 1998: 15]. While T. D. Lee indeed worked at Columbia University, C. N. Yang's "permanent" address at that time was the Institute for Advanced Study at Princeton, New Jersey (see the address bylines in [Lee & Yang 1957]). In fact, Yang has never been associated with Columbia University. A few sentences later, still referring to the work of Yang and Lee, Snow makes another mistake in his sentence, "If there were any serious communication between the two cultures, this *experiment* would have been talked about at every High Table in Cambridge." Lee and Yang's work is purely theoretical, which is to point out that there were no experimental evidence supporting or refuting parity conservation in weak interactions at that time. They went on to propose several experiments to settle this issue without predicting the outcome of these experiments. After this paper, parity nonconservation was indeed discovered in an experiment by C. S. Wu [Wu et al. 1957], a colleague of Lee at Columbia, not by Lee and Yang.

These two errors are minor by themselves and do not affect the rest of the lecture. Yet, they are factual errors that could be easily avoided since Lee and Yang received their Nobel Prize in December 1957, a mere 17 months before Snow delivered his lecture. With all the reporting of the parity nonconservation story in the newspapers and magazines, not to mention the more formal academic publications, only a careless writer like Snow could miss the basic facts. And this is not an isolated incident. In 1932, Snow had to recant publicly his "discovery" of how to produce Vitamin A artificially after his calculation was found faulty. Snow, a trained chemist, decided to leave scientific research completely after this incident and became a novelist [Snow 1998: xx]. He indeed made the correct career move, judging by later developments.

2 The Two Cultures: The Essence

2.1 Emergence of the two cultures

More than 10,000 years ago on Earth, the early *Homo sapiens*, our ancestors, started to wonder about the things around them—things in their immediate surroundings and things in the sky. Curiosity serves not just human needs but for those who figure out how things work from their observations, it is a survival skill via the evolutionary mechanism according to Darwin.

Among these activities, after writing was invented 5,000 years ago, literature is the description of humans' reflection on and understanding of nature. Here, nature include all (human and nonhuman) material systems, such as falling leaves in Autumn, the changing weather and seasons, effect of moonlights on lovers, the way humans treat each other in different spatial and temporal settings, and, quite often, thoughts in one's brain as a function of happenings inside or outside the person's body. When the authors write all these down, they are using their bodily sensors (sighting, touching, smell, hearing, etc.) as the main detectors and their brain as the major information processor. Apart from that, for latecomers, they do benefit from reading what previous writers wrote.

As time went by, the observation and understanding of certain kinds of phenomena progressed faster. For example, how things fall under the influence of gravity can be predicted and measured with high accuracy. This is achieved not because the falling object under study is simple, but because (1) we can approximate it by something simple, and (2) we use detectors and information processors other than those from our own bodies.

For example, a human body falling from a tall building is the same complicated human body described in a piece of literature, but in physics we pretend that it is a point particle (i.e., an idealized particle with zero size) in our calculations. This is an approximation; it works because the size of Earth, the main source of the gravitational force, is much greater than the size of the human body. Furthermore, we can record the positions of the falling body by digital cameras and compare them with our calculations, with the help of calculators or computers. (For smaller falling objects, low-tech devices are used to record the positions at regular time intervals. This is routinely done in freshmen physics labs.)

This branch of study is now called "natural science," which involves mostly nonliving systems even though living systems (such as humans in free fall and other simpler biological bodies) are not excluded. As we just pointed out, natural science presently succeeds because it chooses to deal only with a special subset of phenomena. And literature is stuck with the complicated aspects—such as pride and prejudice—of the complex system called human.

As study deepened, specialization became essential and we were left with two distinctive groups of practitioners, the writers in the literature profession and what Snow called scientists for those working in natural science. Since writers use their own bodies as tools, only those with supreme bodily sensitivity and suitable hard wiring of neurons in their brains can become good writers, while scientists need other types of quality (such as supreme self confidence) to succeed. There is no overlap between these two groups of professionals, as Snow painfully found out for himself.

2.2 Why the gap

The fact that scientists can talk to each other is true only to a certain extent. There is not much to talk about between a particle theorist and a condensed matter physicist if the subject is the Standard Model of particles. But all scientists, be they physicists, biologists or chemists, do share some common knowledge such as the second law of thermodynamics, because this law is a required learning in the college education of these scientists.[4]

Professional activities require high concentration of attention and usually are time consuming, and, especially in the case of science, involve very keen competition. Time is short, for the professionals. Many first-rate scientists do not read books, particularly science books, because what contained in books is usually not fresh enough. Instead, they read research papers that they think might be helpful to their (present or future) work. That is what the scientist had in mind when he, asked by Snow what books he read, replied, "Books? I prefer to use my books as tools." [Snow 1998: 3]. Tools, here, mean something that will help him to do his research. There is in fact a fair chance that literary books will be read by scientists, for relaxing purpose, e.g., when they are in a plane after attending a conference. But these books are not Shakespeare's. The same goes for the literary people. Why should they read any science book if they cannot find anything there that would help them to do their job?

In short, the non-communication between these two groups is not due to the non-overlap of the people involved, but due to the absence of any common language or principle in their trades, at least in the 1950s when Snow delivered his lecture. This is no longer so. Since the 1980s, some general principles arise from the study of simple and complex systems, which are applicable to both the natural and social sciences, and to both living and nonliving objects. Here, we are referring to fractals, chaos and active walks [Lam 1998]. (See Sec. 3 for more.)

2.3 Why close the gap

We want the literary people to learn a little bit basic science and the scientists to read some good literature. The reason is not that we afraid they have nothing to talk to each other in a cocktail party. They can always talk about Ang Lee's *Brokeback Mountain* (2005) or his other movie, *Crouching Tiger, Hidden Dragon* (2000). The movie's storyline is as deep as Shakespeare's, and perhaps more entertaining.

And if the purpose is to make the literary people to appreciate the mental achievement of the humankind via the elegant theories established in science, then learning basic science is not the easiest way to do so and may not even be necessary.[5] In Snow's days, the television and telephone, and if not enough, the two atomic bombs in 1945 and the Sputnik in 1957 should convince every sensible person on Earth the high achievement of science, without the need to know the theories behind them. These days, a cell phone will do.

Yes, knowing some basic science supposedly will help you to cast sensible votes as a citizen on scientific matter, such as laws regarding global warming. But we actually rely on the experts on their professional opinions on these matters. Our rudimentary scientific knowledge is not sufficient for us to make the judgments ourselves, even though it may help us to pick which expert to trust. Unfortunately, this is easier said than done. Sometimes you can't even trust the Nobel laureates. (For example, there is one who believes in astrology, big foot, and that we never went to the moon.) The situation is like picking which financial advisor to help managing your money. You go to the big institutions and also look at their track record. It helps if you have some financial knowledge, but that is not enough.

A good reason for ordinary people to learn some science is to increase their personal safety. Some science knowledge could help people to eat and live healthy and avoid accidents. It could also help them to recognize the crooks when they see one. For example, if someone told you that the Earth had exploded three times in the past and it was he who repaired it (which is so against what we know in science), you could safely ignore what else he told you and should never gave him a dime, nor your other valuable possessions.

But why we want the scientists to read some good literature? [6] This is less clear, not even Snow has anything meaningful to say about this. What is sure is that it is not a bad thing to do, unless you happen to be a young scientist who needs undivided attention to your research. It could help the scientists to meet more interesting people; many of them, at the time this article is written, know only literature but not science.

2.4 How to close the gap

The best time to get the literary people and the scientists to learn something from each other's trade is when they are still students in schools or colleges, when they are forced to attend classes. Apart from requiring the students to take some general education classes in both science and literature before they are allowed to graduate, as is the practice in most American universities, it would be wonderful to teach them something, if exist, that they could use for the rest of their life no matter what profession they end up with. Fortunately, these wonderful things do exist. They are the general principles governing many complex systems [2.3].

And the way to achieve this in the classroom is through educational reform; outside of the classroom, science communication [12]. But first, let us review the essentials of science.

3 Science: The Essentials

3.1 What science is about

Science is about the systematic understanding of nature without introducing God/supernatural considerations [8.1]. And nature, of course, includes all material systems. On the other hand, human beings are biological material bodies called *Homo sapiens*. Consequently, any study related to human beings, literature in particular, should be a part of science [Lam 2002].

Since nature consists of everything in the universe, the two terms science and natural science are identical to each other.[8] In other words, in terms of the objects under study in science, we have

Science = Natural Science

= Science of nonliving systems + Science of living systems (1)

whereas

Living systems = Nonhuman biological systems + Humans (2)

Since we human beings (and not, e.g., the ants) are the ones who do the study and control the research budgets, it is not surprising that a large part of the science activity is related to and is for the benefit of humans. In terms of the disciplines, these human-related studies fall into one of two categories, viz., social science and the humanities. Social science consists of anthropology, business and management, economics, education, environmental science, geography, government policy, law, psychology, social welfare, sociology, and women's studies.[7] Philosophy, religions, languages, literature, art, and music make up the humanities. History, by its very nature, should be part of social science, but is listed in the humanities at some universities.

The aim of literature, music and art in the humanities is to stimulate the human brain—through arrangement of words or colors, sound or speech, or shape of things—to achieve pleasure and beauty, or their opposites, via the neurons and their connecting patterns [Pinker 1997]. The brains, some sort of computer, of the creator and the receiver at the two ends of this process are heavily involved [7.2]. The scientific development of these disciplines as complex systems is at a primitive level, and that is why they are separated from social science, which are at an intermediate level. Linguistic is the study of the tools involved in written words and speeches, supporting the three disciplines mentioned above.

In terms of the disciplines, Eq. (1) could be rewritten as:

$$\begin{aligned} \text{Science} &= \text{Natural Science} \\ &= \text{Physical science} + \text{Social science} + \text{Humanities} \end{aligned} \quad (3)$$

whereas "physical science" includes not just physics, but biology, chemistry, etc.

3.2 Three general principles

There are three established principles that are able to unify many different phenomena found in nature, with examples taken from both the natural and social sciences, and even the humanities. They are [Lam 1998]:

1. **Fractals**—the principle of self-similarity. Self similar means that if you take a small part of an object and blow it up in proportion, it will look similar or identical to the original object. Self-similar objects are called fractals, which quite often have dimension not equal to an integer. A famous example is the Sierpinski gasket [Lam 2004b] [20]. Fractals are everywhere, ranging from the morphology of tree leaves, rock formations, human blood vessels, to the stock market indices and the structure of galaxies. Fractals are even relevant in the corporate culture [Warnecken 1993] and the arts [Barrow 1995].

2. **Chaos**—the common (but not universal) phenomenon that the behavior of many nonlinear systems depends sensitively on their initial conditions. Examples of chaos include leaking faucets, convective liquids, human heartbeats, planet motion in the solar system, etc. The concept is found applicable in psychology, life sciences and literature [Robertson & Combs 1995; Hayles 1991]. A general summary is available [Yorke & Grebogi 1996].

3. **Active walks**—a major principle that Mother Nature uses in self-organization. Active walk is a paradigm introduced by Lam in 1992 to handle complex systems [Lam 2005a, 2006a]. In an active walk, a

particle (the walker) changes a deformable potential—the landscape—as it walks; its next step is influenced by the changed landscape. For example, ants are living active walkers. When an ant moves, it releases chemicals of a certain type and hence changes the spatial distribution of the chemical concentration. It next step is moving towards positions of higher chemical concentration. In this case, the chemical distribution is the deformable landscape. Active walk has been applied successfully to a number of complex systems studied in the natural and social sciences. Examples include pattern formation in physical, chemical and biological systems such as surface-reaction induced filaments and retinal neurons, the formation of fractal surfaces, anomalous ionic transport in glasses, granular matter, population dynamics, bacteria movements and pattern forming, food foraging of ants, spontaneous formation of human trails, oil recovery, river formation, city growth, economic systems, and, most recently, human history [Lam 2002, 2004a, 2006a].

All three principles are an integral part of complexity science, which is becoming important in the understanding of business, governments, and the media.[9]

4 Educational Reform: A Personal Journey

University educational reforms could involve three possible components:

(1) The contents of the courses.
(2) The way of teaching by the instructors.
(3) The learning method of the students.

No matter how it is done, an unavoidable constraint that will crucially affect the success of the reform is usually not mentioned or ignored completely by the reformers; i.e., the reform should *not* increase the teaching load of the instructors.

Item 3: quality of students

The quality of the students taking a course—like the quality of a sample in a physical experiment or the raw material in a factory—is of primary importance; this factor is never emphasized enough. Obviously, with a defective sample, no good experimental result can be expected, no matter how skillful the experimentalist is. This last factor points to the need to start any educational reform from grade one on, or even better, from the kindergartens.

With the constraints understood and resources limited, I tried to do my best as a teacher. There is not much we can do about Item 3. It is very hard for the students to change their learning habit after being wrongfully taught for 12 years before they show up in college, and this is not their fault. I therefore concentrated my effort in the first two Items.

Item 2: change teaching method

On Item 2, I have tried two different methods. The **first** method is quite radically: The class is broken up completely. It is called "MultiTeaching MultiLearning" (MTML) [Lam 1999]. We note that in a physics class, the teacher usually does not have enough time to cover everything. The attention span of a student is supposed to be about 15 minutes. Students in a class have different learning styles. Some students are more advanced than others. Active learning and group learning are good for students.

To overcome these problems in the teaching of two freshmen classes in mechanics and thermodynamics, around 1999, I have tried a zero-budget and low-tech approach. In these classes, we cover about one chapter per week, using *Physics* by Resnick, Halliday and Krane as the textbook. In each course, there are three sessions per week, each 50 minutes long. In the last session of every week, the class is broken up completely. Different "booths" like those in a country fair are set up in several rooms, manned by student volunteers from the class. The rest of the class is free to roam about, like in a real country fair, or like what the professional physicists will do in a large conference with multiple sessions. In this way, we are able to simultaneously offer homework problem solving, challenging tough problems for advanced students, computer exercises, Web site visits, peer instruction, and one-to-one tutoring to the students. The students seemed to enjoy themselves and benefited from it. However, this approach was soon discontinued. It did require a little bit of extra preparation time from the instructor, but more importantly, it did not seem to raise significantly the grades of the students. The "inferior raw material" factor might be at work here.

The **second** method I tried, with better luck this time, is to integrate popular science books into my physics classes [Lam 2000, 2005b, 2008]. This is done by giving extra credits to the students who would buy a popular science (PS) book, read it and write up a report [Lam 2000, 2001: 330-336]. The instructor does not actually teach the books, and hence will not find the teaching load increased. It is like a supplementary reading, a practice commonly used in the English classes but rarely adopted by science instructors. The aim of this practice is (1) to broaden the knowledge base of the students, (2) show them the availability and varieties of PS books in their local book stores, (3) encourage them to go on to buy and read at least one PS book per year for the rest of their life, and (4) become a science informed citizen—a voter and perhaps a legislator who is science friendly. It is about lifetime learning of science matters. This practice is quite successful and is still going on in my classes. This PS book program is not trying to alter the course content per se.

Item 1: change course content

My first attempt in changing course content, Item 1 in educational reform, happened quite early. Soon after I started teaching at San Jose State University (SJSU) in 1987, I created two new graduate courses: Nonlinear Physics and Nonlinear Systems [Lam 1998]. But these two were for physics majors.

In Spring 1997, I established a general-education course called The Real World, opened to upper-division (i.e., third and fourth year in college) students of any major. It results from my many years of research ranging from nonlinear physics to complex systems [Lam 1998]. The description of this course is given in a flyer (right figure, at beginning of article). There were only nine students, majoring in physics, music, philosophy and so on, plus two physics professors sitting in. It was fun. The course stopped after one semester due to nonacademic reasons, falling victim to the sociology of science.

Five years later in Fall 2002, the course was resurrected with the same name but modified to suit incoming freshmen students. It is this general-education freshmen course that will be described in detail in the next session.

5 The Real World: A General-Education Course

In 2001 we have a new provost in campus. This very energetic and ambitious man, Marshall Goodman, wanted to make SJSU distinctive from the other twenty plus campuses of the California State University system. Introducing international programs with a global outlook was his way of doing that. But perhaps more important, with lightning speed as administrative things went, he was able to push through the university senate and actually had 100 brand new freshmen general-education courses set up and running in about half-a-year's time. Each of these courses is limited to no more than 15 incoming freshmen students. The program starting in Fall 2002 is called the MUSE (Metropolitan University Scholar's Experience) program. "MUSE/Phys 10B (Sec. 3): The Real World" was one of the 100. Some details of the course are given below (more in [Lam 2008: 89-118]):

Course Description: To understand how the real world works from the scientific point of view. The course will consist of two parallel parts. (1) The instructor will introduce some general paradigms governing complex systems—fractal, chaos, and active walk—with examples taken from the natural and social sciences, and the humanities. (2) The students will be asked to pick any topic from the newspapers or their daily lives, and investigate what had been done scientifically on that topic, with the help from the Web, library, and experts around the world. Outside speakers and field trips are part of this course.

Learning Objectives and Activities for this Course: This course qualifies as an Area B1 (Physical Sciences) course in the General Education requirements. It is designed to enable the student to achieve the following GE and MUSE learning outcomes. By the end of this course, the student should be able to:

- Use methods of science and knowledge derived from current scientific inquiry in life or physical science to question existing explanations;
- Demonstrate ways in which science influences and is influenced by complex societies, including political and moral issues;
- Recognize methods of science, in which quantitative, analytical reasoning techniques are used;
- Understand the learning process and the student's responsibility and role in it; and
- Know what it means to be a member of a metropolitan university community.

After successfully completing this course, stendents will:

- Realize that there are general paradigms—fractal, chaos, and active walk—governing the functioning of complex systems in the real world, physical and social systems alike.
- What nonlinearity is.
- How "dimension" is defined mathematically.
- The meaning of self-similarity and fractals.
- Recognize and able to evaluate data to show that any physical structure or pattern in the real world is a fractal or not.
- What a chaotic system is.
- Able to distinguish a chaotic behavior from a random behavior given the time series of a system.
- To realize that many complex systems in the real world can be described by Active Walks, and be familiar with a few examples.
- Recognize that there are multiple interpretations or points of view on some ongoing, forefront research topics, and that these interpretations can co-exist until the issue is settled when more accurate data and a good theory become available.
- Know the difference between science and pseudoscience, and the real meaning of the scientific method.
- How scientific research is actually done.

- Able to find out the latest scientific knowledge about any topic of interest in the future.
- Have improved skills in communicating orally and in writing.
- Have increased familiarity with information resources at SJSU and elsewhere.

Course Materials: The following two books are required:

1. Lui Lam, *Nonlinear Physics for Beginners: Fractals, Chaos, Solitons, Pattern Formation, Cellular Automata, and Complex Systems* (World Scientific, Singapore, 1998), paperback (list price: $28). Reading assignments from this book will be announced in class. Additional materials will be provided by the instructor. Other information could be found from the Web, magazines, research journals, and books from the library.
2. *A Spartan Scholar from the Start* (published by SJSU).

6 Conclusion: What Is to be Done?

The "two cultures" issues are clarified. What the literary people and the natural scientists do are similar to each other at the basic level. They both try to understand the world around them. What differentiates them is, roughly speaking, that literary people confine their investigation to using their body as the detector and their own brain as the information processor, while modern scientists use tools other than their own body to do their work (Sec. 2). All these activities could be viewed as parts of a big project—to understand nature (human and nonhuman systems) systematically, except that literature is still doing it empirically and is at a less developed level, scientifically speaking. But this is also the case with the study of many other complex systems, because the problem involved is much harder (Sec. 3).

The gap between the two cultures can never be completely closed, and there is no need to do so. What should be and could be done is to teach everybody the fact that our real world is governed by some unified principles, which are applicable to both the human and nonhuman systems and could be shared beneficially by people in the two cultures, and in fact, in any culture. A general-education course for this purpose has been designed and tried (Sec. 5), which should be taught at school of any level, the earlier the better. Furthermore, to reinforce the effect, lifelong learning through popular science books is strongly urged (Sec. 4). Beyond that, it would be good to make both natural science and literary writings an essential reading for college students of any major. This can be easily done by incorporating a few popular science books—such as James Watson's *The Double Helix*—into the list of required readings in the general education of every student in every university.

It does not help if in science communication of any kind, we keep conveying to the public the wrong impression that natural science, social science, and the humanities are three very different things, without anything in common. A remedy to correct this in the science museums is to show the unifying themes of all natural and social phenomena before the museum exit [Lam 2004c].

Through appropriate effort, the synthesis of natural science and the humanities is possible. Examples include the merging of biology and sociology to form Sociobiology [Wilson 1975], economics and physics for Econophysics [Mantegna & Stanley 2000], sociology and physics for Sociophysics [Galam 2004], and more recently, the creation of a new discipline called Histophysics through the link up of history with physics [Lam 2002, 2004a, 2006].

Notes

1. See "The two Cultures: a second look (1963)" in [Snow 1998].

2. Ssee Stefan Collini's "Introduction" in [Snow 1998].

3. In the famous paper that earned Lee and Yang the Nobel Prize in 1957, the authors' names appears as Lee and Yang [Lee & Yang 1957]. The ordering of the two names in this and other joint papers apparently is not a small matter; it plays an important role in the two men's subsequent total breakup of collaboration and friendship [Yang 1983; Lee 1986; Chiang 2002; Zi et al. 2004].

4. The second law of thermodynamics is the example used by Snow to test the scientific knowledge of the literary people in a social gathering [Snow 1998: 15]. This is in fact quite unfair, because the second law is less universal and useful than people think. It applies only to closed systems and only to their

thermodynamic equilibrium states. It applies neither to humans—an open system and the interest of literary people—nor to the expanding "cosmos" as Snow wrongly claimed [Snow 1998: 74]. The reason is that our universe is ever expanding and is never in an equilibrium state [Lam 2004a]. See [Zhao 2003] for a detailed discussion.

5. Snow, a chemistry major, is mistaken when he writes that asking someone to define "acceleration" is "the scientific equivalent of saying, Can you read?" [Snow 1998: 15].) The definition of acceleration ($a \equiv dv/dt$) involves calculus and the concept of vectors [Halliday & Resnick 1988], and may even be found difficult by some students in a freshmen physics course.

6. In practice, as good literature is concerned, unlike the case in science, there is no unique choice suitable for everybody. Reading Shakespeare or Tang poems/Song lyrics will equally do.

7. http://www.sosig.ac.uk.

8. With this understanding, every possible enquiry undertaken would be about nature. The term "science" in its German sense of Wissenchaft—any systematic body of enquiry—and its use in the English language will then coincide with each other.

9. See http://www.trafficforum.org/budapest for a description of an upcoming conference on "Potentials of complexity science for business, governments, and the media," Budapest, Aug. 3-5, 2006.

References

Barrow, J. D. [1995]. *The Artful Universe: The Cosmic Source of Human Creativity*. New York: Little, Brown and Co.

Chiang, T.-C. [2002]. *Biography of Chen Ning Yang: The Beauty of Gauge and Symmetry*. Taibei: Bookzone.

Galam, S. [2004]. Sociophysics: A personal testimony. Physica A **336**: 49-55.

Halliday, D. & Resnick, R. [1988]. *Fundamentals of Physics*. New York: Wiley.

Hayles, N. K. [1991]. *Order and Chaos: Complex Dynamics in Literature and Science*. Chicago: University of Chicago Press.

Lam, L. [1998]. *Nonlinear Physics for Beginners: Fractals, Chaos, Solitons, Pattern Formation, Cellular Automata and Complex Systems*. Singapore: World Scientific.

Lam, L. [1999]. MultiTeaching MultiLearning: a zero-budget low-tech reform in teaching freshmen physics. Bulletin of the American Physical Society **44**(1): 642.

Lam, L. [2000]. Integrating popular science books into college science teaching. Bulletin of the American Physical Society **45**(1): 117.

Lam, L. [2001]. Raising the scientific literacy of the population: a simple tactic and a global strategy. *Public Understanding of Science*, Editorial Committee (ed.). Hefei: University of Science and Technology of China Press.

Lam, L. [2002]. Histophysics: A new discipline. Modern Physics Letters B **16**: 1163-1176.

Lam, L. [2004a]. *This Pale Blue Dot: Science, History, God.* Tamsui: Tamkang University Press.

Lam, L. [2004b]. A science-and-art interstellar message: The self-similar Sierpinski gasket. Leonardo **37**(1): 37-38.

Lam, L. [2004c]. New concepts for science and technology museums. The Pantaneto Forum, Issue 21, 2006 (http://www.pantaneto.co.uk). See also Lam, L. in *Proceedings of International Forum on Scientific Literacy*, Beijing, July 29-30, 2004.

Lam, L. [2005a]. Active walks: The first twelve years (part I). International Journal of Bifurcation and Chaos **15**: 2317-2348.

Lam, L. [2005b]. Integrating popular science books into college science teaching. The Pantaneto Forum, Issue 19, 2005 (http://www.pantaneto.co.uk).

Lam, L. [2006a]. Active walks: The first twelve years (part II). International Journal of Bifurcation and Chaos **16**: 239-268.

Lam, L. [2006b]. Science communication: What every scientist can do and a physicist's experience. Science Popularization, 2006, No. 2: 36-41. See also Lam, L. in *Proceedings of Beijing PCST Working Symposium*, Beijing, June 22-24, 2005.

Lam, L. [2008]. SciComm, PopSci and The Real World. *Science Matters: Humanities as Complex Systems*, Burguete, M. & Lam, L. (eds.). Singapore: World Scientific.

Lee, T. D. [1986]. *Selected Papers, Vol. 3.* Boston: Birhauser Inc.

Lee, T. D. & C. N. Yang, C. N. [1957]. Question of parity conservation in weak interactions. Physical Review **104**: 254-258.

Mantegna, R. N. & Stanley, H. G. [2000]. *An Introduction to Econophysics.* New York: Cambridge U. P.

Pinker, S. [1997]. *How the Mind Works.* New York: Norton.

Robertson, R. & Combs, A. (eds.) [1995]. *Chaos Theory in Psychology and the Life Sciences.* Mahwah, NJ: Lawrence Erlbaum Associated.

Snow, C. P. [1998]. *The Two Cultures.* Cambridge: Cambridge University Press.

Warnecken, H. J. [1993]. *The Fractal Company: A Revolution in Corporate Culture.* New York: Springer.

Wilson, E. O. [1975]. *Sociobiology.* Cambridge, MA: Harvard University Press.

Wu, C. S., Ambler, E., Hayward, R. W., Hoppes, D. D. & Hudson, R. P. [1957]. Experimental test of parity conservation in beta decay. Physical Review **105**: 1413-1415.

Yang, C. N. [1983]. *Selected Papers 1945-1980 with Commentary*. New York: Freeman.

Yorke, J. A. & Grebogi, C. (eds.) [1996]. *The Impact of Chaos in Science and Society*. Tokyo: United Nation University Press.

Zhao, K.-H. [2003]. The end of the heat death theory. *Philosophical Debates in Modern Science*, Sun, X.-L. (ed.). Beijing: Peking University Press.

Zi, C., Liu, H. Z. & Teng. L. (eds.) [2004]. *Solving the Puzzle of Competing Claims Surrounding the Discovery of Parity Nonconservation: T. D. Lee Answering Questions from Sciencetimes Reporter Yang Xujie and Related Materials*. Lanzhou: Gansu Science and Technology Press.

Original: Lam, L. [2006]. The two cultures and The Real World. The Pantaneto Forum, Issue 24 (2006). Paper presented at *The 9th International Conference on Public Communication of Science and Technology*, Seoul, May 17-19, 2006.

19 First Non-government Visiting Scholars from China to USA

Northwestern University, Evanston, IL, USA.

1 Introduction

People's Republic of China was founded in 1949. After a series of political movements, the country embarked on the road of reform-and-opening up in 1978. It is at this early period, in February of 1979, that a delegation of eight scholars from the Institute of Physics, Chinese Academy of Sciences, was invited to visit and work in the United States. Most of them stayed for two years; six were funded by their USA hosts during their stay—the first group of visiting scholars[1] from China ever receiving this treatment. In other words, they were treated as equals as physicists by their hosts, at Northwestern University, Evanston, IL.

This article is written by one of the eight, telling the story for the first time by reviewing the background, the 1979 visit itself, and the aftereffect of this historical event.

2 Historical Background

2.1 A brief history of the Institute of Physics

In 1928, during the Republic of China period, the Institute of Physics of the National Academia Sinica (国立中央研究所) was established in Shanghai. Next year, an Institute of Physics under the National

Academia Beiping (国立北平研究所) was established in Beiping (Beijing today). In 1948, part of the former institute moved from Shanghai to Nanjing.

Table 1. Timeline of events related to the visit.

Institute of Applied Physics, CAS, established	1950	
Yan Ji-Ci, Director		
Shi Yu-Wei, Director	1957	
Changed name to Inst. of Physics (moved to ZGC)	1958	
	1971	Ping-pong diplomacy
	1972	Nixon visited China
	1976	Hua Guo-Feng succeeded Mao as chairman of CCP
	1977	May, Deng Xiao-Ping resumed working
Jan., Lam came to IoP	1978	
		March, National Science Meeting
Aug.-Dec.., Chia-Wei Woo visited IoP		
		Dec., Third Central Committee of CCP Meeting
	1979	Jan. 1, China and USA established diplomatic relationship
		Jan. 29-Feb. 4, Deng Xiao-Ping visited USA
Feb. 9, IoP delegates left Beijing, starting the journey to Northwestern Univ., USA		
Feb. 15, arriving Northwestern Univ.		
June, Lam returned to Beijing via Hong Kong		
Rest of IoP delegates returned to Beijing	1981	
Guan Wei-Yan, Director		

Soon after the establishment of the People's Republic of China in 1949, the Academia Sinica (中国科学院, called Chinese Academy of Sciences, or CAS, today) was established in Beijing, the capital. Next year, the two institutes of physics were combined to form the Institute of Applied Physics, with Yan Ji-Ci (严济慈, 1900-1996) as the director. Shi Yu-Wei

(施汝为, 1901-1983) became the new director in 1957. The institute changed name to Institute of Physics (IoP) and moved to Zhongguancun (ZGC), the present address, in 1958. In 1981, Shi stepped down (and passed away in 1983), and Guan Wei-Yan (管惟炎, 1928-2003)[2] succeeded as director (Table 1). A detailed history of the IoP can be found in [Zhao et al. 2008].

2.2 China-USA interaction and Chinese politics, 1971-1979

Various political movements planned by the government appeared after 1949. In particular, the Great Cultural Revolution lasted from 1966 to 1976, officially speaking. The political interests of China and the USA, in regard to USSR, overlapped partially in the early 1970s, resulting in the visit of the American ping-pong team in China in 1971 even though there was no diplomatic relationship between the two countries. The next year, President Nixon visited China.

Chairman Mao Ze-Dong (毛泽东) died in 1976 and was succeeded by Hua Guo-Feng (华国锋) immediately. Deng Xiao-Ping (邓小平) resumed working in May 1977. In March 1978, the National Science Meeting was held in Beijing, in which Deng gave the speech that recognized intellectuals as part of the working class and science as the first production force [Luo 2008]. In December of the same year, the most important Third Central Committee Meeting of the Chinese Communist Party (CCP) was held in Beijing, which officially shifted the focus of government work from class struggle to economic developments. This meeting is regarded as the beginning of the reform-and-opening up period of the last 30 years in China's development [Tang 1998].

A few days after this meeting, on January 1, 1979, China and USA established formal diplomatic relationship. Between January 29 to February 4, 1979, Deng visited the United States.

2.3 History of Chinese students going aboard and returned since 1872

Before 1949, there are eight generations of Chinese students going aboard to study, by the Western Returned Scholars Association's counting:[3]

1. First generation (1872-1875): The Qing dynasty government sent out 120 children, aged 12-15, to the USA to study [including the famous Rong Hong (容宏)].
2. Second generation (1877): Nearly 100 navy students sent to Europe in early years of Emperor Kuangxu (光绪).
3. Third generation: Students going to Japan in the early 20th century.
4. Fourth generation: Students going to the USA under the auspices of the Boxer Indemnity.
5. Fifth generation: Students going to France to study and work [including Zhou En-Lai (周恩来), Deng Xiao-Ping…].
6. Sixth generation: Students going to USSR during the 1920s.
7. Seventh generation (1927-1937): Students going abroad (including Yan Ji-Ci, Shi Yu-Wei…).
8. Eighth generation (1938-1948): Students going to Europe and USA [including Chen-Ning Yang (杨振宁), Tsung-Dao Lee (李政道) …].

Continuing with this counting, after 1949, we have three more generations of Chinese students going aboard, namely:

9. Ninth generation (1949-now): Large number of students from Taiwan and Hong Kong, and a few from Macau, went to USA and Europe [including the three Nobel laureates: Yuan-Tseh Lee (李远哲) from Taiwan, Daniel Tsui (崔琦) and Charles Kao (高锟) from Hong Kong].
10. Tenth generation (1950s): Students going to USSR sent by the Chinese government (including Guan Wei-Yan…).
11. Eleventh generation (1978-now): Students from mainland China, going to USA, Europe, etc., sent officially or going privately [Cao 2009].

The history of students returning to China to reside and work is equally interesting. Before 1949, a large number of these students willingly returned to China during the Sino-Japanese war and contributed to the modernization of their motherland. After 1949, there are three generations returning to mainland China voluntarily, called *Haigui* (海归, meaning "overseas returnees") by today's jargon:

1. First generation (early 1950s): Those coming back mainly from USA and Europe, soon after People's Republic of China was established [including Qian Xue-Sen (钱学森) ...].
2. Second generation (mostly1975-1985): Near 100 students of the 9th generation returned to China, the majority of them after the Cultural Revolution.
3. Third generation (after 1980s): These are the 11th generation students returning when the reform-and-opening up process in China is picking up speed.

3 The Visiting-Scholar Delegation

3.1 Lam returned to China to work in January 1978

I was born in Guangdong province and grew up in Hong Kong, where I received my education from grade one on and graduated from the University of Hong Kong with a BS degree, spanning from 1949 to 1965. I then went abroad to Vancouver, Canada, in 1965 and then to New York City in 1966. I received my MS degree from the University of British Columbia and PhD from Columbia University (1973), both in physics. I therefore belong to the 9th generation of Chinese students going aboard.

It was during my graduate student years that the anti-Vietnam War movement erupted, and the students took over the buildings at Columbia University. Chinese overseas students, though present, were mostly outsiders in all these activities.

All these changed suddenly at end of 1970. In that year, after large quantities of potential oil deposits near the group of tiny islands called Diaoyutai (called Diaoyudao in mainland China) were announced by foreign oil companies, both Japan and China reiterated their ownership of Diaoyutai, which, formally speaking, is under the jurisdiction of the Taiwan government. To help keep Diaoyutai under China, overseas

Chinese students in USA started the *Baodiao* (protecting Diaoyutai) movement in December 1970 [The Seventies Monthly 1971]; the first big meeting was held in the basement of the Teachers College, Columbia University, December 1970. Many oversea Chinese, students or otherwise, worldwide were mobilized.[3]

I was a physics graduate student at Columbia University when the Baodiao movement started. I actively participated at the first meeting held at Columbia in December 1970 and subsequent events, including the march in Washington, DC, and the national political congress in Ann Arbor, both in 1971. But since I can speak Cantonese, the major language used in Chinatown in New York City, I chose to go and live in Chinatown for about two years, to "live and learn from the masses" as what was the fashion in China during those Cultural Revolution years. We did earn the trust of the Chinatown masses by living among them and serving their needs, such as selling them vegetables and eggs at wholesale prices which were lower than that in the market. However, in order to concentrate on my thesis research (which I had suspended doing while living in Chinatown) I moved back to live near the Columbia campus. I found an idea to solve the physics problem at hand one day, quite unexpectedly, and finished my thesis quickly [11.11].

Before I earned my PhD (in 1973, formally speaking) I had applied to the Chinese representative to UN office in New York City, asking for permission to come back to China, to serve the country and the people. We were asked to wait because China was not ready to accept us back then. In 1975, my job took me to Europe, first in Belgium and then in West Germany [11.13]. I kept contact with China's embassies there, and eventually, in September 1977 after Chair Mao passed away, I was invited by the embassy to go to China, to participate in the October 1 National day celebration. It was in Beijing that I was informed that my application to return to China was finally granted, and I was assigned to work at the Chinese Academy of Sciences [11.15].

Early January 1978, I, together with my wife and my eight-month old daughter, arrived in Beijing through Hong Kong. And I started to work at the Institute of Physics, CAS [Li 2003]. In other words, I was a member of the 2nd generation haigui; in fact, I was the first haiqui to return to work at CAS after the Cultural Revolution. I was quite welcome by the official press.

3.2 Formation of the delegation

In my first year at IoP, Chia-Wei Woo (吴家玮, Fig. 1) from the Northwestern University, USA, came to our institute as a guest professor in mid-August, staying for about three months. During his stay, he invited the IoP to send a delegation of physicists to go to his department to work as visiting scholars; everything paid by his university once we arrived there.[5] This proposal was very generous, and was approved by the highest level in China, presumable by Hua Guo-Feng himself.[6] (After leaving IoP early December, Woo went to Fudan University as a guest professor for one month.) Woo did his high school in Hong Kong, went to USA in 1955 and obtained his PhD there; he was the chair of his physics department when he visited us. He, like me, belongs to the 9th generation of Chinese students going aboard.

Fig. 1. Chia-Wei Woo.

The IoP at that time was effectively run by Guan Wei-Yan, the deputy director, due to the old age of the director, Shi Yu-Wei. The IoP picked eight members to form the delegation (Fig. 2).[7] The members are: Qian Yong-Jia (钱永嘉), Li Tie-Cheng (李铁成), Zheng Jia-Qi (郑家祺), Shen Jue-Lien (沈觉涟), Wang Ding-Sheng (王鼎盛), Cheng Bing-Ying (程丙英), Gu Shi-Jie (顾世杰) and Lin Lei (林磊).

The youngest one was Cheng Bing-Ying; he was a worker-peasant-soldier graduate of Fudan University, Shanghai. Wang Ding-Sheng studied physics at Peking University (1956-1962) and was a graduate student at IoP (1962-1966); he started working at IoP in 1967. Qian Yong-Jia, a graduate of Fudan University, was the head of the delegation; he was a minor leader in IoP during the Cultural Revolution years [Li 2004: 241].

Fig. 2. The delegates with Chia-Wei Woo in front of the Field Museum of Natural History, Chicago. From left to right: Woo, Shen Jue-Lian, Lin Lei, Zheng Jia-Qi, Gu Shi-Jie, Qian Yong-Jia, Cheng Bing-Ying, Li Tie-Cheng and Wang Ding-Sheng. The visit to the Field Museum was on Feb. 15, 1979, arranged by Northwestern University soon after their arrival in Chicago, and was aired by CBS in its weekly highlight TV program *60 Minutes* on Sunday evening, Mar. 18, 1979.

Some of the delegates were already publishing papers; some examples:

- Li Yin-Yuan (李荫远), Fang Li-Zhi (方励之) & Gu Shi-Jie [1963]. Influence of ferromagnetic defects on spin waves. Wuli Xuebao, 599-612.

- Gu Shi-Jie (translator) [1974]. *Diffraction: Interference in Optics*, by Francon, M. Beijing: Science Press.
- Shen Jue-Lian [1966]. Second order phase transformation of magnetic crystals and magnetic structure of metals of the Lanthanum series. Acta Physica Sinica **22**: 94-110.
- Shen Jue-Lian [1978]. On the theory of second order phase transitions and an exposition on the non-validity of Lifshitz condition. Acta Physica Sinica **27**: 63-84.
- Wang Ding-Sheng & Pu Fu-Cho (蒲富恪) [1964]. Spin wave spectrum and excitation in finite ferro-antiferromagnetic linear chain. Acta Physica Sinica **20**: 1067-1078.
- Wang Ding-Sheng, Chen Guan-Mian (陈冠冕) & Pan, S. T. (潘孝硕) [1973]. Frequency dependence of domain wall creeping in magnetic films. Wuli (Physics) **2**: 169-182.
- Lai Wu-Yan (赖武彦), Wang Ding-Sheng & Pu Fu-Cho [1977]. Dipole-exchange spin waves in a cylindrical ferromagnet. Acta Physica Sinica **26**: 285-291.

4 The 1979 Trip and Stay in USA

4.1 The 1979 Trip

The delegation left Beijing on Feb. 9, 1979, soon after Deng's visit to the United States. We flew from Beijing to Paris by CAAC (the only government airline at that time, now reorganized as Air China), and spent an evening in a hotel in the suburb. Upon arriving, I called up my French friend Roland Ribotta from the Université Paris Sud, and he took several of us for a short sightseeing in Paris.[8] The next day, we took TWA from Paris to New York City. The TWA captain announced a welcome to our delegation to the passengers soon before we landed, and they all applause.

Arriving in New York City on Feb. 11, I suddenly had a good idea. With permission, I spent a few days in New York City and visited my friends in Chinatown there, while the rest of the delegates went on to

Washington, DC. They spent four days there, receiving briefings from the embassy people, and even did some sightseeing. The local patriotic newspaper China Daily News (美洲华侨日报) interviewed me.[9] I then flew to Chicago, and joined my fellow delegates in Evanston.

When our delegation arrived Northwestern University in Evanston on Feb. 15, the first thing that Prof. Woo did was take us to a news conference attended by local newspaper and TV reporters. Of course, that included reporters from the student newspaper. The reception was very warm. All American people were willing and eager to help us, since we came from a China which just pulled herself out of the Cultural Revolution and needed help to rebuild the country. The next thing that Woo did was take us to a closed room in the canteen and taught us how to use fork, knife and spoon to eat American food. A very essential and helpful lesson.

4.2 The Stay in USA

We lived in campus, a few minutes of walking distance from the building in which the physics department was housed. Our daily life involved traveling between two buildings, but on the weekends, we did sometimes venture to the Chicago city and did some sightseeing.

I spent three months at Northwestern University, working with Chia-Wei Woo, a theoretical physicists specialized in many-body problems. Shen Jue-Lian and Li Tie-Cheng did the same. Eventually, after I returned to China, Shen, Woo, Yu Lu (于渌)[10] and I published a joint paper [Shen et al. 1981]. Wang Ding-Sheng worked with Arthur Freeman on band-structure calculations. The rest of the delegates worked on experiments with two professors, George Wong (王克倫) and John Ketterson. In the last few weeks during my stay, I went to quite a number of university campuses to give physics seminars, and general talks on my experience in China (sponsored by the Chinese students there). The reception was very enthusiastic.

Woo then moved to University of California, San Diego, as a provost in summer 1979.[11] He took Shen and Li with him there. All the members, except me, stayed for two years in USA.[12]

5 Back to China

I went back to Beijing, via Hong Kong, in June 1979. In Hong Kong, I gave a seminar at my alma mater, the University of Hong Kong, which led to an interview by the *Wen Wei Po* (文汇报) and was invited to write a popular science article in this newspaper. While still in Hong Kong I wrote up my work on the phase transition of liquid crystals (done at Northwestern) and, with the three diagrams drawn up by my friend Chapman Wong (黄卓民, Fig. 3),[13] I submitted the paper to *Physical Review Letters* [Lin 1979]. This paper turned out to be the first one by mainland-only authors ever to appear in this top journal in the physics profession (Fig. 4).[14]

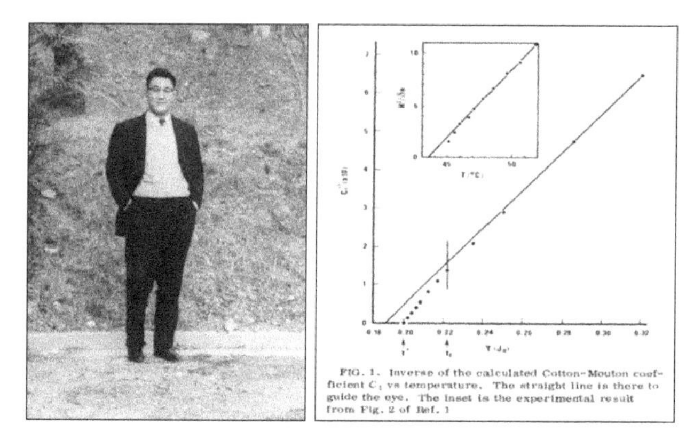

Fig. 3. *Left*: Chapman Wong (c. 1970), who drew up the three diagrams in my first paper in *Physical Review Letters* [Lin 1979]. *Right*: Figure 1 from this paper.

After returning to China in 1981, the other members progressed as physicists over the years. Cheng Bing-Ying became a PhD doctoral advisor at IoP; he passed away due to liver cancer in 1996. Li left IoP later and stayed in Canada, while Gu and his family left China and became residents of USA. Qian moved to Fudan University soon after returning. He and Zheng later worked at the Hong Kong University of Science and Technology; both retired now. Shen stayed on in IoP until he retired at 65. Wang went on to build up an active research group in Beijing; he and his students kept long-term close collaboration with

Arthur Freeman and coauthored many papers (see, e.g., the paper by Wang et al. below). He became an academician of CAS in 2005, and still works at IoP.[15]

Here are some works by these people after 1981:

- Gu Shi-Jie (translator) [1987]. *Principles of Nonlinear Optics*, by Shen, Y. R. Beijing: Science Press.
- Lu Hui-Bing (吕惠宾), Zhou Yue-Liang (周岳亮) & Gu Shi-Jie [1989]. AC power source for pre-ionization gas CW laser. Patent application.
- Qian Yong-Jia (ed.) [1989]. *Selected Papers on High T_c Superconductors Research in Shanghai: 1987-1988*. Shanghai: Fudan University Press.
- Qiu Jing-Wu (邱经武), Zhang Xian-Feng (张先锋), Tang Zhi-Ming (唐志明) & Qian Yong-Jia [1990]. Double-hole RF-SQUID made of high-temperature oxide superconductor. Chinese Journal of Low Temperature Physics, No. 4.
- Yuan Song-Liu (袁松柳), Jin Si-Zhao (金嗣炤), Chen Xiu-Jia (陈兆甲), Cao Ling (曹宁), Zheng Jia-Qi & Quan Wei-Yan [1990]. EPR of high-T_c superconductor BiSrCaCu2OY before the superconducting transition under various temperatures. Chinese Journal of Low Temperature Physics, No. 4.
- Wang Ling (王宁), Chen Kai-Lai (陈凯来) & Wang Ding-Sheng [1986]. Work function of transition-metal surface with submonolayer alkali-metal coverage. Physical Review Letters **56**: 2759-2762.
- Zhou Wei (周薇), Zhang Qi-Ming (张齐鸣), Qu Li-Jia (曲立茄) & Wang Ding-Sheng [1989]. Interaction and charge transfer in the iron nitride Fe(4)N. Physical Review B **40**: 6393-6397.
- Wang Ding-Sheng, Wu Ru-Qian (武汝前) & A. J. Freeman [1993]. State tracking first principles determination of magneto-crystalline anisotropy. Physical Review Letters **70**: 869-872.

VOLUME 43, NUMBER 21 PHYSICAL REVIEW LETTERS 19 NOVEMBER 1979

by numerically integrating the dynamical equations. They include the effects of boundaries, which increases the difficulty and may explain part of the complicated motion they report. Still, the mechanism seems clear: The system evolves toward a local minimum of $\tilde{f}$ at $\beta = 0$ or π; upon reaching the local minimum, $\tilde{f}$ changes, with the local minimum now a local maximum. The external source of the heat current provides the energy dissipated by this process.

The preceding analysis, based on helical solutions, cannot be quantitatively correct, for it is likely that the unstable helices develop into more complicated time-dependent states. Nevertheless, we expect the qualitative behavior to confirm our second basic result that an applied parallel magnetic field $H > H_c$ should induce a marked time-dependent deformation, whose character depends on the nature of the experiment: A persistent current in a torus should lead to a stable wide-angle helix with reversed but diminished supercurrent, whereas heat flow should produce anharmonic but periodic oscillations of the texture.

This work was supported in part by the National Science Foundation Grant No. DMR78-25258.

[1] P. Bhattacharyya, T.-L. Ho, and N. D. Mermin, Phys. Rev. Lett. 39, 1290, 1691(E) (1977).
[2] A. L. Fetter, Phys. Rev. Lett. 40, 1656 (1978).
[3] H. Kleinert, Y. R. Lin-Liu, and K. Maki, Phys. Lett. 70A, 27 (1979).
[4] J. B. McLaughlin and P. C. Martin, Phys. Rev. A 12, 186 (1975).
[5] C. Normand, Y. Pomeau, M. G. Velarde, Rev. Mod. Phys. 49, 581 (1977).
[6] W. F. Brinkman and M. C. Cross, in *Progress in Low Temperature Physics*, edited by D. J. Brewer (North-Holland, Amsterdam, 1978), Vol. VIIA, p. 105.
[7] A. L. Fetter, Phys. Rev. B 20, 303 (1979).
[8] Y. R. Lin-Liu, K. Maki, and D. Vollhardt, J. Phys. (Paris), Lett. 39, 381 (1978), and Phys. Rev. B 20, 159 (1979).
[9] H. Goldstein, *Classical Mechanics* (Addison-Wesley, Palo Alto, 1965), p. 219.
[10] W. M. Saslow and C.-R. Hu, J. Phys. (Paris), Lett. 39, 379 (1978); S. Takagi, Prog. Theor. Phys. 60, 934 (1978).
[11] This situation is analogous to the Eckhaus instability discussed in Ref. 5.
[12] The present dynamical equations are insufficient to calculate the period of the motion because $\partial f/\partial\beta$ vanishes at $\beta = 0$ or π. The uniform state is thus an unstable equilibrium and fluctuations must be invoked for the system to move away.
[13] J. R. Hook and H. R. Hall, J. Phys. C 12, 783 (1979).

Nematic-Isotropic Transition in Liquid Crystals

Lin Lei
Institute of Physics, Academia Sinica, Beijing, People's Republic of China,[a] *and Department of Physics and Astronomy, Northwestern University, Evanston, Illinois 60201*
(Received 6 June 1979)

Correlation functions and the Cotton-Mouton coefficient are calculated for liquid crystals beyond the mean-field approximation. My results in the context of a first-order transition are compared with the recent experiments of Keyes and Shane for N-[*p*-methoxybenzylidine]-*p*-butylaniline (MBBA) connecting with the possible tricritical nature of the nematic-isotropic transition.

Recently, Keyes and Shane[1] measured the gap exponent Δ for the nematic-isotropic (N-I) phase transition in N-[*p*-methoxybenzylidine]-*p*-butylaniline (MBBA) in the isotropic phase. They found $\Delta = 1.26 + 0.10$ which is consistent with the tricritical value $\Delta = 1.25$ but differs from the mean-field prediction $\Delta = 2$, giving the impression that the N-I transition is actually tricritical in nature. In this Letter, among other things, we show that by going beyond the mean-field approximation the so-called gap exponent Δ is not a constant but in general a function of temperature T. Depending on the temperature range under consideration, the effective exponent can deviate from the mean-field value and may be equal to 1.59, for example. Therefore, the measurement of Δ alone is insufficient in determining the critical or tricritical nature of the N-I transition. In addition, the deviation of the inverse of the Cotton-Mouton coefficient from linearity just above T_c is explained.

It has been known for some time that the de Gennes-Landau theory[2] is inapplicable near T_c in the isotropic phase. More recently, contrary to the current belief,[3] Lin and Cai[4] have shown that, quantitatively speaking, the same

Fig. 4. The very first *Physical Review Letters* paper by mainland-only authors from the People's Republic of China. A Texas Instrument calculator (bought by Lui Lam in Hong Kong just before he went back to work in China in Jan. 1978) was used to do a linear fit in Fig. 1 of this paper (Fig. 3, right). This weekly journal was created by American Physical Society in 1958 and quite immediately became the top journal in the physics profession worldwide.

I left China due to family reasons at end of 1983 after a stay of six years [Li 2003]. In Beijing, I trained five graduate students who completed five MS theses and one PhD thesis under my supervision.[16] All of them, except one, are now in the USA, against my wish. Since 1984, I worked at the City University of New York for three-and-half years and moved to San Jose in 1987. In San Jose, I worked on nonlinear and complex systems, and now more on *Scimat*[17] [Lam 2008; Burguete & Lam 2008], in particular, on *Histophysics* (physics of history) [Lam 2002] and literature as part of science [14, 17].

6 Conclusion

The 1979-1981 visit to USA had profound impact on the life and career of my fellow delegates. After all, it was their first foreign trip ever, and it happened after China had been insulated from the West for a long period of time since 1949. Here is how IoP describes the delegation's visit to the USA and its aftereffect[18] (from Major Events in [Sun 2008]):

> Upon the approval by the central government, under the promotion of Professor Woo Chia-Wei of the Northwestern University in USA, an eight-person delegation of our institute consisting of Qian Yong-Jia et al. went to the Northwestern University of the USA to be visiting scholars, signaling the beginning of the wide-range collaboration between our institute and the American physics community which was suspended for many years.

That is absolutely correct. What could be added is that the visit was funded by the American side and helped to set an example for other delegations to come, saving China a lot of money when money was still in short supply. And the visit helped to cement the friendship and promote understanding between the peoples of China and USA, a positive effect advancing the reform-and-opening up movement in China starting 30 years ago.

Acknowledgement

It was always my intention to write down the story of this 1979 visit someday. The fact that it occurred earlier than planned was prompted by Zuoyue Wang (王作跃) on Oct. 14, 2008, who was writing a book on the involvement of Chinese-American scientists in China's reform-and-opening up movement and wanted to know about this story in some detail. To prepare for the writing of this paper, following my usual practice, I gave a talk on this trip in a seminar at Peking University, on Jan. 16, 2009, upon the invitation of Wu Guo-Sheng (吴国盛). I thank both Wang and Wu for their encouragement in this effort. I am also grateful to Wang Ding-Sheng for helpful conversations and insights, and his comments after reading of this manuscript; to Chia-Wei Woo for reading and correcting an early draft of this manuscript, and a conversation in Hong Kong on Mar. 24, 2009; and to all the other members of the 1979 delegate who read the draft of this paper and provided me with corrections and unique insights.

Notes

1. This term "visiting scholar" was coined by Chia-Wei Woo in Oct. 1979 because the visiting Chinese scientists did not have professional titles such as professor or doctoral degrees, so the usual titles of visiting professors or postdocs could not be used. (Chia-Wei Woo, interview in Hong Kong, Mar. 24, 2009; see also [Woo 2007: 36].)
2. Quan Wei-Yan, born Aug. 18, 1928, and died Mar. 20, 2003 (in Taiwan), studied low temperature physics in the 1950s in USSR and returned to China in 1960. He was the president of the University of Science and Technology of China, after serving as the director of the IoP, and was an academician of CAS (since 1980).
3. www.coesa.cn/info/categorymore.shtml?Cid=C01 (Mar. 20, 2009).
4. http://archives.lib.nthu.edu.tw/exhibition/diaoyun/ (April 27, 2009).
5. As it worked out, six of these scholars were supported by Northwestern; Li Tie-Cheng and Cheng Bing-Ying were funded by CAS (at least for the first year). According to Woo (interview in Hong Kong, Mar. 24, 2009), he used the unused part of the Northwestern money to invite two visiting scholars from Fudan University.
6. Wang Dingsheng, interview in Beijing, July 7, 2005.
7. I actually volunteered myself, asking to stay for only three months in the USA.
8. Among the passengers in this small car were Gu Chao-Hao (谷超豪), the mathematician from Fudan University, and C. N. Yang's sister. They happened to be in the Paris airport in transit.

9. I worked as a volunteer in editing this newspaper during my graduate student years.

10. Yu Lu was my colleague in the same group at IoP in 1978. He moved to the newly formed Institute of Theoretical Physics, CAS, in 1979, and is now an academician of CAS.

11. Woo, after four years at UC San Diego, became the president of San Francisco State University in 1983 and, later in 1988, the founding president of Hong Kong University of Science and Technology.

12. Chen Guan-Mian, a physicist at IoP and wife of Wang Ding-Sheng, joined her husband at Northwestern in Sept. 1980 and spent a year there working with L. H. Schwartz in the Materials Research Center. After Woo left Northwestern, some of these seven scholars from IoP repeatedly returned to Northwestern for short visits.

13. Wong, an old friend of mine from Hong Kong, was then a technician in the physics department of the Chinese University of Hong Kong. At that time, there was no software (like Excel or Origin) to draw diagrams; diagrams had to be drawn carefully and professionally on special papers using Indian ink.

14. Many years later after I went back to work in the USA, I met Stanley Liu, the then assistant editor of this journal, for the first time in a physics conference. He still remembered this paper.

15. There is no retirement age for academicians in China.

16. The PhD thesis by Shu Chang-Qing (舒昌清), "Propagating solitons in shearing nematic liquid crystals," was officially in China the first PhD thesis in the field of liquid crystals *physics*.

17. *Scimat* (formerly Science Matters) is a new multidiscipline that treats all research on humans (i.e., the humanities, social science, and medical science) as part of science [Burguete & Lam 2008].

18. The original is written in Chinese: "1978年, 经中央批准, 在美国西北大学吴家玮教授的促成下, 我所钱永嘉等8人赴美国西北大学做访问学者, 标志着中断多年之后我所与美国物理界大规模合作的开始."

References

Burguete, M. & Lam, L. [2008]. *Science Matters: Humanities as Complex Systems*. Singapore: World Scientific.

Cao Cong (曹聪) [2009]. "Brain drain", "brain gain", and "brain circulation" in China. Science & Cultural Review **6**(1): 13-32.

Lam, L. [2002]. Histophysics: A new discipline. Modern Physics Letters B **16**: 1163-1176.

Lam, L. [2008]. Science Matters: The newest and biggest interdicipline. *China Interdisciplinary Science*, Vol. 2, Liu Zhong-Lin (刘仲林) (ed.). Beijing: Science Press.

Li Ya-Ming (李雅明) (ed.) [2004]. *Guan Wei-Yan's Memoir: An Oral History.* Hsinchu: National Tsinghua University.

Li Yuan-Yi (李元逸) [2003]. The unbroken China complex: The life story of a Chinese scientist. Sciencetimes, Aug. 8, 2003.

Lin Lei (Lam, L.) [1979]. Nematic-isotropic transitions in liquid crystals. Physical Review Letters **43**: 1604.

Luo Ping-Han (罗平汉) [2008] *Spring: Chinese Intellectuals in 1978*. Beijing: People's Press.

Shen Juelian, Lin Lei, Yu Lu & Chia-Wei Woo [1981]. Molecular theory of liquid crystals including anisotropic repulsion. Molecular Crystals and Liquid Crystals **70**: 301.

Sun Mu (孙牧) (ed.) [2008]. *IPCAS 80th Anniversary*. Beijing: Institute of Physics.

Tang Ying-Wu (汤应武) [1998]. *Choices: The Road of Chinese Reform Since 1978*. Beijing: Economic Daily Press.

The Seventies Monthly (ed.) [1971]. *Truth Behind the Diaoyutai Incident*. Hong Kong: The Seventies Monthly.

Woo Chia-Wei [2007]. *Establishing Together the Hong Kong University of Science and Technology: Stories and People in the Early Period*. Beijing: Tsinghua University Press.

Zhao Yan (赵岩) et al. [2008]. Searching knowledge for eighty years, glorious physicists: Eighty years anniversary of Institute of Physics. Wuli **37**(6), 363-371.

Original: Lam, L. [2010]. Recall the first delegation of Chinese "non-governmental" visiting scholars to the United States. Science & Culture Review **7**(2): 84-94. Here is the English translation from the Chinese.

20 A Science-and-art Interstellar Message

If extraterrestrial intelligence (ETI) exists, whether carbon-based or not, it is nothing but a particular type of complex system existing in the universe. Being a complex system, it must share the common traits of all (or almost all) known complex systems [2.3, 22]. For example, an ETI, like a human, is very likely an active walker [Lam 1998] that changes its surrounding (physical or virtual) landscape through its every action and is in turn influenced and limited by the changed landscape [13.4].

Like us, being a complex system, an ETI will recognize immediately the property of self-similarity (the principle behind fractals), which is a property shared by many complex systems. Being self-similar is not the same as being identical or similar. An object is self-similar if a small part of it (when blown up in scale) looks similar or identical to the whole. Trees are such an example; in fact, the existence of bonsai (called "plate scenery" by the Chinese) is an example of self-similarity, in which a small branch of a tree may look like the whole tree. In Africa, Ba-ila settlements established before 1944 are arranged in a self-similar pattern

[Eglash 1998], even though the settlers themselves may not be aware of fractals. A self-similar mosaic (a Sierpinski gasket in fact), an art object, actually lies on the floor of a cathedral built in the year 1304 in Agnani, Italy.[1] Finally, the foam-like material structure of our universe is a fractal. Self-similar objects thus exist ubiquitously over many scales on Earth and elsewhere in the universe.

The simplest example of fractals is the Sierpinski gasket (SG), named by Benoit Mandelbrot (1924-2010) after the Polish mathematician Wacław Sierpiński (1882-1969) [Sierpiński 1916]. The SG is obtained by first dividing an equilateral triangle into four equal, smaller equilateral triangles, and then removing the middle one. The process is repeated for the three existing triangles ad infinitum.[2] SG designs are at once simple and beautiful, already existing throughout human civilization, and perhaps also existing in the history of ETI, or is at least recognizable by ETI.

The SG is both a science and an art object; its human-made nature cannot be mistaken by ETI because of its neat design. (In fact, it is more than self-similar; it is self-identical.) Moreover, the SG is extremely easy to draw graphically, and it can be computer-generated in at least four different ways, all with only a few lines of code.[3]

In short, the SG is the ideal science-and-art object to be beamed to any ETI.

Notes

1. Rachel Stanley, at age 10 in 1988, discovered this mosaic in Agnani, Italy. A photo of this mosaic is reproduced in Guyon, E. & Stanley, H. E. [1991]. *Fractal Forms*. Amsterdam: Elsevier.

2. Because of the infinite number of operations, SG is only a mathematical existence. All the graphics obtained after a limited number of operations are not SG itself but can only be called "Pre-fractal" of SG.

3. The code written in Quick Basic has 14 lines [Lam 1998: 234].

References

Eglash, R. [1998]. Fractals in African settlement architecture. Complexity **4**: 21-29.

Lam, L. [1998]. *Nonlinear Physics for Beginners*. Singapore: World Scientific.

Lam, L. [2005]. Active walks: The first twelve years (Part I). International Journal of Bifurcation and Chaos **15**: 2317-2348.

Lam, L. [2006]. Active walks: The first twelve years (Part II). International Journal of Bifurcation and Chaos **16**: 239-268.

Sierpiński, W. [1916]. Sur une courbe cantonrienne qui content une image biunivoquet et continue detoute courbe donne. Comptes Rendus de l'Académie des Sciences, Paris **162**: 629-632.

Original: Lam, L. [2004]. A science-and-art interstellar message: The self-similar Sierpinski gasket. Leonardo **37**(1): 37-38.

21 This Pale Blue Dot

On September 5, 1977, NASA's spacecraft *Voyager 1* set off from Earth. The goal was to fly by Jupiter and Saturn and take pictures, and then fly away from the solar system to interstellar space. The latter mission, accomplished in August 2012, was a first for a man-made spacecraft. Fourty five years since its departure, Voyager 1 is still working, dedicated to its duties, and continues to send messages back to Earth.

On Valentine's Day, February 14, 1990, Voyager 1 was 3.7 billion miles away from Earth. It was just about to leave the solar system. Knowing that this is a one-way trip and feeling sad Voyager 1 wanted to leave a precious Valentine's Day gift for her beloved earthlings. She turned around and took a family portrait of the Sun and its 6 planetary sisters, and beamed it back to Earth. In this photo, Earth appears as a bluish dot (top left and bottom right photos).

Pale blue dot

It was Carl Sagan (1934-1996), a planetary scientist at Cornell University, who contributed to this. He was a member of the Voyager 1 imaging team at the time and felt that the opportunity was so great that he asked to point the spacecraft at Earth and take an image. Later, Sagan said in a Cornell gathering on October 13, 1994:

> We succeeded in taking that picture, and, if you look at it, you see a dot. That's here. That's home. That's us. On it, everyone you ever heard of, every human being who ever lived, lived out their lives.
>
> The aggregate of all our joys and sufferings, thousands of confident religions, ideologies and economic doctrines, every hunter and forager, every hero and coward, every creator and destroyer of civilizations, every king and peasant, every young couple in love, every hopeful child, every mother and father, every inventor and explorer, every teacher of morals, every corrupt politician, every superstar, every supreme leader, every saint and sinner in the history of our species, lived there—on a mote of dust, suspended in a sunbeam.
>
> The Earth is a very small stage in a vast cosmic arena. Think of the rivers of blood spilled by all those generals and emperors so that in glory and in triumph they could become the momentary masters of a fraction of a dot. Think of the endless cruelties visited by the inhabitants of one corner of the dot on scarcely distinguishable inhabitants of some other corner of the dot. How frequent their misunderstandings, how eager they are to kill one

> another, how fervent their hatreds. Our posturings, our imagined self-importance, the delusion that we have some privileged position in the universe, are challenged by this point of pale light …
>
> To my mind, there is perhaps no better demonstration of the folly of human conceits than this distant image of our tiny world. To me, it underscores our responsibility to deal more kindly and compassionately with one another and to preserve and cherish that pale blue dot, the only home we've ever known.

His book *Pale Blue Dot* (bottom left figure) was published in 1994. On December 20, 1996, Sagan died at the age of 62. NASA built a Carl Sagan Center in the Bay Area of California in 2001, which combines scientific research and popularization.

This pale blue dot

Tamkang University, located in Tamsui, is a private university co-founded by Clement Chang (1929-2018) and his father. It was established in 1950 and assumed its present name in 1980. Tamkang University has a futurology research institute.

On August 12-16, 2001, the Foundation For the Future in Bellevue, Washington held a "Future of Humanities: Seminar 3" in Seattle. I gave an invited talk at the meeting: Modeling history and predicting the future: the active walk approach. Chang, who attended the meeting, invited me to his university to be the speaker of the Tamkang Chair Lectures. He paid the round-trip air ticket and a speaker fee.

As a result, on December 9-11, 2003, as an "internationally renowned physicist and futurist," I gave three talks at Tamkang University: Why is the world so complicated? Does God exist? How to model history and predict the future? These reports are collected in *This Pale Blue Dot* (Tamkang University Press, 2004; top right figure), which is my first popular science book. This book is not for sale and is in limited print run. On the back cover of the book, there are a few paragraphs written by me:

The year was 1990. When NASA's *Voyager 1* space probe was leaving the Solar System, it turned its camera around and took one last picture of Earth. In this rare picture, our dear Earth appears as a pale blue dot. The blue is reflection from the seas while the white comes from the clouds. It is this pale blue dot we share every day. It is on this pale blue dot our joy and sorrow come and go. We are curious about the real world happening on this pale blue dot and beyond. We are curious about the trees, ants, sunset, and the stars up in the sky. We are curious about the fate of us humans—past, present, and future. And we keep on wondering whether there is a God out there.

These questions were raised systematically about 2,600 years ago by the Greeks in the West and the Chinese in the East. The complete answer did not come, not even today. However, in the past 400 years since Galileo, modern science has prospered, and we know much more. We even have the answer to some of the big questions raised by our ancestors.

Our understanding of this pale blue dot and its inhabitants, we humans among them, comes from all branches of science, but especially from the study of complex systems in the last four decades as well as the century-long development of evolutionary science and neuroscience.

To understand where we came from, why we humans behave the way we do, and how we can make the world better tomorrow, we must look back 13.7 billion years in the past—a long lookback in time, just like the Voyager 1's long lookback in space. In this book, some of these understandings are presented, spanning from science to art and human history, and from philosophy to the God question.

Original: Lam, L. [2022]. This pale blue dot. *This Pale Blue Dot*, Lam, L. San Jose: Yingshi Workshop. English translation here.

22 Why the World Is So Complex

For most of us the world looks very complex. Why? At the societal level, the world is sometimes unnecessarily complex because people have conflicting interests and some of them are selfishly motivated, according to some observation. What this means is that the internal states of a human body—thinking, mood, memory, background, and all that—play an important role and are the culprit. But what about the inanimate world? Our world (or the universe) is nonlinear, quite often nonsymmetrical, and always in nonequilibrium. Furthermore, mistakes frequently occur, and *emergent* properties appear often. What do these concepts mean and how relevant are they? Here, the deep reasons driving the complexity of our world will be discussed. Interestingly, it turns out that out of complexity, simple mechanisms or laws can be discerned. The real question then becomes: Why is the world so simple? And, why is our world at all comprehensible?

1 Introduction

There is a feeling that our world is very complex. According to the Merriam-Webster dictionary:

- **Complex**: composed of two or more parts…hard to separate, analyze, or solve.
- Complex suggests the unavoidable result of a necessary combining and does not imply a fault or failure.
- **Complicated** applies to what offers great difficulty in understanding, solving, or explaining.

In short, Complex = inevitable complicated, and Complicated = avoidable complex.

Scientifically, the mechanisms driving the complexity of our world are more than what the dictionary suggests; it is not purely due to the existence of two or many parts. And, in practice, complex systems are hard to analyze, but progress has been made in the last few decades.

Science is about the study of nature, which includes any material systems. Humans are (biological) material bodies called *Homo sapiens*, which fall naturally into the domain of science. Thus, the societal world and the inner world of individual psyche are also parts of science.

The societal world is deemed complex, a fact usually attributed to the human brain, meaning that humans can think "freely," have conflicting interests, mood swings, and diverse cultural backgrounds. In other words, there are too many internal states, too many degrees of freedom, within a human body. But a water molecule also has many internal states—discrete states associated with the vibrations and rotations of the two hydrogen atoms and the oxygen atom that make up the molecule, and the infinite number of continuum states due to the movement of the molecule as a whole.

The way that physics deal with these internal states is to ignore most of them and keeping only those few that are most relevant, depending on what aspects of the molecules are under study. For example, in understanding the flow of water the discrete internal states of the water molecules are dropped completely. The same approach can be applied to human studies.

And how about the inanimate world? We still see a lot of complexity there. Beautiful, intriguing spatial-temporal patterns show up everywhere, such as clouds in the sky, swirling smoke columns arising from the chimneys, trees and flowers in various shapes, and many, many others.

About 300 years ago, in the *deterministic*, clockwork world of Isaac Newton (1642-1727), everything is predetermined. The world is governed by a few eternal laws; only the initial conditions vary. (And, according to Newton, even that was set and adjusted by God, from time to time.) The world could be complex, but boring. This rigid world was replaced by the *probabilistic* world with the advent of quantum mechanics, about 100 years ago, but it concerns the small-scale world only, at the atomic level, with a few exceptions.

The exceptions showing up in the large-scale or macroscopic world are superconductivity and superfluidity, manifestations of the so-called Bose-Einstein condensation. (Incidentally, the 2003 Nobel Prize in physics was awarded to works done on superconductivity and superfluidity.) Apart from that, quantum effects are still relevant to our real world. The electrons running in the electric wires or in the silicon chips in our computers are governed by quantum mechanics. A radioactive nucleus may or may not decay within a certain time, and if that nucleus is linked to a shot gun that could shoot at a confined dog, then the fate of the dog would depend on whether the nucleus actually decayed or not—an example of uncertainty in the microscopic world showing up in the macroscopic world, as the dog is concerned. But all these quantum effects are kind of remote, if one wants to understand the complexity in our daily world, the world we can see, smell, and feel.

What then makes our real world so complex?

2 What Is Complexity?

It would be nice if we can quantitatively define *complexity* and use it to compare different situations. Unfortunately, complexity is like love, you know it when you encounter it, but it is hard to pin it down. In fact, people came up with more than 30 definitions of complexity, all different. For our purpose, a working definition is enough: Complexity is the length of the shortest description of a system.

For example, let us look at three sequences of 1's and 0's.

A. 11111111111111111111111111111111

B. 11001010001100000110010101101001 10

C. 10010111001010111000000110101000 11

Sequence A can be described by "All 1's." Sequence B is generated randomly, and so is described by "1's and 0's generated randomly." There is no pattern in sequence C, and the description has to be a recitation of the whole sequence, and so, C is the most complex among the three.

Such a definition is, of course, problematic. If we are not informed that sequence B is generated randomly, we will not know by examining it, and we will assign it the same complexity as for C. The definition is also subjective. It could be we are not smart enough to identify the rhythm or pattern in C, or we may be able to identify it a month later.

A similar experience occurred historically in the dispute between Galileo (1564-1642) and the Vatican. Galileo confirmed Nicolaus Copernicus' Sun-centered theory that by putting the Sun at the center of the solar system the paths of the planets appear as closed loops (ellipses in fact, according to Johannes Kepler)—a description much simpler than that with the Earth at the center—and the solar system appears less complex.

Well, let us keep these limitations in mind and move forward.

3 The World Is Nonlinear

A system is *nonlinear* if the output from the system is not proportional to the input. Mathematically, the signature of a nonlinear system is the breakdown of the *Superposition Principle*, which states that the sum of two different solutions of the equation describing the system is again a solution. For example, the sound wave equation:

$$\partial^2 y/\partial x^2 = \partial^2 y/\partial t^2 \tag{1}$$

is a *linear* equation, where y(x,t) is the displacement in the sound wave. The simple-pendulum equation:

$$d^2\theta/dt^2 + (g/L)\sin\theta = 0 \qquad (2)$$

is a *nonlinear* equation, where θ is the displacement angle (Fig. 1). In Eq. (2), sin θ is the nonlinear term; it causes the failure of the superposition principle since $\sin\theta_1 + \sin\theta_2 \neq \sin(\theta_1 + \theta_2)$.

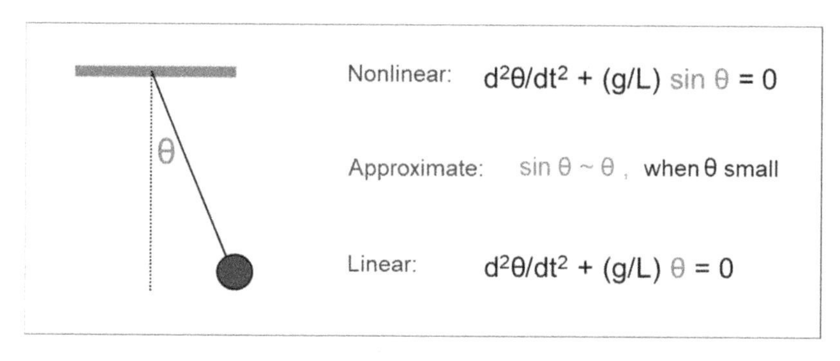

Fig. 1. Simple pendulum and its equation. L = pendulum length; g = acceleration due to gravity.

Apart from sound waves, electromagnetic waves also obey linear equations. And these are important practical cases. Radio communications are based on the propagation of electromagnetic waves and constitute an important part of our information technology. However, while the linear world is relatively simple, like "1 + 1 = 2," the nonlinear world is not.

An important advance in science in the last few decades is the study of *Chaos*. In the daily language, chaos means "a state of utter confusion." In the realm of science, chaos is a technical word representing the phenomenon that the behavior of *some* nonlinear systems depends sensitively on the initial conditions. Not every nonlinear system is chaotic, but chaos does occur in many mathematical and real systems such as a dripping faucet, a waterwheel, thermal convection of liquids, electronic circuits, chemical reactions, and heart beats.

Chaos can happen in very simple, deterministic systems, too. The simplest example is the *Logistic Map*:

$$x_{n+1} = r\,(x_n - x_n^2) \quad (3)$$

where $n = 0, 1, 2\ldots$; $0 \le x_n \le 1$ and $0 \le r \le 4$; x_n^2 is the nonlinear term in this iteration equation. In one application, x_n represents the population of moths in a certain habitats in its nth generation. The last term mimics the negative effect if the population gets too large since food will be in short supply. (In fact, to make the mathematics tidy, x_n is the population divided by the maximum population.)

For a fixed parameter r and starting from an initial x_0, Eq. (3) is used to generate a sequence of numbers $\{x_n\} = \{x_0, x_1, x_2, x_3 \ldots\}$. As r increases from zero, there is a certain range of r that the sequence behaves periodically with period 1 for n large ($\{x_n\} = \{\ldots$ a, a, a $\ldots\}$); then another range, the period jumps to 2 ($\{x_n\} = \{\ldots$ b, c, b, c $\ldots\}$); then to 4, to 8, etc. This is called *period doubling*, a behavior quite unexpected (Fig. 2). But more surprisingly, some very different, real systems like a dripping faucet and a thermally convecting liquid actually behave like this—an example of *universality* in nature.

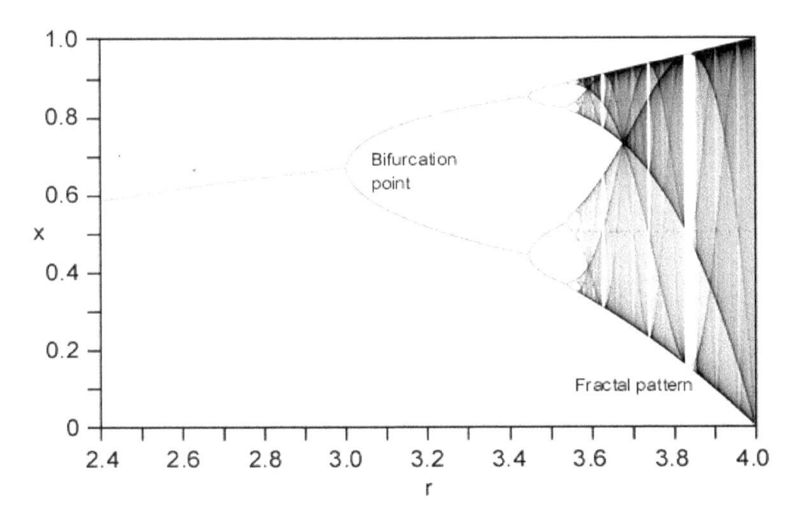

Fig. 2. Period doubling effect of the Logistic Map. x is x_n when n is large enough.

For r ≥ 3.5699456…, the sequence looks very random. In this so-called chaotic regime, two sequences with slightly different initial x_0 soon diverge from each other (Fig. 3). Since initial conditions in the *real* world cannot be controlled to a 100% precision, the outcome of a chaotic system is unpredictable *in practice*, even though the system itself is classical (i.e., non-quantum mechanical) and deterministic. Thus, an unavoidable, very slight change of the initial condition due to a falling autumn leaf in San Jose could change the weather in Tamsui. (In reality, this effect is not observable due to other, larger disturbances.) No wonder our world is so **complex**!

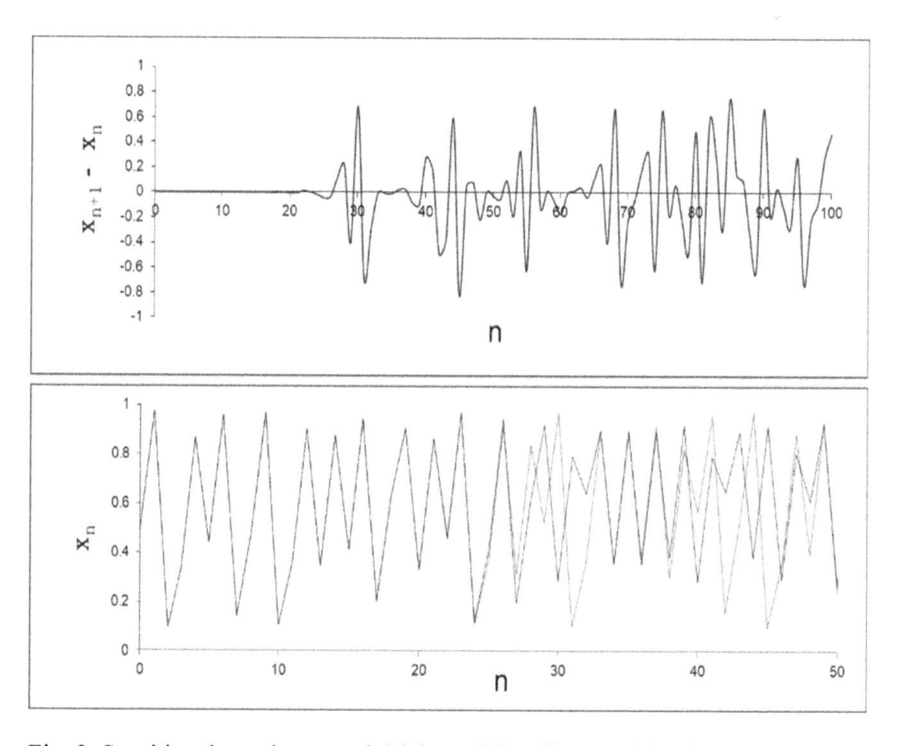

Fig. 3. Sensitive dependence on initial condition illustrated by the Logistic Map. r = 3.9. For the two sequences x_0 = 0.5 and 0.5001, respectively. The two sequences are almost identical for n < 22 but diverge for n > 22 (*top*), as can be seen more clearly in the bottom figure.

4 The World Is Nonsymmetrical

If you look around yourself or look out of the window, you will see that all symmetrical objects are man-made (Fig. 4, left). Nature is mostly nonsymmetrical (the prominent exception being crystal structures that are beyond our naked eyes). A nonsymmetrical figure is more complex than a symmetrical figure because it takes longer to describe a nonsymmetical shape (Fig. 4, right.)

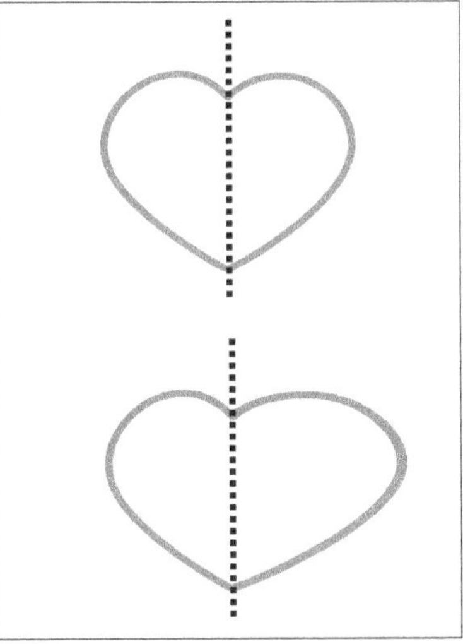

Fig. 4. *Left*: A street in Paris (Photo/Lui Lam). *Right*: An asymmetrical heart (bottom right) is more complex than a symmetrical heart (top right) because the former requires more words to describe its shape.

There is a mechanism adopted by nature that is called *spontaneous symmetry breaking*, which means that the physical state of a system tends to be more asymmetric than what theory allows or people expect. An example is the symmetry of molecular arrangements (lattices) in crystals, which is weaker than the spatial-translation symmetry embedded in theory. Two simple examples of symmetry breaking are presented in Fig. 5.

Symmetry breaking, spontaneous or not, makes the world more **complex**!

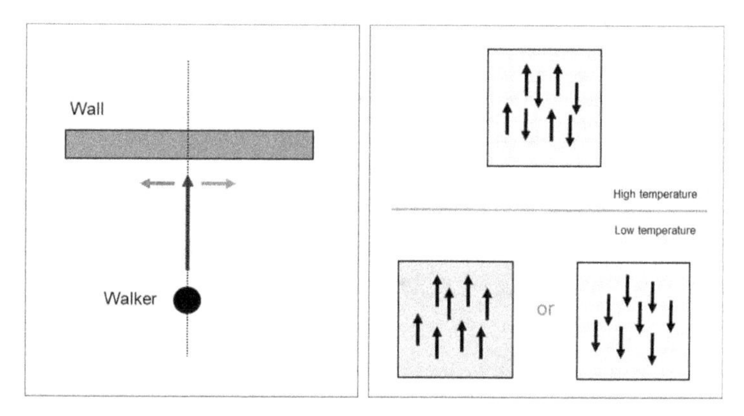

Fig. 5. Symmetry breaking. *Left*: Walk toward a brick wall. The left-right symmetry is broken when the walker makes a turn toward either left or right before hitting the wall. *Right*: A magnet. The up-down symmetry is broken as the magnet is cooled.

5 The World Is in Nonequilibrium

To improve on the efficiency of steam engines, thermodynamics was ushered in by the industrial revolution about 200 years ago, similar to the more recent case that nanotechnology is driven by the need of our information era. Thermodynamics represents one of the great triumphs of theoretical physics, peaked with the Second Law.

In its innocent form, the second law of thermodynamics says simply: Heat *by itself* cannot go from a cold place to a hot place. Nothing dramatic; it says something we already know from our daily experience. Expressed in another form, it says: For any *closed* system, when the system moves from an *equilibrium* state to another equilibrium state, the entropy of the new state is larger or equal to that of the old state. Still, interesting but not alarming. The second law becomes a bombshell once we understand that entropy is a measure of disorder, through the effort of Ludwig Boltzmann (1844-1906) a century ago.

Pretty soon, scientists and laypeople alike apply the second law to our universe, a closed system, and conclude that everything will disintegrate, and our universe will eventually end in a heat death, in complete disorder. How sad, for the universe and us. Luckily, they are plainly wrong. The universe is expanding all the time since the big bang and is never in an equilibrium state, and so the second law does not apply.

A second common mistake in applying the second law is by ignoring the prerequisite, a closed system. People look around themselves and find the increase of order everywhere, and they are puzzled. The fact is, most of the interesting systems in our world, including humans, are not closed systems. For example, a living human body intakes air and food; it is not isolated from its environment. Our body temperature is higher than the room temperature; it continuously exchanges heat with the outside world. For open systems, the second law does not apply, and ordering can indeed increase as time goes by. That is why human bodies and plants and other complex living things can grow. Furthermore, these systems, like our universe, are not in equilibrium states. A human body is in equilibrium only when it is dead.

However, it should be pointed out that there do exist important closed systems in our world which are governed by the second law, such as chemical solutions in the test tube, and crystals when they are properly isolated. But these are very exceptional cases.

Systems far from equilibrium are nonlinear; its dynamical behavior has bifurcation points; order can be produced from disorder. Nonequilibrium systems are *complex systems*. Examples are thermal convection in fluids (the collective name for liquids and gases), turbulence, snowflake formation, hurricanes, and most biological systems.

Thermal convection could happen in a layer of liquid. As the temperature from below the layer is increased, patterns of less and less symmetry appears, one after the other, before chaos appears. Similarly, *electro-*convection could happen in a thin layer of liquid crystals under the influence of a uniform vertical electric field: As the vertical voltage is increased, a sequence of patterns called normal rolls, zig-zag, skewed varicose, and bimodal—each one more complex than its predecessor—appears before chaos sets in [Ribotta et al. 1986] (Fig. 6).

Even bacteria, a living system, know how to play this game. The colonies of *Bacillus subtilis* develop more complex patterns as the growth conditions become more adverse (Fig. 7).

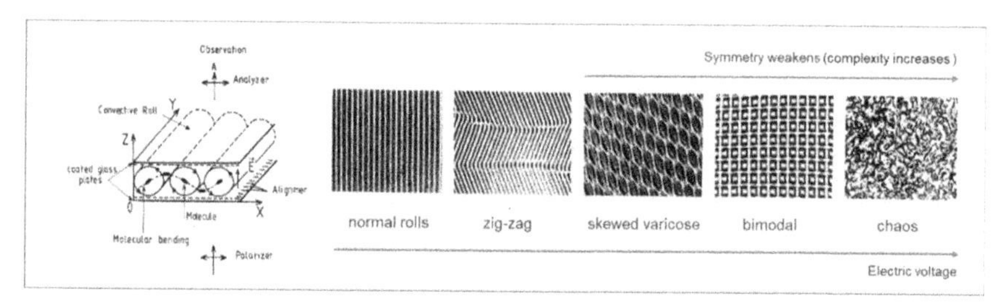

Fig. 6. Pattern formation in electroconvection of liquid crystals.

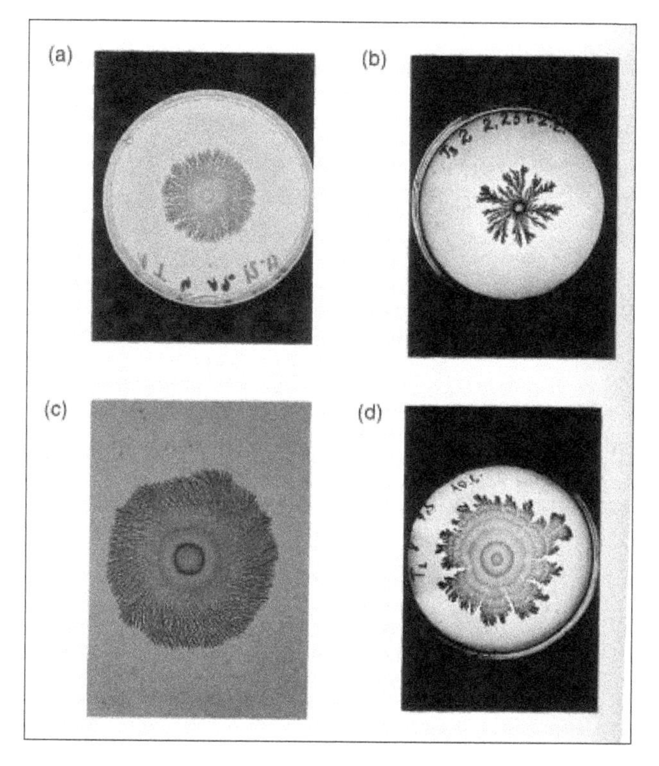

Fig. 7. The collective arrangement of bacteria *Bacillus subtilis* under different living conditions—an example of self-organization. The complexity of the pattern varies, resulting from the collective effort of bacteria to survive.

6 The World Is Full of Emergent Properties

The real world is made up of various interconnecting hierarchies, or levels. Going up in size, above the level of molecules, we have the level of condensed matter, cells, organisms, humans, and human society. Going down, we have atoms, nucleons and electrons, and quarks.

At each level, new properties emerge which are quite impossible to guess a priorily from the knowledge of the level under it. For example, knowing water are made of molecules and knowing all we can know about the molecules, no one can guess that water can flow. The *fluidity* of water is an *emergent* property. Emergent properties are surprises. And the world is full of many levels and a lot of emergent properties, making the world very **complex**!

Even though the characteristics found at one level can be explained, in principle, from the properties of the constituents at the next level below it, in *practice*, working from the bottom up—called the reduction method—is very difficult, if not impossible, in most cases. In the case of water, after observing that water can flow, a phenomenological equation, the Navier-Stokes equations, are first written down. The molecular nature of water at the lower level is then used to calculate the viscosity coefficients contained in the equations.

It should be noted that the knowledge of the lower levels beyond the adjacent lower one will have no immediate effect on the level under study. In other words, the fundamental laws governing elementary particles, even if they are eventually established, will have zero effect on the understanding of human affairs, since humans are many levels far above the level of elementary particles.

7 Mistakes Are Made

Mistakes are unintentional chance events (Fig. 8). They are usually bad, but could turn out to be good depending on the circumstances and your viewing angle. And they provide this world with surprises and make it more **complex**!

Fig. 8. A man asks God why we are living in this very imperfect world. God replied: "The Lord God of the universe take full responsibility. Mistakes are made."

Many famous discoveries in science were discovered when mistakes were made. In fact, we humans are here precisely because error in the copying of DNA molecules enables different versions of a species to exist, which then compete in evolutionary survival á la Charles Darwin (1809-1882), allowing more complex species to emerge and eventually leading to the existence of humans. In order words, without biological mistakes we would not even be here!

8 Simplicity Out of Complexity

Out of the many complexity in the real world, the surprising thing is that simple rules or laws can be found. In the large-scale world, we are blessed with Newton's three laws of motion, which govern the movement of ants, humans, planets and even water. Out of the apparent randomness of a dissipative chaotic system, there exists orderly structure in the state space called *strange attractors* [Lam 1998].

To generate the large variety of structures and patterns efficiently, it seems that Mother Nature adopts a few simple organizing principles. The two extreme cases are the Principle of Complete Order and the Principle of Complete Disorder, leading, for example, to crystals and gases, respectively. A more interesting scheme is the Principle of Self-

Similarity, giving rise to *fractals* (Fig. 9). However, not every structure in nature is a fractal; there must exists at least one more scheme, and that is the Principle of Active Walk [Lam 2005].

Fig. 9. Examples of fractals. From left to right: Sierpiński gasket, distribution of tree branches, and distribution of clouds.

Active Walk (AW) was introduced by Lam in 1992 as a paradigm of self-organization and pattern formation in nonlinear complex systems. In an AW, the walker changes the *deformable* landscape as it walks and is influenced by the deformed landscape in choosing its next step. For example, ants are living active walkers. When an ant moves, it releases pheromone of a certain type and hence changes the spatial distribution of the pheromone concentration. Its next step is moving towards positions of higher pheromone concentration. In this case, the pheromone distribution is the deformable landscape.

By leaving the landscaping rule and the stepping rule of the walker open, to be specified according to the particular phenomenon under modeling, the AW scheme is very simple and flexible, and hence very powerful.

Active walk has been applied successfully to various complex systems from both the natural and social sciences (Fig. 10). Examples include pattern formation in physical, biological, and chemical systems such as surface-reaction induced filaments and retinal neurons; the formation of fractal surfaces; anomalous ionic transport in glasses; population dynamics; bacteria movements and pattern forming; food foraging of ants; spontaneous formation of human trails; localization-delocalization transitions; oil recovery; granular matter; economics; and positive-feedback systems. More recently, AW has been used to model human history [Lam 2002, 2008].

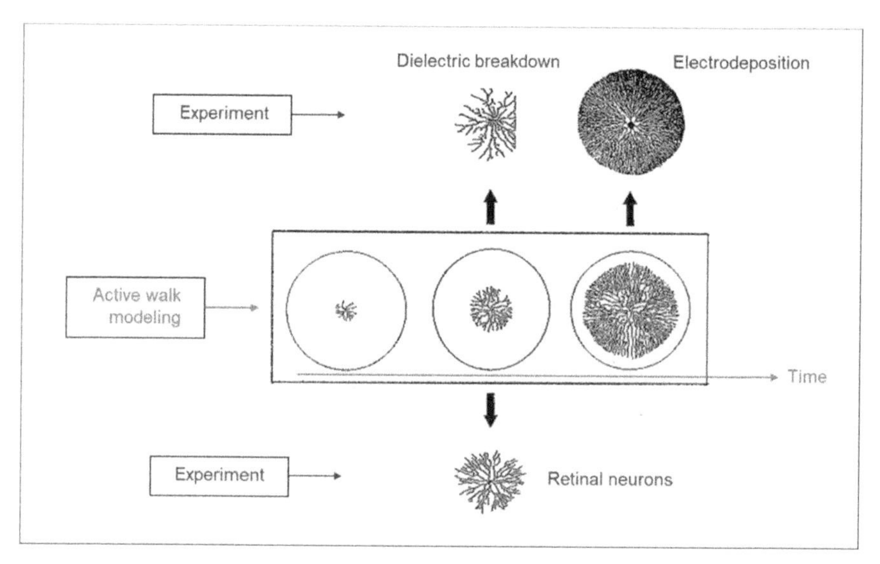

Fig. 10. A specific result of the active walk simulation [Lam 1998]. The resulting complex patterns are quite consistent with the three patterns in real systems, indicating that the Principle of Active Walk is indeed one of the organizing principles existing in nature.

9 Conclusion

The world is complex for the following reasons:

1. Quantum mechanics removes determinism at small scales, and chaos does that at large scales.
2. The world is nonlinear, nonsymmetrical, and in nonequilibrium.
3. Unexpected emergent properties appear at every level.
4. Mistakes are made often.

In short, the world results from the interplay of chance and necessity. The world is not just complex, but is creative. It is splendid, beautiful, and full of surprises.

Perhaps the question is not why the world is so complex, but why it is so simple. The more important ultimate question is: Why is the world understandable?

References

Lam, L. [1998]. *Nonlinear Physics for Beginners: Fractals, Chaos, Solitons, Pattern Formation, Cellular Automata and Complex Systems*. Singapore: World Scientific. (This book bridges the gap between popular science accounts and serious studies in the field of nonlinear science.)

Lam, L. [2002]. Histophysics: A new discipline. Modern Physics Letters B **16**: 1163-1176.

Lam, L. [2005]. Active walks: The first twelve years (Part I). International Journal of Bifurcation and Chaos **15**: 2317-2348.

Lam, L. [2008]. Human history: A science matter. *Science Matters: Humanities as Complex Systems*, Burguete, M. & Lam, L. (eds.). Singapore: World Scientific.

Ribotta, R., Joets, A. & Lin, L. (Lam, L.) [1986]. Oblique roll instability in an electroconvective anisotropic fluid. Phys. Rev. Lett. **56**: 1595; **56**: 2335 (E).

Trinh Xuan Thuan [1998]. *Le Chaos et L'Harmonie: La Fabrication du Réel*. Paris: Librairie Arthème Fayard. (This book is better than most English books on similar topics.)

Original: Lam, L. [2004]. Why the world is so complex. *This Pale Blue Dot*, Lam, L. Tamsui: Tamkang University Press. Updated here.

23 Science and Religion: Does God exist?

correspondence

Leading scientists still reject God

Sir — The question of religious belief among US scientists has been debated since early in the century. Our latest survey finds that, among the top natural scientists, disbelief is greater than ever — almost total.

Research on this topic began with the eminent US psychologist James H. Leuba and his landmark survey of 1914. He found that 58% of 1,000 randomly selected US scientists expressed disbelief or doubt in the existence of God, and that this figure rose to near 70% among the 400 "greater" scientists within his sample[1]. Leuba repeated his survey in somewhat different form 20 years later, and found that these percentages had increased to 67 and 85, respectively[2].

In 1996, we repeated Leuba's 1914 survey and reported our results in *Nature*[3]. We found little change from 1914 for American scientists generally, with 60.7% expressing disbelief or doubt. This year, we closely imitated the second phase of Leuba's 1914 survey to gauge belief among "greater" scientists, and find the rate of belief lower than ever — a mere 7% of respondents.

Leuba attributed the higher level of disbelief and doubt among "greater" scientists to their "superior knowledge, understanding, and experience"[2]. Similarly, Oxford University scientist Peter Atkins commented on our 1996 survey, "You clearly can be a scientist and have religious beliefs. But I don't think you can be a real scientist in the deepest sense of the word because they are such alien categories of knowledge."[4] Such comments led us to repeat the second phase of Leuba's study for an up-to-date comparison of the religious beliefs of "greater" and "lesser" scientists.

Our chosen group of "greater" scientists were members of the National Academy of Sciences (NAS). Our survey found near universal rejection of the transcendent by NAS natural scientists. Disbelief in God and immortality among NAS biological scientists was 65.2% and 69.0%, respectively, and among NAS physical scientists it was 79.0% and 76.3%. Most of the rest were agnostics on both issues, with few believers. We found the highest percentage of belief among NAS mathematicians (14.3% in God, 15.0% in immortality). Biological scientists had the lowest rate of belief (5.5% in God, 7.1% in immortality), with physicists and astronomers slightly higher (7.5% in God, 7.5% in immortality). Overall comparison figures for the 1914, 1933 and 1998 surveys appear in Table 1.

Table 1 Comparison of survey answers among "greater" scientists

Belief in personal God	*1914*	*1933*	*1998*
Personal belief	27.7	15	7.0
Personal disbelief	52.7	68	72.2
Doubt or agnosticism	20.9	17	20.8
Belief in human immortality	*1914*	*1933*	*1998*
Personal belief	35.2	18	7.9
Personal disbelief	25.4	53	76.7
Doubt or agnosticism	43.7	29	23.3

Figures are percentages.

Repeating Leuba's methods presented challenges. For his general surveys, he randomly polled scientists listed in the standard reference work, *American Men of Science* (*AMS*). We used the current edition. In Leuba's day, *AMS* editors designated the "great scientists" among their entries, and Leuba used these to identify his "greater" scientists[1,2]. The *AMS* no longer makes these designations, so we chose as our "greater" scientists members of the NAS, a status that once assured designation as "great scientists" in the early *AMS*. Our method surely generated a more elite sample than Leuba's method, which (if the quoted comments by Leuba and Atkins are correct) may explain the extremely low level of belief among our respondents.

For the 1914 survey, Leuba mailed his brief questionnaire to a random sample of 400 *AMS* "great scientists". It asked about the respondent's belief in "a God in intellectual and affective communication with humankind" and in "personal immortality". Respondents had the options of affirming belief, disbelief or agnosticism on each question[1]. Our survey contained precisely the same questions and also asked for anonymous responses.

Leuba sent the 1914 survey to 400 "biological and physical scientists", with the latter group including mathematicians as well as physicists and astronomers[1]. Because of the relatively small size of NAS membership, we sent our survey to all 517 NAS members in those core disciplines. Leuba obtained a return rate of about 70% in 1914 and more than 75% in 1933 whereas our returns stood at about 60% for the 1996 survey and slightly over 50% from NAS members[1,2].

As we compiled our findings, the NAS issued a booklet encouraging the teaching of evolution in public schools, an ongoing source of friction between the scientific community and some conservative Christians in the United States. The booklet assures readers, "Whether God exists or not is a question about which science is neutral"[5]. NAS president Bruce Alberts said: "There are many very outstanding members of this academy who are very religious people, people who believe in evolution, many of them biologists." Our survey suggests otherwise.

Edward J. Larson
Department of History, University of Georgia, Athens, Georgia 30602-6012, USA
e-mail: edlarson@uga.edu
Larry Witham
3816 Lansdale Court, Burtonsville, Maryland 20866, USA

1. Leuba, J. H. *The Belief in God and Immortality. A Psychological, Anthropological and Statistical Study* (Sherman, French & Co., Boston, 1916).
2. Leuba, J. H. *Harper's Magazine* **169**, 291–300 (1934).
3. Larson, E. J. & Witham, L. *Nature* **386**, 435–436 (1997).
4. Highfield, R. *The Daily Telegraph* 3 April, p. 4 (1997).
5. National Academy of Sciences *Teaching About Evolution and the Nature of Science* (Natl Acad. Press, Washington DC, 1998).

"Does God exist?" is a question that almost everyone has asked at least once in one's lifetime. According to survey, about 40% of American scientists answered this question in the affirmative. Why? Do they really know? Can one ever know? Here, we discuss these questions and provide the answers, from the broader perspective of the real nature of science and that of religion, and the interaction between these two pillars of Western civilization.

1 Introduction

My interest in the God question became serious in 1996 when a friend of mine, who has a bachelor's degree in the humanities from a top Taiwan university, became a loyal follower of a certain "religious" figure residing in Taiwan. With good intention, my friend tried to explain it to me. But from what I could gather, the critical thinking that my friend supposed to pick up from the university training was gone.

Soon, an article by Edward Larson and Larry Witham published in *Nature* caught my attention [Larson & Witham 1997]. It turned out that in 1996 they repeated a survey taken 82 years ago and found that about the same percentage of American scientists—close to 40%—believed in a personal god.

With this data, the God question becomes a *scientific* question. Why 40 %? Why there is no change of this figure in a span of 82 years, in spite of the tremendous progress made in science since the year 1914 when the first survey was taken? Isn't it that there is intrinsic conflict between science and religion, and so how can those 40% scientists maintain their faith in God and still doing research in science every day? Do they actually know what they are doing? In other words, for me, the God question suddenly turned into a science and religion question.

My preliminary findings and thinking were presented in two physics seminars at my university: Does God Exist? April 21, 1998; God, Science, Scientist, December 7, 1999.

With intense interests and enthusiastic responses from the audience, a public lecture *series* "God, Science, Scientist" was established by me in December 1999. The series is a vehicle to popularize science, not religion; God in the title is meant to get people's attention. The first three speakers in this series are Michael Shermer, Eugenie Scott, and Charles Townes (1915-2015)—the Nobel laureate and inventor of laser.

Here I will present my current understanding of this important issue.

2 Why Religion? Why Science?

Religion

In the beginning, many thousands of years ago, religion was human's pursuit for an explanation of *everything* that was observed in the sky and on earth. The basic drive is human's *curiosity*. Curiosity could be an advantage for evolutionary survival à la Darwin: Those who are curious about their surroundings may see dangers coming and find ways to avoid them, and hence survive longer and better.

In the absence of any scientific knowledge at that time, it is natural to explain everything by using analogies and lessons learned in the daily lives. Consequently, the rise and disappearance of the sun everyday over the sky was explained by the effort of a sun god—called Apollo by the Greeks—and similarly for the moon. When there are enough gods, all natural and human phenomena can be explained, but they cannot be predicted (Fig. 1).

When one was sick, without the benefit of any medical knowledge, an easy explanation was that a certain god was offended. To get well again, the god's anger had to be removed, and that could be done by bribery—in the form of animal or even human sacrifice—in the same way that it worked with humans themselves. With enough number of gods, this "theory of everything" worked. There was no conflict with existing evidence, at least for a long time.

The next step was the reduction of the number of gods from many to one, reflecting human's desire for *simplification*. Monotheism appeared in Egypt more than 3,000 years ago and then disappeared. It was not until the Hebrews of Babylon adopted it more than 2,000 years ago that monotheism was consolidated in the world.

This principle of simplification remains in the core of modern science and is called Occam's Razor: What can be done with fewer is done in vain with more. That is: the simpler the better (TSTB).

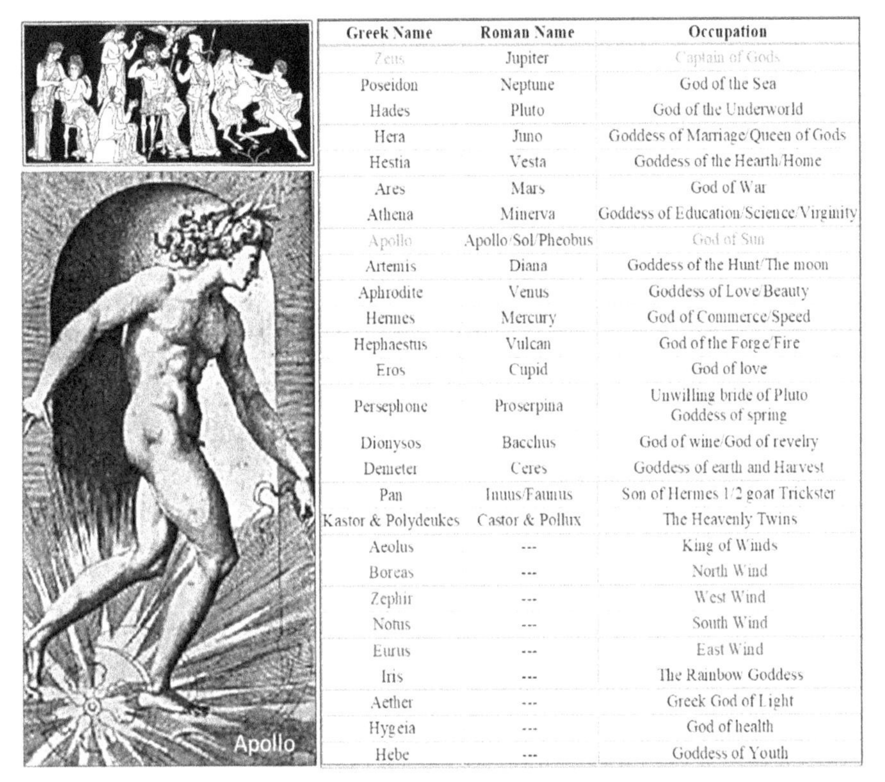

Greek Name	Roman Name	Occupation
Zeus	Jupiter	Captain of Gods
Poseidon	Neptune	God of the Sea
Hades	Pluto	God of the Underworld
Hera	Juno	Goddess of Marriage/Queen of Gods
Hestia	Vesta	Goddess of the Hearth/Home
Ares	Mars	God of War
Athena	Minerva	Goddess of Education/Science/Virginity
Apollo	Apollo/Sol/Pheobus	God of Sun
Artemis	Diana	Goddess of the Hunt/The moon
Aphrodite	Venus	Goddess of Love/Beauty
Hermes	Mercury	God of Commerce/Speed
Hephaestus	Vulcan	God of the Forge/Fire
Eros	Cupid	God of love
Persephone	Proserpina	Unwilling bride of Pluto Goddess of spring
Dionysos	Bacchus	God of wine/God of revelry
Demeter	Ceres	Goddess of earth and Harvest
Pan	Inuus/Faunus	Son of Hermes 1/2 goat Trickster
Kastor & Polydeukes	Castor & Pollux	The Heavenly Twins
Aeolus	---	King of Winds
Boreas	---	North Wind
Zephir	---	West Wind
Notus	---	South Wind
Eurus	---	East Wind
Iris	---	The Rainbow Goddess
Aether	---	Greek God of Light
Hygeia	---	God of health
Hebe	---	Goddess of Youth

Fig. 1. Gods in Greek mythology. There are many and fine division of labor, e.g., south, north, east, and west winds are each controlled by one god. Each god is multitasking. For example, the male god Apollo, in addition to managing the sun, makes mankind aware of their sins and purified them, presides over religious and civil laws, and predicts the future, and is also the god of crops and cattle. The goddess Athena is in charge of education, science, and chastity at the same time.

Science

Religion existed 9,000 years ago. Science, invented by the ancient Greeks, appeared much later. Specifically, science began 2,600 years ago with Thales (c.624-c.546 BC)—the Father of Science. How did this happen?

It happened in 1867, in England. In that year, the present understanding of science that it is human's effort to understand nature without supernatural considerations—a conscious decision to *decouple* science from religion—happened. People then look back in history and find Thales' theory of "all things are matter" is the earliest theory that fits this definition. For more, see [8.1].

Interestingly, the term *scientist* was coined earlier, in 1840, reflecting the birth of a new *profession.*

3 Are Science and Religion in Conflict?

Since science had voluntarily decoupled from religion in 1867, implying that science had no position on the existence of God, how do some people still get the impression that science and religion are in conflict?

Conflict

First, as long as religious descriptions of nature contradict known scientific findings, the two are in conflict. (For example, the Bible says the universe was created in six days, while scientists say it took 13.7 billion years to evolve from the big bang.) Otherwise, there would be no contradiction between science and religion.

In other words, when science moves forward and religion is reluctant to take a step back, conflict will occur. A famous example is Galileo Galilei (1564-1642), a Roman Catholic. He was punished by the church for refusing to abandon his conclusion that the center of the solar system is the Sun and not the Earth.

Also, when science advanced quickly in the modern-science era, started by Galileo 400 years ago, religion has removed natural phenomena from its concerns, but not quickly or completely enough. For example, the Vatican still maintains its own observatory.

Second, the 1867 definition of science is not widely known or deeply rooted in the hearts of the people, including many scientists; and historians, philosophers, and popularizers of science. For example, according to the June 1999 issue of *APS News*, the American Physical Society described science as follows:

> Science is the systematic enterprise of gathering knowledge about the world and organizing and condensing that knowledge into

testable laws and theories. The success and credibility of science anchored in the willingness of scientists to:

- Expose their ideas and results to independent testing and replication by other scientists. This requires the complete and open exchange of data, procedures, and materials.
- Abandon or modify accepted conclusions when confronted with more complete and reliable experimental evidence.

Adherence to these principles provides a mechanism for self-correction that is the foundation of the credibility of science.

Sounds good, but it does not mention that science's premise is it cannot introduce the supernatural, probably because it has been practiced for so long since 1867 and so is taken for granted. This is problematic. For example, Newton's long-term effort to study the Bible carefully and *systematically* came to the prediction that the world will be destroyed by 2060—a falsifiable prediction if one waits patiently. Then, according to the APS's notion, it should be a *scientific* prediction, too. Yet, this conclusion is obviously not believed, because no one seems to take it seriously and prepare for it.

Dialogue

Although there are occasional conflicts, religion and science feel the need to communicate with each other for different reasons, and there is dialogue between the two from time to time. On the religious side, religion is an enterprise at the organizational level, often a global enterprise. Many religions offer membership (called believers). Like any business, even a non-profit organization must meet the requirements of the people it serves—members or customers [10.1].

In the United States, more than 90% of the population believes in a personal God, and those are taxpayers who can influence the government funding of science. On the other hand, they owe modern science for their comfortable living (for example, air conditioning, electrical lighting and cell phones) and, for most of them, even their jobs. These people would like to see their religion more in harmony with science.

Seeing mutual benefits, both the National Academy of Sciences and the American Association for the Advancement of Science have set up projects to promote a dialogue between science and religion. After all, with science on one end and religion on the other, there lies a spectrum of common enemies—pseudoscience, antiscience, antireligion and pseudo-religion—between them, the two greatest enterprises on earth.

4 What Is the Difference Between Science and Religion?

In the early days, religion had the same interest as science and wanted to be the interpreter of all things in nature. At present, religion's interest in nature has shrunk but its interest in the supernatural has not diminished. In any case, the two do not understand the world or the universe in the same way (Table 1).

Among them, science requires the Reality Check, i.e., "confirmation" through experiment or practice (within the margin of error), or at least be consistent with established data [8.3]. In particular, nonhuman systems can generally be experimented, because electrons and the like will not complain, and experiments involving animals are subject to scientific ethics. When it is within the scope of ethical permission, there are also experiments used in human-system research, such as questionnaires or brain scans in psychological studies, and the use of contrast group of healthy people in clinical trials of vaccine developments. While the verification of economic theories may take years, sociological theories will take longer—sometimes tens to hundreds of years. And so, social theories are difficult to establish scientifically, which should always be used with caution, otherwise countless deaths involving the life of millions or hundreds of millions of people could happen (see Epilogue).

For more about science and religion, see [10.1].

Table 1. Current contrast between science and religion.

Science	Religion
Object: Nature (including nonhuman systems and human system—the humanities, social science, medical science)	**Object**: Soul (and unexplained phenomena in nature, such as the origin of big bang)
Correctness: derived from "rational" thinking and Reality Check	**Correctness**: derived from revelation
Objective: same for all people	**Subjective**: depends on culture and individual understanding
Universal: any trained person can do it	**Non-universal**: only a few selected ones can do it
Truth: tentative	**Truth**: absolute
Proof: Reality Check (observation or experiment)	**Proof**: none (needs only faith)
Prediction: mature theory can predict and be tested	**Prediction**: none that can be confirmed

5 Do Scientists Know Better?

Science is about the natural world, which includes all material systems. Since humans are (biological) material bodies called *Homo sapiens*, the societal world of humans (social science) and the inner world of individual psyche (the humanities) fall naturally into the domain of science.

Science cannot provide absolute proof of the existence or nonexistence of God. Reason: God, the creator of the universe, is a supernatural while science (as defined in 1867) is about nature only, not the supernatural.

In recent centuries, some scientists have discovered that it is possible to understand nature satisfactorily without introducing God. One example: Galileo's astronomical study 400 years ago; another example: Laplace's celestial mechanics 200 years ago. In 1799 Pierre-Simon de Laplace (1749-1827) published his *Celestial Mechanics*. Napoleon Bonaparte (1769-1821) asked Laplace what part God played in his mechanics. Laplace answered: I have no need of that hypothesis. Today many scientists share the same sentiment, but others hold the opposite view. Why?

In the eyes of the public, scientists are usually considered to be people who think logically and act rationally. Therefore, if 40% of them think there is a personal God, it could very well be true. But is it so?

To answer these two questions, let us first look at how scientists are trained.

Scientific method

Scientists are supposed to be trained in the use of the so-called "scientific method." What then is the scientific method? According to the Merriam-Webster dictionary:

> **Scientific method** (1854): principles and procedures for the systematic pursuit of knowledge involving the recognition and formulation of a problem, the collection of data through observation and experiment, and the formulation and testing of hypotheses.

This is fine. But how is it actually done? The steps in a scientific method are sketched in Fig. 2. There are three paths that really involve "reasoning," which has three different types: those based on *experience* (by analogy/guessing), by induction, and by deduction. The is only one path involving prediction; here, only guessing, induction or deduction is used. Among all these thought processes, only induction and deduction involve logic.

And in some real cases, the final result is guessed first, before induction or reduction is used to confirm it. In science, guessing the right result is called *intuition*, which is highly valued in the profession. You could get a Nobel Prize by obtaining the right result based on wrong reasoning.

In short, scientists are *not* trained 100% in logical thinking. And being humans, they are *not* always rational and objective in making decisions. You cannot always trust them [10.8]. Moreover, a few scientists do have a dark side. Generally speaking, natural scientists receive more training in scientific skills than social scientists, with historians among the worst trained.

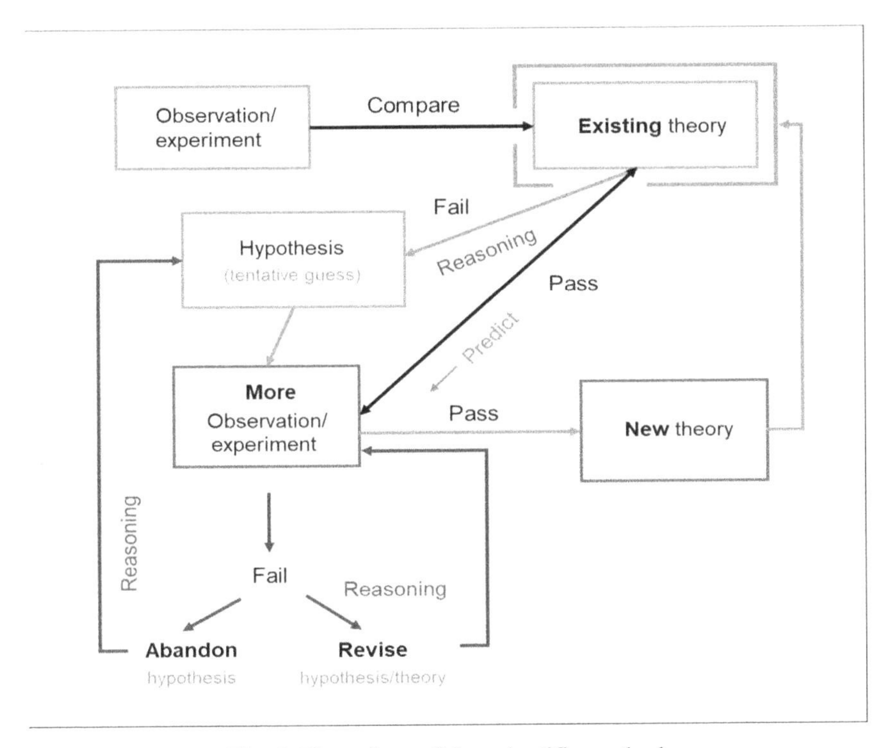

Fig. 2. Flow chart of the scientific method.

It should be pointed out that the so-called "scientific method" above is only a conventional understanding summarized from many cases and cannot be taken seriously, because scientific breakthroughs are actually breakthroughs in exceptional cases. Strictly speaking, there is no step-by-step scientific method like a recipe, otherwise there would be more than a few scientific powerhouses in the world. In fact, science does not care about methods, but there are two premises and two follow-ups. Two premises: pick a thing (a new phenomenon, say) and ask why. Two follow-ups: guess the answer and seek evidence. "Ask why" is the first step to becoming a scientist, and "guess the answer" refers to answers that do not contain supernatural [Lam 2022a]. In fact, Thales became the Father of Science for precisely these two reasons.

6 What Do Scientists Say and How Do They Cope?

Let us start with the great Isaac **Newton** (1642-1727), a profoundly religious man. After formulating successfully the three laws of motion and the law of universal gravity, he tried to reconcile the Bible with his own science. To explain how the world could be made in six days according to the Bible, he first pointed out that the biblical Book of Genesis says that God created the Earth on the first day and the Sun on the fourth day. Newton followed by asking: What is a day? His answer: Only *after* the Sun is there, and the Earth rotates once against its own axis, that is a day. Thus, if there is no Sun yet, you cannot define a day. And so, the first three days in the Genesis are not the three days that we generally understand; it could be N years or as long as you want it to be. Accordingly, the Bible does not conflict with science.

Moreover, to explain why the stars in the sky could remain separated from each other in spite of the gravitational forces among them, Newton invoked God again. He says: God uses his hands to separate the stars from time to time.

In other words, God's hand is needed in setting the initial condition of the differential equations, with which the laws are expressed. This practice of Newton calling God in at the edge of science is called the "God of the gaps." The trouble with this approach is obvious. Every time when a gap is filled with new science, God has to move out of the gap. And history has shown that God had to keep on moving. But this line of thought will never exhaust itself, since new gaps in our knowledge keep on appearing when old gaps are closed, at least for a long time to come.

Albert **Einstein** (1879-1955; Nobel Prize 1921), perhaps the greatest scientist after Newton, is sometimes quoted to be a religious man. This is *not* true. In his own words:

- A knowledge of the existence of something we cannot penetrate, of the manifestations of the profoundest reason and the most radiant beauty, which are only accessible to our reason in their most elementary forms—it is this knowledge and this emotion that constitute the truly religious attitude; in this sense, and in this sense alone, I am a deeply religious man. I cannot conceive of a God who rewards and punishes his creatures, or has a will of the type of which we are conscious in ourselves.

- The ideals which have lighted me on my way and time after time given me new courage to face life cheerfully, has been Truth, Goodness, and Beauty.

Thus, even though Einstein's worldview is sometimes referred to as "cosmic religiousness," I do not think this description is correct either. Einstein was an *atheist* at heart, and the God he said is synonymous with Mother Nature. Perhaps the confusion arises from the occasional practice of physicists taking an ordinary English word and giving it a different meaning. Some examples are given in Table 2.

Table 2. Different meanings of words used by laypeople and scientists.

Word	Laypeople	Scientists
Work	Labor	$\int F \cdot ds$ (≈ force × distance)
Chaos	Confusion	Sensitive dependence on initial conditions
God	Creator and ruler of the universe	Mother Nature (Einstein & others)

Max **Planck** (1858-1947; Nobel Prize 1918), the father of quantum, had this to say: "There can never be any real opposition between religion and science; for the one is the complement of the other."

As for contemporary scientists, the two extreme positions are respectively occupied by Charles Townes (1915-2015; Nobel Prize 1964) and Steven Weinberg (1933-2021; Nobel Prize 1979). All other scientists are somewhere in between.

Townes invented the laser and is a lifelong churchgoer—he prays daily. He "regards religion and science as two somewhat different approaches to the same problem, namely that of understanding ourselves and our universe." As for the origin of the universe, he believes in the big bang;

for origins of *Homo sapiens*, the evolutionary theory. In other words, he ignores the Bible when it is contradicted by science.

Weinberg is a particle theorist. He is famous for his saying at the conclusion of his book *The First Three Minutes*: "The more the universe seems comprehensible, the more it also seems pointless." Later he clarified, "I'm not taking that line back, but I did add that people can grant significance to life by loving each other, investigating the universe, and doing other worthwhile things."

The uncompromising position of an atheist is exemplified by Richard **Dawkins**, an evolutionary theorist at Oxford University and author of famous popular science books such as *The Selfish Gene* (1976). He said:

- The universe we observe has precisely the properties we should expect if there is at bottom, no design, no purpose, no evil and no good, nothing but pointless indifference.
- Only the scientifically illiterate accept the "why" question where living creatures are concerned.

He now finds religion "very boring and not worth talking about."

To conclude, let us see what the smart Richard **Feynman** (1918-1988; Nobel Prize 1965; Fig. 3) has to say. Feynman observed that Western civilization stands on two great heritages and said on May 2, 1956, in his presentation of "The relation of science to religion" at Caltech YMCA:

> One is the scientific spirit of adventure into the unknown...The other...is Christian ethics....These two heritages are logically, thoroughly consistent....How can we draw inspiration to support these two pillars of Western civilization so that they may stand together in full vigor, mutually unafraid? This is the central problem of our time.

He also pointed out that the consistency between science and religion is difficult to attain on a *personal* level, because in science, it is imperative to doubt, while religion demands absolute faith. As far as we know, Feynman is not a religious person.

To sum up: The full spectrum of opinions and positions are taken up by our top scientists. What they say are educated but personal opinions. In other words, one cannot draw conclusion on the God question with the help of the scientists.

Fig. 3. Richard Feynman (Photo/Lui Lam).

7 Can One Really Know?

If faith is not enough and empirical evidence is demanded in showing God's existence, then it is very difficult and in *practice* quite impossible. Any required empirical evidence will involve the physical world and its interaction with the observer's mind. And there is the problem.

Miracle

One particular kind of empirical "evidence" is *miracle*. Miracles are those happenings that cannot be explained by humans, at a certain moment in time. Many miracles are destroyed later under closer scrutiny when a physical explanation is found—another example of the "God of the gaps" situation. For example, quite often the observation of Madonna's face on some window surfaces turned out to be an optical illusion.

And even something cannot be explained immediately does not mean that a miracle is called for. The explanation of some phenomenon simply takes a long time to come by. Scientists are used to this and learn to be patient, while many laypeople are not. They demand an answer within their lifetime, even for important and difficult questions like the existence of God and the meaning of life. They are not alone. The famous philosopher René Descartes (1596-1650) also thought that a rational understanding of the world should be possible for one generation or one man.

To show how patient scientists are let us mention a quite recent example, the case of *superconductivity*. The phenomenon was discovered in 1911 but the explanation did not come until 1957 [8.4]. No single scientist who died within the 46 years in between called superconductivity a miracle; no one demanded the answer within her or his lifespan.

Some miracles can be easily explained by *probability* theory. You chatted with someone in a cocktail party and found that you two have the same birthday. You might call it God's will. The truth is that, according to probability calculations, it takes only 23 people in a room to have more than 50% chance for that to happen (Fig. 4.)

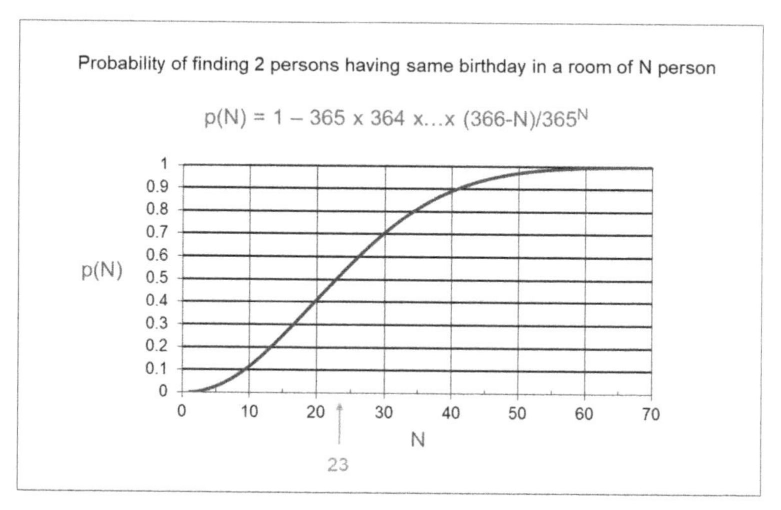

Fig. 4. Probability of finding two persons in a room (of N persons) having the same birthday.

Eyewitness

Another type of empirical "evidence" comes from **eyewitness**. One would say, "I saw it with my own eyes; it must be true." Unfortunately, you cannot always trust your eyes. Magic is based on this principle. There is a more fundamental reason behind this: Life evolves so slowly while modern society is changing so rapidly that the "software" in the brain does not keep up with external developments, and it interprets what it observes incorrectly.

This is illustrated very well in an experiment from the first-year college physics lab. In this lab, a ball is rotating uniformly in a circle with frequency f_0—a fact kept secret from the students. The room is completely dark. A light source flashing at frequency f (called a stroboscope) points towards the object. The instructor can tune f at will.

(a) When $f = f_0$, the students will see the ball at *rest*.

(b) For $f < f_0$, the students see the ball moving *forward* in a circle.

(c) For $f > f_0$, the students see the ball moving *backward* in a circle.

And in all these three cases, the students really believe what they see, while the reality is that the ball is uniformly moving in a circle forward all the time (Fig. 5).

Why? What happened is that light signals reflected from the moving ball enter the student's eyes, triggering the brain to think. Take the $f > f_0$ case, for example. A good brain should interpret from the signals that there are two possibilities:

1. The ball is rotating uniformly, while the flashes has a frequency greater than that of the object.
2. The ball is moving backward.

But our brain always gives us the second answer, the wrong answer. The reason is that the stroboscope was invented only recently in history, and our brain did not have enough time to adapt and is caught here off guard.

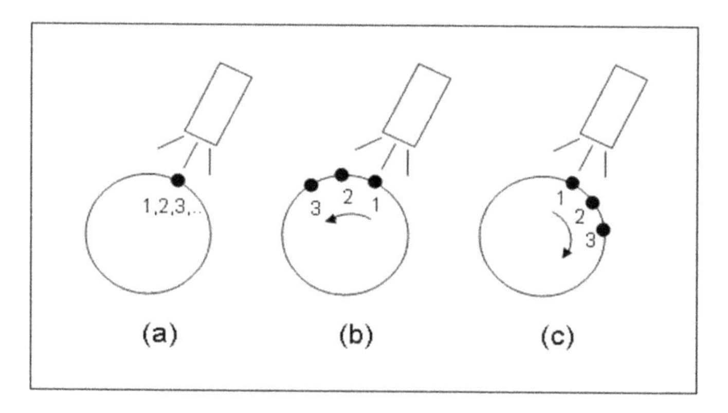

Fig. 5. Our faulty brain. A ball, represented by a black dot, is physically moving forward, anticlockwise in a circle, with frequency f_0 in a dark room. The light box flashes at frequency f. (a) $f = f_0$: the ball is seen at the same position every time when it can be seen (i.e., when the light is on). Our brain tells us the ball is at rest. (b) $f < f_0$: the ball appears at positions 1, 2, 3 sequentially, and our brain tells us the ball is moving forward. (c) $f > f_0$: the ball appears at positions 1, 2, 3 sequentially, and our brain tells us the ball is moving backward.

The alien problem

What if God really exists and shows up suddenly in front of us one day, in the form some people conceive him to be, with full beard and long hair and all that. Can we tell it is really he?

I don't think so. How do we know it is not a highly intelligent extraterrestrial (ET) enjoying itself on its Halloween day? Even if this being is able to recite your personal secrets only you know about—like in the movie *Ghost* (1990)—does it prove that this being is God himself? No way! It could be the ET is picking your brain, collecting data on your thoughts and memory, with wi-fi technology. What I want to say is that, for any physical communication involving the capacity of your mind or body, there is no way to tell whether it is coming from God or from a technologically advanced extraterrestrial.

While scientists cannot help and current science cannot offer absolute proof either way, the problem is compounded by the impossibility of recognizing God through physical communications. Faith is the last resort. And we are back to square one.

8 Does God Exist?

Not everything in the world is knowable. For example, consider someone alone in a room who wrote "10 = 7 + 3" on the blackboard and then erased "7 + 3" before he had a heart attack and dropped dead. Someone else came to the room would have no way to know what the two erased numbers were. That is a limitation of our knowledge at the practical level. More interestingly, there may be a limitation at a fundamental level. Cats will never be able to understand quantum theory. Similarly, there may be some knowledge that is always beyond humans.

When everything is considered, the sensible and honest answer to the question "Does God exist?" is "Maybe." Here, maybe means a probability ranging from 0 to 100%; the exact number is your pick.

For some people, religion does serve some practical purposes. If you cannot sleep well without knowing the answer, or an answer, to the question "Why am I here?" religion is a way out. Nothing wrong. It is a personal choice according to your personal taste. It is like tea or coffee. Some people need it; some people don't. It won't hurt you as long as you don't over drink it; it may even be good for your health. After all, as the American philosopher William James (1842-1910) pointed out long time ago in 1890, people believe in what makes them feel good.

No matter what, remember these two rules: 1. Don't hurt others. 2. Don't hurt yourself.

9 Afterthought

In ancient Greece, philosophy is the branch of human enquiry that asks questions about *everything* without invoking the Greek gods. Later, the non-supernatural part of it was posthumously recognized as a science [Lam 2022b]. The difference between religion and science is that the latter does not introduce God/supernatural into its reasoning.

Philosophers did ask some good questions, but because they refused to introduce the empirical method into their explorations, many of their topics were hijacked by the physicists [Morris 2002].

Judging by the TSTB principle, religion is obviously the winner over science. No answer can be simpler than saying "God did it." Meanwhile, science is still struggling to find the "theory of everything" and the superstring theory (or M theory) may or may not be the answer. And even if the superstring theory succeeds, all the emerging phenomena in complex systems remain to be tackled separately [22].

Yes, the scientific approach is clumsier when compared to religion. Yes, in science, faith like "the world is understandable" is indeed involved. But in fact, for a scientist, at the working level, only the faith that "this problem under study is understandable" is needed—not a big deal! More importantly, science "delivers." All the comfort in modern living, and all the medical cures invented are due to science. No other branch of enquiry can claim the same.

In case you are still pondering the God question, don't forget to take some time out to enjoy the splendid world around you. To help your thinking, take a good physics course. The survey of Larson and Witham shows that among the *top* scientists, only 7 % believe in God [Larson & Witham 1998]. In other words, a good mind, rigorous critical thinking, and an independent and confident personality will reduce superstition. Science education or science popularization alone would not be effective or is not enough.

Finally, a question remains. Why 40%? Why not 20% or 45%? That is a scientific question, and should be answerable.

References

Cunningham, A. & Williams, P. [1993]. De-centring the "big picture": The origins of modern science and the modern origins of science. The British J. History of Science **26**: 407-432.

Edis, T. [2002]. *The Ghost in the Universe: God in Light of Modern Science*. Amherst, NY: Prometheus. (On science and religion, written by a physics professor; comprehensive, with useful references.)

Lam, L. [2008]. Science Matters: A unified perspective. *Science Matters: Humanities as Complex Systems*, Burguete, M. & Lam, L. (eds.). Singapore: World Scientific.

Lam, L. [2014]. About science 1: Basics—knowledge, Nature, science and scimat. *All About Science: Philosophy, History, Sociology &*

Communication, Burguete, M. & Lam, L. (eds.). Singapore: World Scientific.

Lam, L. (林磊) [2022a]. *Science and Scientist*. San Jose: Yingshi Workshop.

Lam, L. [2022b]. *New Humanities*. San Jose: Yingshi Workshop.

Larson, E. J. & Witham, L. [1997]. Scientists are still keeping the faith. Nature **386**: 435-436. (https://doi.org/10.1038/386435a0)

Larson, E. J. & Witham, L. [1998]. Leading scientists still reject God. Nature **394**: 313. (https://doi.org/10.1038/28478)

Morris, R. [2002]. *The Big Questions: Probing the Promise and Limits of Science*. New York: Henry Holt. (What modern science can say about the big questions asked by philosophers, including the God question.)

Pigliucci, M. [2002]. Science and Religion. *The Skeptic Encyclopedia of Pseudoscience*, Shermer, M. (ed.). Santa Barbara: ABC-CLIO. (A concise introduction to the issues by an atheist.)

Williams, R. [1983]. *Keywords: A Vocabulary of Culture and Society*. New York: Oxford University Press.

Original: Lam, L. [2004]. Does God exist? *This Pale Blue Dot*, Lam, L. Tamsui: Tamkang University Press.

Epilogue: From Two Cultures to Bettering Humanity

I have been curious about the human condition more than why the stars are bright and shine. I always wonder why people behave the way they do and how the world has become like this as we see it today. And I kept wondering how the world can be a better place tomorrow.

In my research, I happened to spend the first 30 years on natural science (physics, chemistry, complex systems) and the last 20 years in the humanities (history, art, philosophy). This transition is motivated by my long-held interest on human affairs. The bridge that allowed me to cross from the former to the latter is the Active Walk, a paradigm in complex systems I invented in 1992.

My work in the humanities started with the invention of Histophysics (physics of history) in 2002. After that, I want to tackle the ultimate question concerning the entire humanities: Why the overall research level of the humanities has been stagnant for such a long time—2,400 years, in fact, since the time of Plato. My entry point to this project is a paper on the *two cultures* presented in Seoul in 2006, which led to the establishment of a new multidiscipline in 2007—Scimat (science of human).

The idea of scimat is very simple: Humans are material bodies made up of atoms, the same atoms that make up everything else in the universe. Thus, any study of human is part of science since science is the study of all material systems (without supernatural considerations). So, the three categories of human studies—the humanities, social science, and medical science—should be considered as a whole and put under one umbrella, which I called Scimat, implying that they are all science matters.

In fact, in the last 400 years or so since Galileo, the study of nonhuman systems under the name of "natural science" or modern science did enlighten deeply our understanding of nature (e.g., big bang), make our living easier (cell phone) and help to prolong our lives (for good or bad).

But that is not enough as the future of humanity is concerned, as the so-called "revolt against science" tried very hard to remind everybody. It is the humanities that determine our quality of life (e.g., to pollute or not to pollute) and bring us genuine happiness (human relationships, art). While the study in "natural science" should be continued, it is time for us to return to the Aristotle tradition of treating the human system and nonhuman systems as equally important in our search for knowledge even though, unfortunately, this tradition was interrupted by the phenomenal success of modern science.

Furthermore, the humanities hold the key to world peace. The reasons are:

1. There are two types of tragedies: natural and human-make, but the casualties of human-make disasters outnumber greatly those of natural disasters. For example, death toll of the Indonesia Tsunami of 2004 is 170,000; the Tangshan earthquake of 1976, 700,000. In contrast, death tool in 1975-1979 under the communist regime of Cambodia is 2 million; the holocaust in Germany during WWII, 6 million. No comparison.
2. Human disasters are caused by erroneous decision makings by undereducated dictators while decision making is a discipline among the humanities. (For example, Hitler is a high-school dropout; Stalin is a high-school graduate.)

Therefore, the way to make the world better is to raise the scientific level of the humanities and make sure that the future dictators are well versed in decision making.

But we do not know who the future dictators are and some of them may not even finish middle school. We thus recommend that the three basic scientific knowledge about human spelled out in Secs. 1.1-1.3 be taught in every grade-1 class over the world (and reinforced through later schooling years). Hopefully, it will make everybody understand that we are all related to each other in blood, belonging to one family as descendants of fish. And humans, being recycled stardust with a written history of merely a few thousand years compared to the 13.7 billion years of the universe, will become humble and be nice to each other.

Finally, deepening the humanities' research and taking it to the next level do not require large increase of the budget. No smashing machines needed to be built. What is needed is a change of our concept of science, our perception of priority, and shifting our focus from simple systems to complex systems, from nonhuman systems to the human system. All these could be started immediately.

The humanities were important for the Greeks and the Chinese in ancient times. It was the frontier for them then but is the new frontier for the rest of us now.

A1 Lui Lam's Academic Life

1949-1955	All Saints Primary School (Hong Kong)
1955-1961	Clementi Middle School (Hong Kong)
1961	Ranked second in All Hong Kong Chinese High School Examination
1961-1962	King's College (Hong Kong), Upper Six
1962-1965	Undergrad student in physics and math, University of Hong Kong
1965	BS (First Class Honor), HKU
1965-1966	PhD student in physics, University of British Columbia, Canada (MS, 1968)
1966-1972	PhD student in physics, Columbia University, USA
1972	Finish PhD thesis (after publication in 1974,[1] there are Lam-Platzman Theorem [2] and Lam-Platzman Correction [3] in the physics literature)
1972-1975	Postdoc at Physics Department, City College, City University of New York (created theory of dissipation function for complex materials [4])
1973	PhD, Columbia University
1975-1976	Research associate, Universitaire Instelling Antwerpen, Belgium
1976-1977	Research associate, Universitat des Saarlandes, West Germany
1978-1983	Associate professor/researcher, Institute of Physics, Chinese Academy of Sciences, Beijing
1979	Publish first paper by mainland-only authors in *Physical Review Letters* [5]
1980	Cofound Chinese Liquid Crystal Society
1982	Invent Bowlic liquid crystals (published in *Wuli* [6]); pioneer study of solitons in (shearing) liquid crystals [7]
1984-	Guest professor, Institute of Physics, CAS, Beijing
1984-1987	Associate professor, Queens Community College, CUNY; adjunct professor, City College, CUNY

1987-1990	Associate professor, San Jose State University, California
1988	Invent bowlic polymers and predict bowlic room-temperature superconductors [8]
1990	Initiate and establish International Liquid Crystal Society [9]
1990-2018	Professor, San Jose State University, California
1992	Invent Active Walk [10]; founder and editor-in-chief, Partially Ordered Systems book series (Springer)
2001-	Guest professor, China Research Institute of Science Popularization, China Association for Science and Technology
2002	Establish the new discipline Histophysics (physics of history) [11]
2006	Discover Bilinear Effect in complex systems [12]
2007	Establish the new multidiscipline Scimat [13]; propose the third definition of Science [13,14]
2008	Founder and editor-in-chief, Science Matters book series (World Scientific); publish first scimat book *Science Matters: Humanities as Complex Systems*
2011	Publish new theory on origin and nature of art [15]
2013	*Renke* (人科), Chinese translation of *Science Matters*, published by China Renmin University Press
2017	Distinguished Service Award, San Jose State University, CA [16]
2018	Publish Chinese paper on innovation [17]
2019-	Professor emeritus, San Jose State University, CA
2021	Publish 2 Chinese books: *Wenliren* (文理人), *Lam Lectures: New Humanities, Science, Hawking* (林磊演讲录：新文科、科学、霍金)
2022	Publish 6 Chinese books: *New Humanities* (新文科), *Science and Scientist* (科学与科学家), *Research and Innovation* (研究与创新), *Being Human* (做人做事), *This Pale Blue Dot* (这淡蓝一点), *China Complex* (中国情结)
2023	Publish 2 English books: *Humanities, Science, Scimat*; *Scimat Anthology: Histophysics, Art, Philosophy, Science*

Notes

1. Lam, L. & Platzman, P.M. [1974]. Momentum density and Compton profile of the inhomogeneous interacting electronic system. I. Formalism. Phys. Rev. B **9**: 5122.
2. Bauer, G. E. W. [1983]. General operator ground-state expectation values in the Hohenberg-Kohn-Sham density functional formalism. Phys. Rev. B **27**: 5912-5918.
3. Callaway, J. & March, N. H. [1984]. Density functional methods: Theory and applications. Solid State Physics **38**: 135-220; Papanicolaou, N. I., Bacalis, N. C. & Papaconstantopoulos, D. A. [1991]. *Handbook of Calculated Electron Momentum Distributions, Compton Profile, and X-ray Form Factors of Elemental Solids.* Boston: CRC Press; Blass, C., Redinger, J., Manninen, S., Honkimäki, V., Hämäläinen & Suortti, P. [1995]. High resolution Compton scattering in Fermi surface studies: Application to FeAl. Phys. Rev. Lett. **75**: 1984-1987.
4. Lam, L. [1977]. Dissipation function and conservation laws of molecular liquids and solids. Z. Physik B **27**: 101; Reciprocal relations of transport coefficients in simple materials. Z. Physik B **27**: 273; Constraints, dissipation functions and cholesteric liquid crystal. Z. Physik B **27**: 349. This dissipative function theory was later used for biaxial nematic phase liquid crystals [Das, P. & Schwartz, W. H., Mol. Cryst. Liq. Cryst. **239**: 27-54 (1994)]. In fact, it remains the best and easiest way to deal with the fluid dynamics of any new thermoviscous solid or molecular fluids with microscopic structures, including liquid crystals.
5. Lin Lei (林磊, Lam, L.) [1979]. Nematic-isotropic transitions in liquid crystals. Phys. Rev. Lett. **43**: 1604-1607.
6. Lin Lei [1982]. Liquid crystals phases and "dimensionality" of molecules. Wuli (Physics) **11**: 171-178.
7. Lin Lei, Shu Changqing, Shen Juelian, Lam, P.M. & Huang Yun [1982]. Soliton propagation in liquid crystals. Phys. Rev. Lett. **49**: 1335-1338; Lam, L. & Prost, J. [1992]. *Solitons in Liquid Crystals*. New York: Springer.
8. Lam, L. [1988]. Bowlic and polar liquid crystal polymers. Mol. Cryst. Liq. Cryst. **155**: 531-538.
9. Lam, L. [2017]. Prehistory of International Liquid Crystal Society, 1978-1990: A personal account. Mol. Cryst. Liq. Cryst. **647**: 351-372.
10. Lam, L. [1995]. Active walker models for complex systems. Chaos Solitons Fractals **6**: 267.
11. Lam, L. [2002]. Histophysics: A new discipline. Mod. Phys. Lett. B **16**: 1163-1176.
12. Lam, L. [2006] Active Walks: The first twelve years (Part II). Int. J. Bifurcation and Chaos **16**: 239-268; Lam, L., Bellavia, D. C., Han, X.-P. Liu, C.-H. A. , Shu, C.-Q., Wei, Z., Zhou, T. & Zhu, J. [2010]. Bilinear effect in complex systems. EPL (Europhys. Lett.) **91**: 68004.

13. Lin Lei [2008]. Science Matters: The newest and biggest interdiscipline. China Interdisciplinary Science, Vol. 2, Liu Zhong-Lin (刘仲林) (ed.). Beijing: Science Press; Lam, L. [2008]. Science matters: A unified perspective. *Science Matters: Humanities as Complex Systems*, Burguete, M. and Lam, L. (eds.). Singapore: World Scientific. pp 1-38.

14. Lam, L. [2014]. About science 1: Basics—knowledge, Nature, science and Scimat. *All About Science: Philosophy, History, Sociology & Communication*, Burguete, M. and Lam, L. (eds.). Singapore: World Scientific. pp 1-49.

15. Lam, L. [2011]. Arts: A science matter. *Arts: A Science Matter*, Burguete, M. and Lam, L. (eds). Singapore: World Scientific. pp 1-32.

16. Jackson, J. H. [2017]. Lui Lam: 2017 Distinguished Service Award. *SJSU Washington Square*. https://blogs.sjsu.edu/wsq/2017/02/21/lui-lam-2017-distinguished-service-award/.

17. Lin Lei, Liu Li (刘立) & Sun Nan (孙楠) [2018]. Make full use of domestic journals to obtain priority in publications. Science and Technology China (科技中国), July 2018, Issue 7: 48-50.

A2 Bowlics: A Chinese Innovation Story

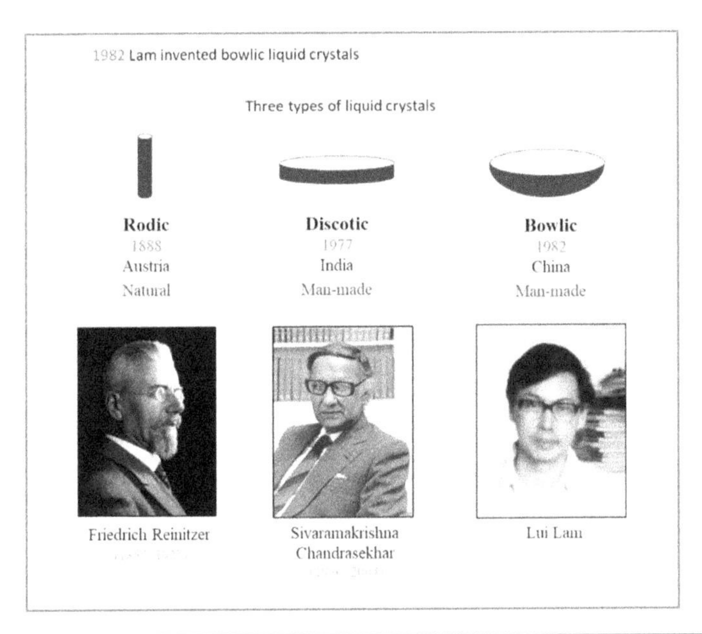

1982	Bowlics proposed by Lin Lei (Lui Lam), while working at the Institute of Physics, Chinese Academy of Sciences; published in *Wuli*, a journal of the Chinese Physical Society [7]
1985	Bowlic monomers synthesized in Europe (one experiment in Paris, another by a France-Germany-Israel group [9,16])
1986	First review on bowlics given by Lam, as invited talk at the 11th International Liquid Crystal Conference, Berkeley, USA [8]
1988	A new kind of bowlics synthesized at Tsinghua University, Beijing [12]
1988	Lam proposed bowlic polymers and predicted bowlic room-temperature superconductors [5]
1992	Bock, Helfrich and Heppke in Europe applied for a patent related to bowlics [2]; 1993, Swager rand Xu in USA applied for another patent [11]

1994	Review of bowlic monomers and polymers published by Lam in a research monograph [6]
1996	The Glenn Brown Award for best PhD thesis was awarded to the postdoc at Harvard University, Bing Xu (done at University of Pennsylvania), for his work on bowlics, presented at the 16th International Liquid Crystal Conference, Kent, Ohio (Xu was a student from China)
1999	Bowlic polymers was synthesized in USA—prompted by the 1994 review article—by Zeng Er-Man (曾尔曼), a PhD student from China at Georgia Institute of Technology [15]
2012	Switchable (ferroelectric) bowlic columnars were realized in the laboratory
2017	A comprehensive review of bowlics' history, advancement and applications was published [13]

Nomenclature: The word Bowlic coined by Lam is used by others in the title of their liquid crystal papers (such as Refs. 4, 10, 12 and 14) and is recognized officially [1] by the IUPAC and formally [3] in *Handbook of Liquid Crystals*
Summary: The story of bowlics represents an interesting story of innovation in China. Bowlics belong to the domain of new and strategic materials, and have potential applications in ultrafast LCD, room-temperature superconductor and other areas—the latter is a Nobel class work when done

References

1. Barón, M. & Stepto, R. F. T. [2002]. Definitions of basic terms relating to polymer liquid crystals. Pure Appl. Chem. **74**: 493-509.

2. Bock, H., Helfrich, W. & Heppke, G. [1992]. Switchable columnar liquid crystalline systems. European Patent EP0529439B1 (filing date: 08/14/1992; publication date: 02/14/1996).

3. Demus, D., Goodby, J. W., Gray, G. W., Spiess, H. W. & Vill, V. [2008]. *Handbook of Liquid Crystals*. New York: Wiley.

4. Dong Yan-Ming, Chen Dan-Mei, Zeng Er-Man, Hu Xiao-Lan & Zeng Zhi-Qun [2009]. Disclination and molecular director studies on bowlic columnar nematic phase using mosaic-like morphology decoration method. Science in China Series B: Chemistry **52**: 986-999.

5. Lam, L. [1988]. Bowlic and polar liquid crystal polymers. Mol. Cryst. Liq. Cryst. **155**: 531-538.

6. Lam, L. [1994]. Bowlics. *Liquid Crystalline and Mesomorphic Polymers*, Shibaev, V. P. & Lam, L. (eds.). New York: Springer. pp 324-353.

7. Lin Lei (Lam, L.) [1982]. Liquid crystals phases and "dimensionality" of molecules. Wuli (Physics) **11**: 171-178.

8. Lin Lei [1987]. Bowlic liquid crystals. Mol. Cryst. Liq. Cryst. **146**: 41.

9. Malthête, J. & Collet A. [1985]. Nouv. J. Chemie **99**: 151.

10. Miyajima, D., Araoka, F., Takezoe, H. & Aida, T. [2009]. Synthesis of new bowlic liquid crystals responsive to electric field. 12th International Conference on Ferroelectric Liquid Crystals (FLC'09).

11. Swager, T. M. & Xu, Bing [1993]. Calixarene-based transition metal complexes and photonic devices comprising the same. United States Patent 5453220 (filing date: 01/29/1993; publication date: 09/26/1995).

12. Wang Liang-Yu (王良御) & Pei Xue-Feng (裴学锋) [1988]. Bowlic liquid crystals. J. Tsinghua University (Natural Science) **28**, Supplement 4: 80-84.

13. Ling Wang (王玲), Dali Huang (黄达理), Lui Lam & Zhengdong Cheng (成正东) [2017]. Bowlics: History, advances and applications. Liquid Crystals Today **26**(4): 85-111. DOI:10.1080/1358314X.2017.1398307.

14. Xu, B. & Swager, T. M. [1993].Rigid bowlic liquid crystals based on tungsten-oxo CaliM41arenes: Host-Guest effects and head-to-tail organization. J. Am. Chem. Soc. **115**: 1159-1160.

15. Zeng Er-Man (曾尔曼) [2001]. Design, synthesis and characterization of columnar discotic and bowlic liquid crystals. PhD thesis, Georgia Institute of Technology, Atlanta, USA.

16. Zimmermann, H., Roupko, R., Luz, Z. & Billard, J. [1985]. Naturforsch. **40a**: 149.

Index

D

E

F

G

H

M

Q

R

S

Z

Printed in the USA
CPSIA information can be obtained
at www.ICGtesting.com
LVHW011306080524
779188LV00003B/9

9 789811 284397